CAMBRIDGE LIBRARY COLLECTION

Books of enduring scholarly value

Monographs of the Palaeontographical Society

The Palaeontographical Society was established in 1847, and is the oldest Society devoted to study of palaeontology worldwide. Its primary role is to promote the description and illustration of the British fossil flora and fauna, via publication of an authoritative monograph series. These monographs cover a wide range of taxonomic groups, from microfossils, trilobites and ammonites through to Coal Measure plants, mammals and reptiles, and from all ages from Cambrian to Pleistocene. They form a benchmark for understanding the past life of the British Isles and many include the original descriptions of numerous key species. The first monograph (on the Crag Mollusca) was published in March 1848 and the Society still continues this work today. Notable authors in the series include Charles Darwin (fossil barnacles) and Richard Owen (dinosaurs and other extinct reptiles). Beginning in 2014, the Cambridge Library Collection and the Society are collaborating to reissue the earlier publications, focusing on monographs completed between 1848 and 1918.

A Monograph of British Trilobites from the Cambrian, Silurian, and Devonian Formations

At the time of his death, John William Salter (1820–69) was dubbed 'the prince of palaeontologists'. A true naturalist with a particular interest in Trilobita, he started publishing this monograph in 1864, and three further parts were issued in 1865–7. But his heath failed and he died, leaving around half of the trilobites then known from Britain untreated; the work was wound up in 1883. Though unfinished, Salter's monograph has thirty excellent plates which make it an essential starting point for the study of British Devonian, Silurian and Ordovician trilobites. The monograph is also frequently cited because Salter's 'preliminary classification of trilobites', though brief, was the best devised up to that time, and has since served as a framework for the classifications used today. Salter introduced four orders of trilobites, three of which – Phacopida, Asaphida, Agnostida – are generally recognised today; only his group 'Ampycini' is not now used.

Cambridge University Press has long been a pioneer in the reissuing of out-of-print titles from its own backlist, producing digital reprints of books that are still sought after by scholars and students but could not be reprinted economically using traditional technology. The Cambridge Library Collection extends this activity to a wider range of books which are still of importance to researchers and professionals, either for the source material they contain, or as landmarks in the history of their academic discipline.

Drawing from the world-renowned collections in the Cambridge University Library and other partner libraries, and guided by the advice of experts in each subject area, Cambridge University Press is using state-of-the-art scanning machines in its own Printing House to capture the content of each book selected for inclusion. The files are processed to give a consistently clear, crisp image, and the books finished to the high quality standard for which the Press is recognised around the world. The latest print-on-demand technology ensures that the books will remain available indefinitely, and that orders for single or multiple copies can quickly be supplied.

The Cambridge Library Collection brings back to life books of enduring scholarly value (including out-of-copyright works originally issued by other publishers) across a wide range of disciplines in the humanities and social sciences and in science and technology.

A Monograph of British Trilobites
from the Cambrian, Silurian, and Devonian Formations

John William Salter

CAMBRIDGE
UNIVERSITY PRESS

University Printing House, Cambridge, CB2 8BS, United Kingdom

Cambridge University Press is part of the University of Cambridge.
It furthers the University's mission by disseminating knowledge in the pursuit of education, learning and research at the highest international levels of excellence.

www.cambridge.org
Information on this title: www.cambridge.org/9781108081221

© in this compilation Cambridge University Press 2015

This edition first published 1864–83
This digitally printed version 2015

ISBN 978-1-108-08122-1 Paperback

This book reproduces the text of the original edition. The content and language reflect the beliefs, practices and terminology of their time, and have not been updated.

Cambridge University Press wishes to make clear that the book, unless originally published by Cambridge, is not being republished by, in association or collaboration with, or with the endorsement or approval of, the original publisher or its successors in title.

THE

PALÆONTOGRAPHICAL SOCIETY.

INSTITUTED MDCCCXLVII.

LONDON:

MDCCCLXIV—MDCCCLXXXIII.

THE CAMBRIAN, SILURIAN, AND DEVONIAN TRILOBITES.

DIRECTIONS TO THE BINDER.

The Monograph on the Cambrian, Silurian, and Devonian formations will be found in the Volumes of the Palæontographical Society for the years 1862, 1863, 1864, 1866, and 1883.

Cancel Title-pages affixed to the Parts issued in the Palæontographical Volumes for the years 1862, 1863, 1864, and 1866, and *substitute* the General Title-page provided in the Volume for 1883. *Cancel* pages 81, 82, and 83 in the Volume for 1862 (these pages will be found reprinted in the Volume for 1863). *Cancel* pages 215 and 216 in the Volume for 1866 ; and place the sheets and plates in the order indicated below.

ORDER OF BINDING AND DATES OF PUBLICATION.

PAGES	PLATES	ISSUED IN VOL. FOR YEAR	PUBLISHED
Title-page.	—	1883	October, 1883
1—80	I—VI	1862	August, 1864
81—128	VII—XIV	1863	June, 1865
129—176	XV—XXV	1864	April, 1866
177—214	XXV*—XXX	1866	June, 1867
215—224	—	1883	October, 1883
"Corrigenda"	—	1862	August, 1864
"Errata"	—	1863	June, 1865
"Addenda et Corrigenda"	—	1864	April, 1866

A MONOGRAPH

OF THE

BRITISH TRILOBITES

FROM THE

CAMBRIAN, SILURIAN, AND DEVONIAN FORMATIONS.

BY THE LATE

J. W. SALTER, A.L.S., F.G.S.

LONDON:

PRINTED FOR THE PALÆONTOGRAPHICAL SOCIETY

1864—1883.

NOTICE.

THE 'Monograph on the British Trilobites,' as originally proposed by the late Mr. J. W. Salter, was intended to comprise descriptions of the Group from the first appearance of the forms in the Cambrian strata to their final cessation in the beds of the Carboniferous series Unfortunately death stayed the ready pen of the author in 1869; and the work, begun in 1864, was not resumed after 1867. Very many British Trilobites remain to be described. The author at the time of his decease had not reached the Carboniferous genera, and, indeed, had not had the opportunity of illustrating more than about one half of the total number of the Cambrian, Silurian, and Devonian species.

T. W.

6TH SEPTEMBER, 1883.

A MONOGRAPH

OF

BRITISH TRILOBITES.

A MONOGRAPH of the British species of the Trilobite group has long been desirable, and would, in accordance with the wish of many friends, have been prepared by me before now, had public engagements permitted it.

The materials for such a work have hitherto been much scattered; nevertheless, of late years, they have become very numerous, whilst the publication of such sterling works as those of Barrande, Burmeister, Beyrich, and Lovén, have disposed naturalists, as well as geologists and collectors, to pay increased attention to the group,

I think that a true and natural arrangement of the Trilobites is yet to be given; and therefore prefer at present to publish the materials in hand, that in the mean time systematic zoologists may have the means of forming their judgment. The group is a very complete one, and no artificial scheme is admissible for its classification. Dr. Emmerich's Essay, of 1845, seems to me to be the nearest approach to a natural grouping of the families; and I shall follow it meanwhile, in figuring the species,[1] with such additions as are proposed on the next page. The dichotomous arrangement here given is proposed merely as temporary; but I believe it to be, in the main, a natural one.

And I trust that, by the time we have arrived at the conclusion of the work, some new and clearer light may be thrown on the affinities and arrangement of this order of Crustacea, as unique in character as it is well defined in geological place.

[1] Our friends will much oblige by forwarding us their best specimens, including any illustrative fragments, in the order of the families as given on page 2. In that way all the materials for a genus will be under the eye together, and no long detention of the specimens need take place. They may be sent to my care, at Mr. SOWERBY's Natural History Offices, 45, Great Russell Street, London.

PRELIMINARY CLASSIFICATION OF TRILOBITES.

I have here given chiefly British genera as examples. There are, of course, many others. It is evident that the lowest and most rudimentary group of all is that of the *Agnostidæ,* which, therefore, may be at once placed at the base; and there is not likely to be much difference of opinion as to the place of the *Phacopidæ,* as the typical and most perfect group of the order. The other genera fall more or less naturally into an intermediate place, but are developed along two distinct lines, as follows:

A.—Trilobites with facial suture ending on the external margin. Eyes well developed, usually facetted externally. (Phacopini.)

1. **PHACOPIDÆ.**
Phacops. Trimerocephalus. Acaste. Chasmops. Odontochile. Cryphœus.

2. **CHEIRURIDÆ.**
Cheirurus. Staurocephalus. ? Cybele. Sphærexochus. Deiphon. ? Encrinurus. Amphion. ? Zethus.

B.—Facial suture ending on the posterior margin. Eyes (usually) moderately developed, smooth. (Asaphini.)

3. **ACIDASPIDÆ.**
Acidaspis, &c.

4. **LICHADÆ.**
Lichas.

5. **CYPHASPIDÆ.**
Cyphaspis. Aulacopleura, &c.

6. **HARPEDIDÆ.**
Harpes.

7. **CALYMENIDÆ.**
Calymene. Homalonotus.

8. **CONOCEPHALIDÆ.**
Conocoryphe. Sao. Solenopleura. ? Centropleura. Holocephalina. Angelina.

9. **OLENIDÆ.**
Olenus. Sphærophthalmus. Paradoxides. Anopolenus. Remopleurides. Triarthrus. Tiresias. Cyphoniscus. Hydrocephalus.

Number of body-rings considerable. Caudal portion reduced.

12. **PROETIDÆ.**
Brachymetopus. Griffithides. Phillipsia. Proetus. Phaeton.

11. **BRONTEIDÆ.**
Bronteus.

10. **ASAPHIDÆ.**
Illænus. Nileus. Stygina. Isotelus. Asaphus. Ogygia. Barrandia. Basilicus. Niobe. Psilocephalus. Æglina.

Number of body-rings reduced. Caudal portion proportionally enlarged.

C.—Facial sutures obscure, or submarginal, or none. Eyes often absent. (Ampycini).

13. **TRINUCLEIDÆ.**
Dionide. Trinucleus. Ampyx.

D.—Without eyes or facial suture. (Agnostini.)

14. **AGNOSTIDÆ.**
Agnostus. Trinodus.

March, 1864. J. W. S.

BRITISH TRILOBITES.

History.—Professor Burmeister has given a very full account of the authors who have written upon this subject, in his celebrated treatise on the 'Organisation of Trilobites,' published at Berlin in 1843, and translated for the Ray Society by Prof. Thomas Bell and the late Prof. Edward Forbes.

A short summary of the Professor's observations will be sufficient for our purpose, for the book is widely circulated, and is indeed the only summary to which we can turn for full information. He has enriched it with a full bibliography collected by himself, and traced the work done by successive naturalists and geologists to the time of Emmerich's scientific treatise in 1839,[1] and Milne-Edwards' Catalogue, in his volumes on the 'Crustacea.'

It would seem that our own countryman Dr. Llhwyd was really the first to call attention to the "regularly figured stones lately found by him," but though he referred them to "the sceletons of some flat fish," he took good care to explain that "not these or any other marine terrestrial bodies were really parts of animals," and he only marvelled that the "*Piscis Icon*" should be raised above the surface of the stone "*ac si verus piscis esset.*" This view of the fossils was about as correct as most of those which followed till the time of Linnæus, although Mortimer and Da Costa had declared in favour of their crustacean origin. It was Walch's 'Natural History of Petrifactions,' published at Nuremberg in 1771, which first brought into a focus the scattered information regarding the group, and established the views of Da Costa, Linnæus, Wilckens, and other writers who had regarded Trilobites as entomostraca. Walch first gave them the name of Trilobites, and Parkinson's figures and the descriptions of several German authors pointed out the necessity of distinguishing the species.

[1] Reproduced with improvements in 1845. This is the earliest scientific arrangement of the tribe: See Leonhard und Bronn's 'Jahrbuch,' 1845.

In 1821, two distinguished French naturalists entered this field. Latreille, whose fame as a student of the *Articulata* needs no illustration, strangely enough overlooked the manifest characters which placed the Trilobites among Crustacea, and pronounced for their affinity with *Chiton.* Audouin, on the contrary, compared them with the Isopod Crustacea, declared they had no feet, but appendages for breathing organs, and, in short, led the way, as Burmeister admits, for all subsequent research in the same direction.

Then followed Wahlenberg's work, and Brongniart's 'Histoire Naturelle des Crustacés,' a work in which the genera of Trilobites were first defined, and seventeen species described. Schlotheim soon raised these to twenty-six, and Dalman's complete treatise on the group in 1826 gave a new importance to the subject; while the very perfect state of the Swedish specimens enabled him to present better figures than had before appeared. He called the group *Paleadæ,* and altered some of the generic names without much reason, but in this he has not been followed.

De Kay in America, and Count Sternberg in Germany, meanwhile, described many forms; and Eichwald and Razumousky, in Russia, prepared the way for Dr. Pander's careful treatise, in which the labrum, first observed by our countryman Charles Stokes in American specimens, was fully described.

Dr. Green's monograph, in 1832, was only valuable for the casts which illustrated it; and up to the time of the publication of the 'Silurian System,' in 1837, but very little was known about Trilobites in England. That work, however, had a wide fame, and deservedly so, and by costly illustrations of the best specimens procurable, and by the philosophic remarks contributed by Dr. McLeay, gave a new impetus to the collection and description of the species. Dr. Buckland, a year before, had illustrated some of the more common kinds in his 'Bridgwater Treatise.'

But the period was now coming for the scientific arrangement of the facts collected. Illustrations were abundant both in England and on the Continent, especially in Sweden and Norway, where Sars and Boeck, Esmark, Hisinger, and other authors, had done their best to make the Scandinavian species known. About 1837 Prof. Burmeister began to turn his attention to the group, and Dr. Quenstedt, of the Mineralogical Museum of Berlin, published some important observations with regard to the number of rings in the body, which Burmeister justly regards as of great consequence. It is doubtful whether the honour belongs to Burmeister or Quenstedt of first calling attention to this, the chief means of distinguishing the various genera, as well as a point of great importance in determining the affinity. Dr. Burmeister, at all events, suggested it to Quenstedt, and afterwards worked it out, considering that Trilobites differ from all other Entomostraca in having no definite fundamental number of segments to the thorax, while the living Entomostraca and Malacostraca are ruled by different but always definite numbers.

While these investigations were going on, Dr. Emmerich succeeded Dr. Quenstedt in the care of the Berlin Museum, and followed out his researches by a complete and beautiful essay, well known as the 'Dissertatio Inauguralis,' Berlin, 1839. In this work the author, while ranging over the more recondite portions of the study, threw for the first time the whole Trilobite group into a series of natural families. And these must be, I conceive, the basis of any true classification of the order *Trilobita.* The separation of the large-eyed Trilobites with eleven body-rings, was not indeed due to Emmerich, but to Quenstedt; but Dr. Emmerich's essay confirmed this important view, named the group Phacops—our first described one, and went on to apply the principle thus gained to the arrangement of the whole. The species are carefully made out, the synonyms collected, and a model set for all succeeding works.

Other observers were not idle. Von Buch, Bronn, Green, Goldfuss, and Münster were figuring the new species with various merit; and Milne-Edwards had compiled all the known synonyms in his great work on the Crustacea. In 1843, three most important works appeared, in one of which Dr. Burmeister placed before the German reader all the facts regarding the history, structure, and affinities of the group, while Dr. Goldfuss gave a systematic arrangement of Trilobites and description of new species in the 'Jahrbuch' for 1843. In England, the profound and careful work of the late Gen. Portlock first called attention to all the new discoveries which had been making abroad, while the many new forms which he described and illustrated[1] have given his work the very highest rank. The American species began now to be figured by Hall, and Emmons, and Vanuxem; Lovén was producing his classic descriptions in the Transactions of the Swedish Academy. (Ofvers. Kongl. Vetensk. Akad., 1844, &c.); Emmerich repeated and improved his classification in the 'Neues Jahrbuch' for 1845; Dr. Beyrich, in 1846, was giving us his accurate descriptions of species; and in the same year appeared the 'Notice Préliminaire' of M. Barrande, the herald of a work which has thrown nearly all other works on trilobites into the shade.

This remarkable sketch by M. Barrande, the fruit of thirty years of labour, indicated at once the commencement of a new era for the group. A supplement published by M. Barrande the same year raised the number of described Bohemian species alone to 152. The rich "terrain" which M. Barrande has so emphatically made his own still furnished abundant work for Beyrich; and M. Corda, the keeper of the National Museum at Prague, even attempted to snatch the "spolia opima" from the hand that had won them. The 'Prodrom einer Monographie der Bohmischen Trilobiten,' while it attests the rapid industry of the well-known botanist of Prague, shows how little is gained by hasty generalisation, and especially "appropriation" in natural history. It was a melancholy failure. While M. Barrande had patiently traced the metamorphosis of some thirty different kinds of Trilobites, and was preparing for their illustration, these young and undeveloped Trilobites were figured by Corda as so many distinct genera and species. A certain number of new forms were doubtless named, and a few errors of nomenclature corrected; but the absurd

[1] His illustrations do not do him justice. Owing to a misfortune of the printers, the whole of the beautiful plates drawn for the work by Mr. G. V. Dunoyer had to be hastily transferred by an employé, and the character is greatly lost in the transfer.

mistake of grouping the Trilobites according to the pattern of their tail-fringes, instead of following Emmerich's natural arrangement, was committed in presence of such materials as Emmerich probably never saw!

Angelin's first instalment of the Swedish Trilobites, appeared in 1852, after many memoirs by Beyrich, and Lovén, and Kutorga, and Volborth, on the Swedish or Russian forms; but it is impossible to notice all the works that followed Emmerich's essay. M'Coy, and Fletcher, and Prof. Wyville Thomson, and myself, have done our best, as opportunity offered, to illustrate the British forms after Portlock's model in 1843. The Decades of the survey by E. Forbes and J. W. Salter appeared in 1849 and 1853. In the last-named year, M. Barrande's long-expected volume made its appearance, and in its illustrations and descriptions of 250 Trilobites is a work without a rival. He is now preparing the second volume, which will add some fifty or sixty more. But as he will be referred to in every chapter that follows, it is not necessary to say more of his work here. On one point only has the author left the field open. He has honestly and modestly stated that he does not profess to *classify* the Trilobites; and on this point we are therefore free to follow Dr. Emmerich as before.

The years 1855 to 1863 have seen great additions to our knowledge of Trilobites made from all quarters: the primordial zone has yielded up its treasures to the search of English and American geologists, and has received additions from all parts of North Europe.

A formation utterly unknown to science till Barrande established its history has proved to be a most extensive and rich repository for Trilobites. Barrande in Bohemia and Spain, Angelin in Sweden, myself in Britain,—Logan and Dale Owen, and Billings, and more lately James Hall, in the New World, have contributed materials from this most ancient zone of life—the Cambrian. And if of late years systematic classification has not kept pace with the description of the forms, the fault has not been with the palæontologist in neglecting to supply the data.

Geological Place.—The geological history of the Trilobite group is very clear and succinct. Though not the oldest animal forms known, they meet us in the earliest formation in which we have any abundant traces of marine life, viz., the Lingula-flags (Cambrian of *Sedgwick*, Upper Cambrian of *Lyell*, Lowest Silurian of *Murchison*).

In this their commencement we have some of the smallest and most rudimentary, as well as some of the largest forms; but the group did not attain its maximum, nor rise to its most perfect forms, till the period of the Llandeilo and Caradoc formations—the typical Lower Silurian deposits.

Above this point few new types were introduced; and though individuals were numerous and species most abundant, there were fewer *genera* in the Upper than in the Lower Silurian. In the Devonian the reduction went still further. In the Carboniferous Limestone they were reduced to three genera; and the group was extinct before the later portion of the Coal-period.

As the Trilobites will be constantly referred to the special groups of beds in which they occur, it may be well to give, in a short tabular form, the Palæozoic arrangement followed in this work, with the distribution of the Trilobites as to maximum and minimum in these zones.

The Palæozoic groups are nearly those adopted of late years by our leading geologists. For the determination of the lower ones we are almost equally indebted to Murchison and Sedgwick, and their labours since 1831 have now become the common property of the scientific world.

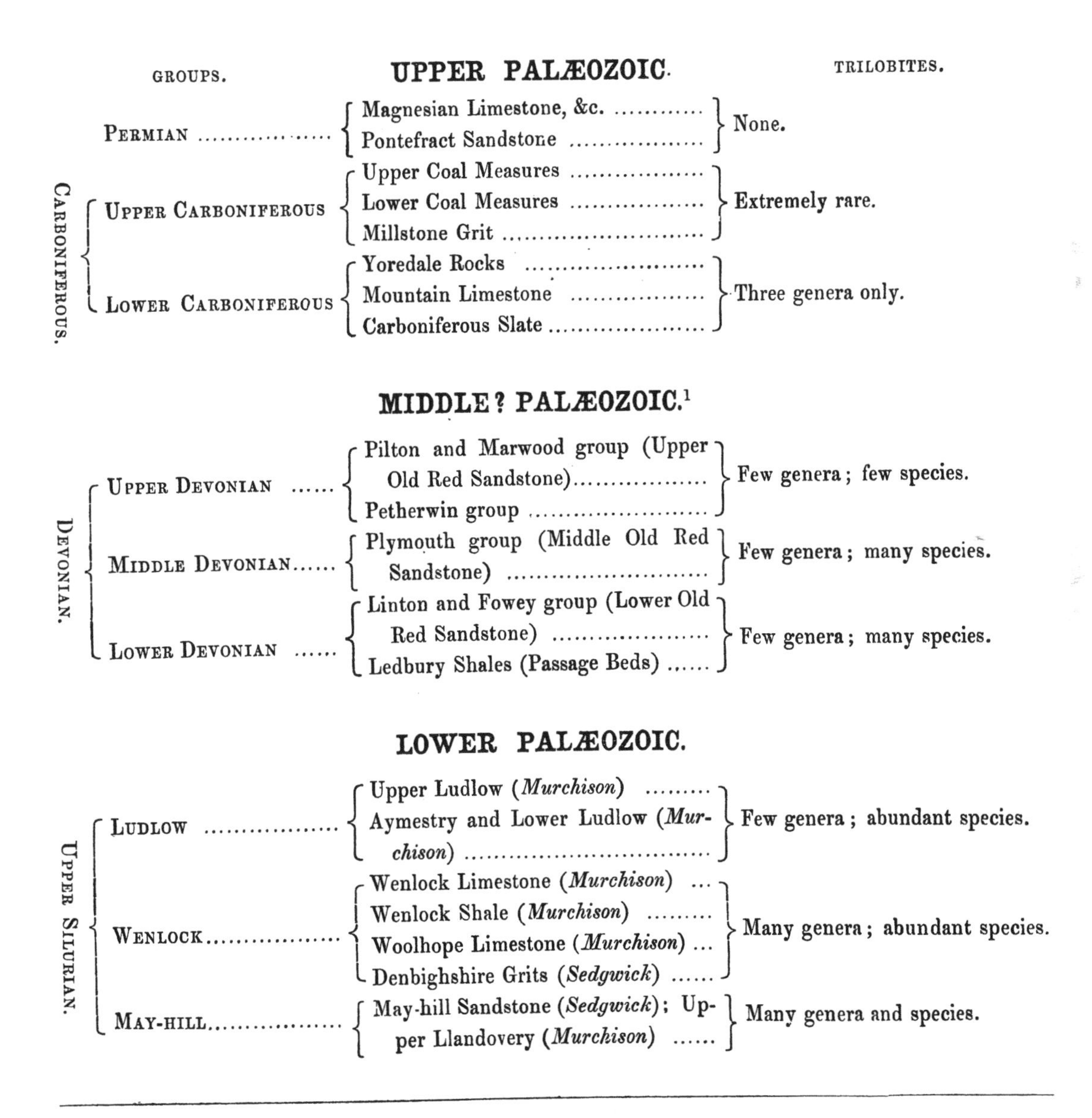

UPPER PALÆOZOIC.

	GROUPS.		TRILOBITES.
	PERMIAN	Magnesian Limestone, &c.; Pontefract Sandstone	None.
CARBONIFEROUS.	UPPER CARBONIFEROUS	Upper Coal Measures; Lower Coal Measures; Millstone Grit	Extremely rare.
CARBONIFEROUS.	LOWER CARBONIFEROUS	Yoredale Rocks; Mountain Limestone; Carboniferous Slate	Three genera only.

MIDDLE? PALÆOZOIC.[1]

	GROUPS.		TRILOBITES.
DEVONIAN.	UPPER DEVONIAN	Pilton and Marwood group (Upper Old Red Sandstone); Petherwin group	Few genera; few species.
DEVONIAN.	MIDDLE DEVONIAN	Plymouth group (Middle Old Red Sandstone)	Few genera; many species.
DEVONIAN.	LOWER DEVONIAN	Linton and Fowey group (Lower Old Red Sandstone); Ledbury Shales (Passage Beds)	Few genera; many species.

LOWER PALÆOZOIC.

	GROUPS.		TRILOBITES.
UPPER SILURIAN.	LUDLOW	Upper Ludlow (*Murchison*); Aymestry and Lower Ludlow (*Murchison*)	Few genera; abundant species.
UPPER SILURIAN.	WENLOCK	Wenlock Limestone (*Murchison*); Wenlock Shale (*Murchison*); Woolhope Limestone (*Murchison*); Denbighshire Grits (*Sedgwick*)	Many genera; abundant species.
UPPER SILURIAN.	MAY-HILL	May-hill Sandstone (*Sedgwick*); Upper Llandovery (*Murchison*)	Many genera and species.

1 This term is introduced in deference to published authorities. But the Upper and Middle Palæozoic groups are together only equivalent to the Lower, and might well be all included under the one term "Upper Palæozoic."

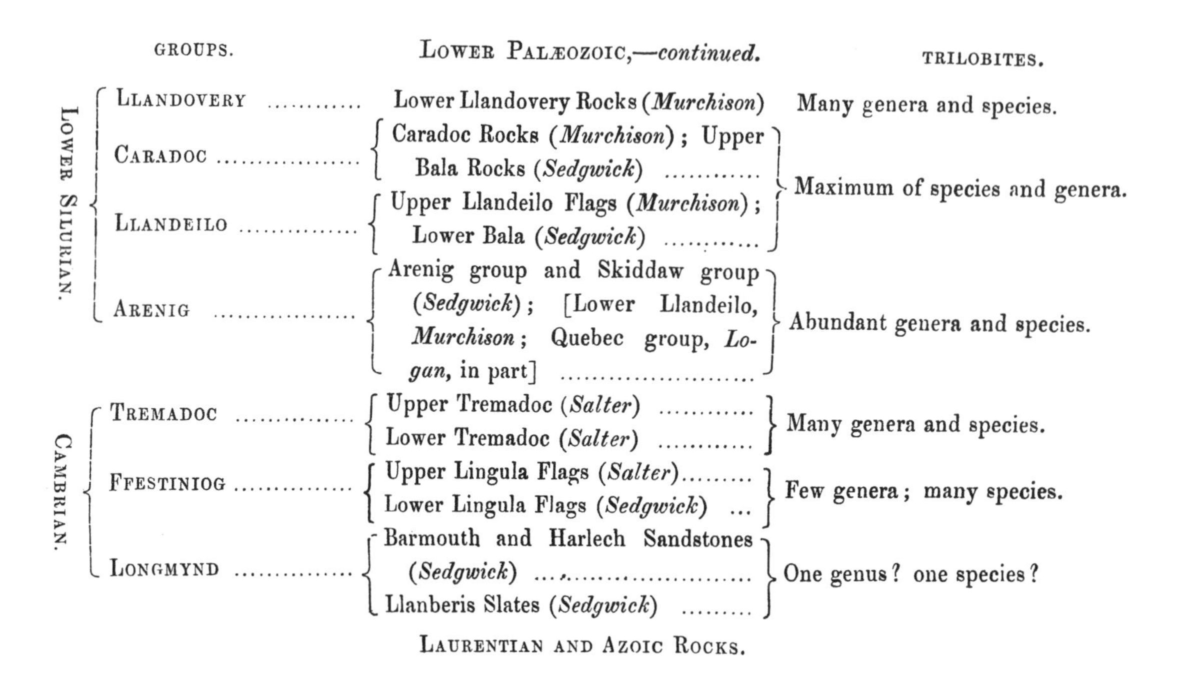

	Groups.	Lower Palæozoic,—*continued.*	Trilobites.
Lower Silurian.	Llandovery	Lower Llandovery Rocks (*Murchison*)	Many genera and species.
	Caradoc	Caradoc Rocks (*Murchison*); Upper Bala Rocks (*Sedgwick*)	Maximum of species and genera.
	Llandeilo	Upper Llandeilo Flags (*Murchison*); Lower Bala (*Sedgwick*)	
	Arenig	Arenig group and Skiddaw group (*Sedgwick*); [Lower Llandeilo, *Murchison*; Quebec group, *Logan*, in part]	Abundant genera and species.
Cambrian.	Tremadoc	Upper Tremadoc (*Salter*) Lower Tremadoc (*Salter*)	Many genera and species.
	Ffestiniog	Upper Lingula Flags (*Salter*) Lower Lingula Flags (*Sedgwick*)	Few genera; many species.
	Longmynd	Barmouth and Harlech Sandstones (*Sedgwick*) Llanberis Slates (*Sedgwick*)	One genus? one species?
		Laurentian and Azoic Rocks.	

Structure and Habits.—Of their natural history I do not intend to say much here, as I think it foreign to the purpose of this work to discuss such points at length, the chief object of the Palæontographical Society being to collect accurate figures and descriptions of the species of British fossils. I shall, at all events for the present, content myself with very few remarks; and chiefly for the use of the general reader.

I need only say, then, that Prof. Burmeister contends that Trilobites belong to the Entomostraca, or lower division of the Crustacea, from the numerous (not definite) number of the segments of the body, and is disposed to place them among the Phyllopoda. The learned McLeay, in his discussion of their affinities in the 'Silurian System,' gave them a higher position, intermediate between the Isopodous group and the Phyllopods;[1] and it is pretty certain that they form a distinct order, and do not belong to any modern group of Crustacea.

Every author who has written on Trilobites has more or less perceived their analogy with the Limulus or King-crab, to which tribe there is, indeed, a good deal of external

[1] McLeay was, of course, not aware of the metamorphosis of the Trilobite, since made out by M. Barrande. He characterises them as having the "head distinct, without antennæ; the feet rudimentary, soft, and almost useless." His judgment, as to their habits, is suggestive, and, at the same time very cautious. He evidently thought it likely they had no feet at all; and I give his remarks in brief. "Whether they moved (they were probably to a certain degree sedentary) by soft, rudimentary feet, by undulation of setigerous segments, as the worm moves, or by an undulation of the lower soft surface, as in *Chiton*, is uncertain. They probably adhered in masses, as *Chitons* do; and, as the mouth is like that of *Apus*, they were probably carnivorous, as is that genus, and may have fed on *Acrita, Annelida*, or naked *Mollusca*."

resemblance. But this resemblance totally fails when we examine the under side of the animal; for all the researches hitherto made (and they are many) fail to detect the slightest trace of limbs in the Trilobite. It is impossible, seeing the state of preservation in which they occur, to suppose that in every case,—in fine shale, in limestone, in arenaceous mud,—all traces of these organs should have been lost, had they ever existed.

We are compelled to conclude that Trilobites had not even membranaceous feet, and that the ventral surface was destitute of appendages. It is of course difficult to prove this. And almost all naturalists are disposed to allow them soft gills, attached to the under side. I do not see that the Trilobite had any need of appendages, further than what might be necessary as breathing organs. In this I have the concurrence of Prof. Wyville Thompson, who has given some thought to the affinities of the group. If gills existed at all, they were probably quite minute.

There is some reason to believe that, like its predecessor, the Annelide, the habit of the Trilobite was to gorge itself with the carbonaceous mud, and extract from it the nutritive portions. Such material has, indeed, been found in the straight intestinal canal of the Trilobite. Barrande has figured a specimen in which this viscus is preserved, a natural cast being taken of the interior by the sabulous matter swallowed by the animal.[1] What the nature of this sabulous matter was originally may be matter of conjecture; but it was solid enough to retain the stomach and intestine in a dilated form, while the surrounding matrix was solidified. It must, therefore, have consisted of a hard food, such, for instance, as the shells of Lingulæ, or, if the habit was carnivorous, of the contents of the bodies of the worms devoured,—or, lastly, of the silty mud among which the creatures lived. Against the former supposition we have strong reason to conclude, for Trilobites certainly possessed no hard jaws capable of comminuting shells or corallines, which we know existed during the same period. I see no likelihood of the carnivorous habit, and venture the latter suggestion. Mr. Spence Bate also thinks the mouth was contractile.

The only hard portion of the under side is the immoveable upper lip or labrum; and this may have been the instrument by which the food was scraped together. The absence of feet, and the presumed nature of the food would give me reason for believing that the Trilobite did not swim, as supposed by many authors, but crawled along the bottom. The shape was fitted for this; all Trilobites, whatever be their ornament on the upper side, present an even contour round the margin, which would apply itself to a flat surface accurately, while the under side was no doubt smooth and soft.[2] Probably, in many cases, the Trilobite lay half-buried in the silt, as is the frequent habit of the large Limulus, or King-crab.

[1] In the genus *Trinucleus*, vol. i, pl. 30, fig. 38, of the 'Système Silurien de Bohême,' par Joachim Barrande, 1852, Prague and Paris, vol. i. A magnificent and costly work, of which only the Trilobites are yet published.

[2] The analogy with *Chiton*, perceived by some of the old writers, is not altogether fanciful. At least, the habit must have been very similar, though of course there is no direct relation.

I am indebted to Mr. Spence Bate for some friendly criticisms of these views, and shall wait with interest for his promised memoir on the 'Homologies of the Trilobite, and its Habits.'

The general structure of the animal will be best perceived by referring to the woodcuts a little further on; and while there is, in the greater part of this structure, a sufficient resemblance to the ordinary Crustacea, there are one or two points in which the Trilobite differs from all other groups, and they happen to be obvious ones.

The curious so-called *facial suture*, a line of division which is only faintly indicated in the Limulus, and which has, perhaps, no other representative[1] in the whole Crustacean class, sufficiently distinguishes the Trilobite. It divides the head into two portions, an anterior one that bears the eye, and a posterior that covers the stomach. The latter segment is much larger than the former, and may be formed of several rings.

And then there is the "trilobation." Whatever tendency some of the higher Crustacea may show to this, and in whatever degree a few of the Trilobites may lose it, it is the conspicuous character of the whole order, and has, doubtless, an important meaning. Limulus also shows a trace of this trilobation; but it is accidental, rather than characteristic, in other groups.

I give here a copy of the original figure in the 'Memoirs of the Geol. Survey,' vol. ii, Part I, p. 334. It is not drawn from any particular Trilobite, but is a general expression of the structure.

And I have added one or two terms from Barrande's more complete figures.

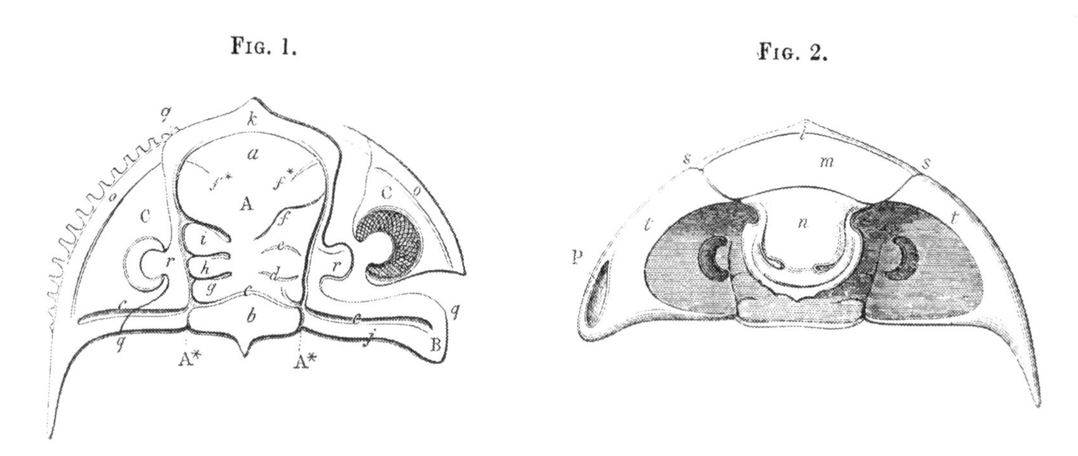

Fig. 1. Fig. 2.

Beginning with the HEAD, or carapace (Fig. 1, upper side; Fig. 2, under side), we recognize the following portions:

GLABELLA (A), bounded by the axal furrows (A*), and including the neck-furrow (*c*);

[1] I am bound to state that Mr. S. Bate believes he has discovered the analogue of this suture on the under side of the crab and lobster; he finds it too in *Argulus*, &c. But his views are not yet fully published.

the basal furrows (*d*); middle or ocular furrow (*e*); upper or frontal furrow (*f*); with an occasional pair of frontal furrows (*f**).

These furrows bound and include the various lobes, designated thus: the frontal lobe (*a*); upper or third lobe (*i*); middle lobe (*h*); basal lobe (*g*); neck-lobe (*b*).

The sides or CHEEKS consist of a portion fixed to the glabella (B *j r k*); and separated from the free or moveable cheeks (*o* C) by the FACIAL SUTURE (*q q*). These moveable cheeks bear the EYES; the fixed cheeks include the eye-lobe (*r*), the neck-furrow (*c*), and in some cases the posterior angles (B), often produced into spines.

The posterior margin (*j*) is usually divided by the facial suture, which, in other cases, cuts the outer margin at *q*.

The front margin of the head is sometimes produced into a point (*k*), and on its under side (fig. 2) shows the ROSTRAL SHIELD (*m*), with its suture (*l*) (rostral suture); the inferior branches of the facial suture (*s s*), the incurved under margin (*t*), the LABRUM or EPISTOMA[1] (*n*), and sometimes a cavity (P) for the reception of the ends of the pleuræ in rolling up.

Each THORAX-RING (fig. 3), in like manner, consists of the AXIS (A), with its articular portion (*b*), divided by the axal furrows (A*) from the side-lobes or pleuræ. These last, whatever be their nature, are in one piece with the axis, never articulated with it.

The PLEURÆ consist of a posterior portion (*f*), a fulcral or anterior portion (*e*), separated by the deep pleural groove (*d*). At *c* the fulcral points are seen, beyond which in most genera the segment is facetted (*a a*), for rolling up. These facets (*a a*) are always smooth, and slide under the preceding joint in the act of rolling.

Sometimes the terminal portion is produced into spines, (*g g*).

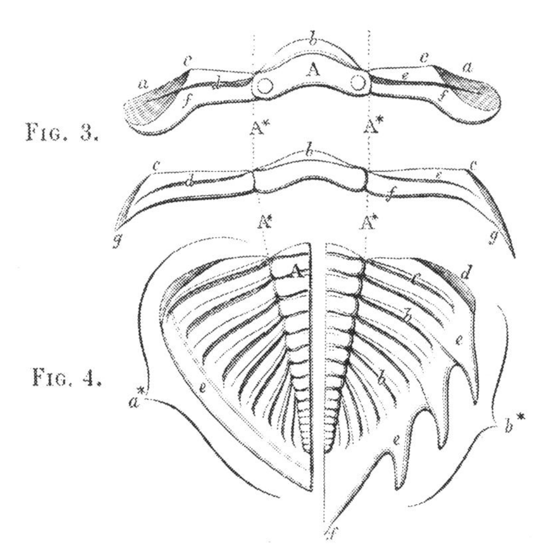

The TAIL, or PYGIDIUM (fig. 4), consists also of an AXIS (A), with its articular portion, as in the thorax; the LIMB or lateral portion is either entire, as at (*a**), or shows its component

[1] The exact nature of this piece is not quite certain. I use '*labrum*' in the descriptions.

pleuræ as at (b^*). The furrows (b b) correspond to the pleural grooves of the thoracic joints, and the finer intermediate lines mark the sutures between the several segments of which the tail is compounded. The anterior groove (A) is generally the strongest. The margin (e) is mostly smooth and even, and often concave. At (d) the facet is seen on the front edge; (f) is the mucro, either short or prolonged into a spine, and often absent altogether.

These are not all the points of structure visible on the crust of the Trilobite. But it is desirable not to multiply terms. I prefer the ordinary appellations *head* and *tail* to the more correct designations, *carapace* and *post-abdomen*. The *axis* is a convenient and well understood term for the middle portion, and is better than *tergum;* and the terms *side-lobes, limb,* or *lateral portions* of the tail; *free* and *fixed cheeks,* &c. to the head, are sufficiently intelligible and well known to render it unnecessary to give the more technical designations "*epimera,*" *&c.*, even if we were quite sure of the correctness of all of these.

It is indeed quite possible that the free cheeks, with the connecting portion in front, constitute the first or ocular ring, and the rostral shield would form the sternal portion of the same ring. But there is not absolute certainty of this, and some naturalists disbelieve it. The rostral shield *may* be the only representative of a small anterior ring, and in that case we should be only inventing prematurely a new term, in deciding to call the first segment, as McCoy has done, the ocular ring. Nor would it be so convenient in description.

Again, the glabella certainly covers the region of the stomach, and the glabella-furrows doubtless mark the attachment of the muscles of the several segments proper to the head, and covered by the expanded carapace. There are always three (and in Ogygia and some other genera, four) of these lateral furrows; and if they indicate the number of segments combined in the carapace, instead of the number of thorax-segments covered by it, we should still be at a loss whether to call the neck-segment, which is always combined with the carapace, the fifth or sixth segment in the general plan. But by adhering to the terms "free cheeks," "facial" and "rostral sutures," "rostral shield," "glabella-furrows," "neck-segment," "tail," &c., while there is no false assumption in the terms, so neither are they too far removed from a scientific and technical nomenclature to be useful.

It may before long be practicable to adopt a more complete terminology. But in these descriptions I adhere to Dalman's formula, modernized a little to adapt it to our more advanced knowledge of the group.

FAMILY—PHACOPIDÆ. *Emmerich, Corda, Salter, &c.*

Eyes largely facetted, the cornea convex over each facet, forming a granulated, not a smooth eye. Facial suture ending posteriorly on the outer margin of the cheek. Thorax with eleven rings. Includes at present only one genus, PHACOPS. (*Sub-genera.*—TRIMEROCEPHALUS, PHACOPS, ACASTE, CHASMOPS, ODONTOCHILE or DALMANIA, CRYPHŒUS.)

Beginning, then, with the family of the Phacopidæ, as being on the whole the most typical and highly organized, I may remark that there is less difference between the various groups into which this natural family is divisible than between the various members of the neighbouring families. So much is this the case, that palæontologists have as yet been generally unwilling to break up this group into genera, or to consider its divisions as more than sub-genera of the great genus *Phacops*. Or, if they divided it, they have been obliged to include a greater variety of forms in some of the divisions than in others. *Dalmania*, or *Dalmanites*, is an example of this. It was intended by its author, Emmerich, to include only the broad expanded forms of the genus; but M. Barrande, whose authority has much weight, has widened its meaning so as to include all the forms which have distinct lobes to the glabella; thus including in *Dalmanites* both convex and flat forms, in fact, four-fifths of all the species; while the original term, *Phacops*, includes the rest.

Prof. Goldfuss had previously taken the same view, but applied the term *Acaste* to the larger group, leaving only the species with inflated lobeless glabella in *Phacops*. This view is a consistent one, for Dr. Emmerich, in founding his genus, gave these last as the type of it. But in the 'Neues Jahrbuch' for 1845, Dr. Emmerich objected to this plan, preferring to unite all the more compact and convex forms in *Phacops*, whether with lobed or lobeless glabella, separating only the more expanded forms, as *Dalmania*, a term which, though in general use, had unfortunately been in previous employ for a group of insects. Prof. Burmeister did not attempt to divide the group at all, and Prof. McCoy included all under *Phacops*, while he recognized truly most of the subgeneric groups.

It will be seen that there is a considerable diversity of opinion as to the value of the subdivisions; and this arises, I think, from the fact before noticed, that the various sub-genera in this, the highest group, differ only by characters of proportion and degree of development of the different parts, while the main features of the group remain constant.

The great characters pointed out first by Quenstedt, viz., that the Phacopidæ have eleven, and only eleven, rings to the thorax, while other groups are variable in this respect, and that all have the strongly facetted eyes, have in the eyes of naturalists overruled the minor distinctions, and disposed them to undervalue the real differences of proportion which exist. In the Decades of the Geological Survey, I have endeavoured to do justice to all the above distinguished authors; and, retaining the name *Phacops* for the whole, pointed out the several natural sub-genera. I believe still we shall best consult the convenience of students by retaining the common name.

No doubt, if we had the living animal, we should attach greater value to what in the fossil appear subordinate characters. The degree of development of the eye, for instance, should surely be a point of much importance in any group. The expanded form and large size of one division, contrasted with the contracted dimensions and compact habit of another, is certainly of consequence, and may well afford generic characters.

And when we find the caudal margin in one form even and compact, while another has the component pleuræ free at their extremities, simulating the character of neighbouring genera, we cannot doubt that we are looking at a group of genera combined by common characters into a natural family.

If I express my own opinion, that the sub-generic groups here given are natural genera in the ordinary sense of the word, and will be hereafter used as such—while for convenience sake, and till the sub-divisions are completely established, we adopt the common term, it will perhaps be all that in the present state of our knowledge is advisable.

Genus—PHACOPS, *Emmerich.* Characters those of the family.

Range. Lowest Silurian—to Uppermost Devonian.

Subgenus I. TRIMEROCEPHALUS, *M'Coy,* 'Annals Nat. Hist.,' 2nd series, vol. iv, 1849.

FIG. 5.

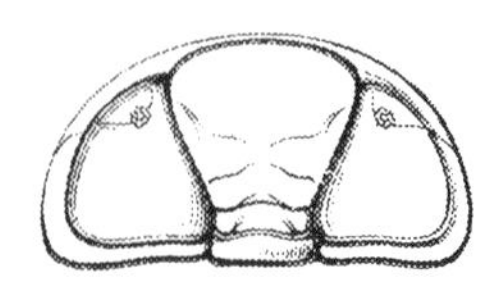

P. (*Trimeroc.*) *Volborthi,* Barr., Wenlock rocks of Bohemia.

Form compact, glabella inflated and expanded in front; the lobes, except the basal ones, obscure. Eyes small, of few large lenses (often ost by abrasion). Head-angles not spinous. Pleuræ all rounded. Tail small, of few segments, with even border, and not at all produced. Ex. *P. lævis, P. Volborthi, P. micromma,* &c.

Range. Upper Silurian—to Upper Devonian.

Subgenus II. PHACOPS, *Emmerich,* 1845.

Form compact, glabella inflated and expanded in front, the two front pairs of furrows obscure. Eyes large and well developed, of numerous lenses. Head-angles not spinous. Pleuræ all rounded. Tail moderate, of few (often coalesced) segments, with an even border, never produced.

Range. Upper Silurian—to Uppermost Devonian.

Subgenus III. ACASTE, *Goldfuss,* 1845.

Form compact, or at least not expanded. Glabella not inflated, nor much expanded in front; all the furrows distinct. Eyes well developed, of numerous lenses. Head-angles spinous. Pleuræ rounded or truncate, not produced into spines. Tail moderate, of less than eleven segments, with an even border, but often mucronate.

Range. Lower and Upper Silurian.

Subgenus IV. CHASMOPS, *McCoy*, 1849.

Form rather large and depressed. Glabella greatly expanded in front, the lobes unequal, the hinder ones being contracted, and almost obsolete; the front pair greatly expanded, and overlapping the others. Head-angles spinous. Pleuræ truncate. Tail large, of few or many segments, not dentate, and seldom mucronate.

Subgenus V. ODONTOCHILE, *Corda*, 1847. (DALMANIA, *Emmerich*, 1845.)

Form large, depressed. Glabella depressed, not much expanded in front; all the lobes distinct, the front ones not enlarged at the expense of the others. Head-angles long-spined. Pleuræ truncate, and the hinder ones often produced. Tail large, of more than eleven segments, with an even border,—often mucronate.

Range.—Lower Silurian, rare; Upper Silurian, common.

Subgenus VI. CRYPHŒUS, *Green*, 1837.

Form of moderate size, depressed. Glabella depressed, not much expanded in front; all the lobes distinct, the front ones not greatly enlarged. Head-angles long-spined. Pleuræ truncate, and the hinder ones often produced, into spines. Tail large, of many segments; the margin spinose

I believe there are more subgenera of *Phacops* to be discovered and described yet. For instance, there is probably more than one included here under *Trimerocephalus*, and there is a South African form of *Cryphœus*, which has the characters of *Acaste* in the compact habit and convex tail. All the other subgenera also contain species which it would be difficult to assign rightly to one or the other.

It will be observed that we have here a wide range of characters, and in regular gradation, from the most expanded, flattened forms, with glabella deeply lobed, and the segments of the large tail almost free, to the most compact and rounded form, with the caudal extremity reduced in size, of even contour, and with the segments most coalesced. We begin with the subgenus Trimerocephalus.

Subgenus—TRIMEROCEPHALUS, as above.

1. PHACOPS (TRIMEROCEPHALUS) LÆVIS, *Münst.*, sp. Pl. I, figs. 5, 6, 7.

ASAPHUS or TRINUCLEUS, *Sow.* Geol. Trans., 2nd series, vol. v, pl. 57, fig. 30, 1840.
TRINUCLEUS LÆVIS, *Münster.* Beiträge, Heft 5, t. x, fig. 6, 1842.
CALYMENE LÆVIS, *Phillips.* Palæozoic Fossils, pl. 55, fig. 250 (1841).
TRIMEROCEPHALUS LÆVIS, *M'Coy.* Ann. Nat. Hist., vol. iv, p. 404, woodcut, 1849.
— — *Ibid.* Synopsis Woodw. Mus., 1851, p. 178.

Diagnosis.—P. (*Trim.*) *late ovatus, lævis, glabellâ latissimâ, brevi, genas subsphæricas trigonas impendente, lobis basalibus minutis distinctis. Axis thoracis angustus, fulcro pleurarum axin approximato, sulcoque brevi. Cauda latissima brevis, axi longo* 5-*annulato, lateribus* 4-*sulcosis, margine nullo.*

This was first figured in England from very imperfect specimens, which came from the only English locality yet known, viz., the Knowl Hill, near Newton Bushel. They are mostly distorted, and, as first noticed by Mr. Pengelly, the head is usually disjoined from the body and inverted, as if the animal had habitually kept it bent under, and been preserved in the slate in that position. In that condition it is often difficult to distinguish the parts correctly.

But the number of specimens already collected enable us to restore the animal pretty completely; and there is not much doubt that it is identified rightly with the *Trinucleus lævis* of Münster.[1] Both Mr. Sowerby and Professor M'Coy have recognized it for this fossil, and Dr. Sandberger has quoted it as a synonym, only he mixes up two or three other species with it,—among others the (*Calymene*) *Phacops lævis* of Münster.

It is clear, however, that it is distinct from the small trilobite originally named *Calymene lævis* by Münster, and next described; as the following description will show.

And should it prove that the fossil called *Cal. lævis* by Münster is a real species, we should have to find a new name for this. Dr. Sandberger wishes to connect it with the *Phacops cryptophthalmus* of Emmerich, a fossil also occurring in the same Devonian formation. But though much disposed to believe that our fossil has true but superficial eyes, which are not easily preserved, I cannot think that this is a true identification. The eye of *P. cryptophthalmus* is very much more distinct, and I have figured what I believe to be that species, with the eye, a little further on. *P. cryptophthàlmus*, too, is described and figured by Dr. Sandberger as having a larger semicircular tail, while ours has a very short one. I may now describe the Knowl Hill fossil.

[1] From the red shales of Guttendorf. The head only is figured by Münster.

General form broad-oval, not very convex. Length occasionally $1\frac{3}{4}$ inch. Head smooth, semicircular, deeply trilobed, the lateral angles rounded. Glabella very broad in front, spherical-triangular, occupying much more than one third the width of the head; convex, but not gibbous, and overhanging the front; neck-furrow strong, basal lobe distinct, and with two lateral tubercles, the rest of the lobes obsolete. Cheeks triangular, evenly convex, with a narrow margin, which is strong at the rounded angles, and lost in front of the glabella; the neck-furrow strong. (Eyes absent in our English specimens, but probably present in perfect individuals.)

Thorax of eleven segments, with convex narrow axis and rounded pleuræ; the segments of the axis tuberculate at the sides; the pleuræ not much bent back, rounded at the end, the groove narrow and short, the fulcrum placed at less than half way out from the axis; facet rather large. Tail short, transverse, flattened, arched in front, straighter behind, about as long as the axis of the thorax is broad, and more than twice as wide as long, of few joints, the axis conical, and reaching nearly to the margin, blunt at the tip, and with four or five rings. The sides wide, with not above four furrows, which do not reach the margin, and are faintly interlined with other furrows.

Localities.—UPPER DEVONIAN. Knowl Hill, Newton Bushell; specimens figured from Mr. Pengelly's cabinet (figs. 5, 6), and Mr. Vicary's (fig. 7). I distrust the other South Devon localities given in the "Pal. Foss.," viz., Mudstone Bay and Durlstone. But it is probable the species occurs at Brushford, North Devon, as quoted by Professor Phillips.

P. CRYPTOPHTHALMUS, *Emmerich?* Pl. I, fig. 8.

PHACOPS CRYPTOPHTHALMUS, *Emmr.*, in Leonhard und Bronn's Jahrbuch., 1845, pp. 27, 40, &c.
— — *Roemer.* Palæontographica, vol. iii, pl. vi, p. 14 (bad figure), 1854.
— — *Sandberger.* Verst. Rheinisch. Schicht. Syst., t. i, fig. 6 (exclude his synonyms, as he includes several species, among others, the *P. lævis* figured above), 1850.
PHACOPS LIMBATUS, *Richter?* (*fide* Sandberger).

"*P. capite semiorbiculari, lateribus frontis rectilineis, ad angulum acutum convergentibus. Annulus fere rectilineus. Oculi parum evexi. Thorax latus. Pygidium breve obrotundatum, ex articulis* 8, *pseudopleuris* 5, *compositum. Superficies subtilissimé granulata.*" Sandberger.

I suppose this to be the species given in Sandberger's beautiful plates. The character, "eyes but little prominent," well agrees with this species. He figures the eyes as lunate, and with fewer lenses than our Newton specimen. But the shape of the glabella is the same, and I do not see that there is much room for doubt.

Locality.—UPPER DEVONIAN: Newton Bushell (Mus. P. Geology).

Subgenus—PHACOPS, as above.

2. P. (PHACOPS) GRANULATUS, *Münster*. Pl. I, figs. 1—4.

CALYMENE, sp., *Sow.* Geol. Trans., 2nd series, vol. v, pl. 54, figs. 23, 24, 1840.
— GRANULATA, *Münst.* Beitr., Heft 5, t. v, fig. 3, 1842.
— LÆVIS, *Id.* Ib., t. v, fig. 4.
— GRANULATA, *Phillips.* Pal. Foss., fig. 248 (exclude figs. *m, n, o, p*), 1841.
PORTLOCKIA GRANULATA, *M'Coy.* Synopsis Woodw. Foss., p. 177, 1851.

Not an inch long. General form broad-oval. Head semicircular, very convex, covered with granules, the sides much bent downward. Glabella fully half the width of the head, very tumid, slightly pointed in front and overhanging the front margin. Its greatest width exceeds its length, even including the neck-segment. Sides converging at an angle of 90°, the base narrow; a small basal lobe, with a tubercle on each side, but no trace of upper furrows. Cheeks moderate, with a strong border. Eyes large, prominent, with few, about thirty-six lenses, five in a row. (M'Coy.)

Thorax ?—

Tail semicircular, small, convex, with a smooth, declining, distinct margin, and prominent conical axis, reaching rather more than two thirds down, tapering and almost pointed at the extreme end, which fades into the limb, marked with six or seven rings, obscure at the tip. The lateral lobes have five or six strongly duplicate furrows.

There seems no good reason for separating Münster's *C. lævis* from the above. It is merely a decorticated specimen, the surface-granulation consequently absent. The tail, as figured by Münster, has too many ribs, and too long an axis, but is otherwise like ours; and Münster's figures are not fully to be trusted.

Locality.—UPPER DEVONIAN. Petherwin, Cornwall. (Our best specimen, fig. 1, is from Mr. Pengelly's cabinet; the others from the Mus. P. Geology.)

P. (PHACOPS) LATIFRONS, *Bronn.* Pl. I, figs. 9—16.

CALYMENE LATIFRONS, and C. SCHLOTHEIMII, *Bronn.*, in Leonhard's Zeitschr. f. d. Miner., 317, t. ii, figs. 1—8, 1825.
CALYMENE LATIFRONS, *Bronn.* Lethæa. Geogn., t. ix, fig. 4, 1835.
— TUBERCULATA, *Murch.* Sil. Syst., pl. xiv, fig. 4, 1837.
— LATREILLII, *Steininger.* Mémoires Soc. Géol. France, vol. i, pl. ii, 1834.
— SCHLOTHEIMII, C. BRONGNIARTII. Ibid., p. 350, 351.
— LATREILLII, *Phillips.* Pal. Foss., fig. 249, 1841.
— ACCIPITRINA. Ibid., p. 128.
PORTLOCKIA LATIFRONS, *M'Coy.* Synops. Woodw. Mus., 1851, p. 177.
PHACOPS LATIFRONS, *Sandberger.* Verstein. Rheinisch. Schichten-Syst. Nassau, t. i, fig. 7, 1850.
— — *Roemer.* Palæontographica, vol. iii, t. ix, figs. 24, 25, 1854-1855.

General form.—A large species; foreign specimens often attaining a length of $2\frac{1}{2}$ inches, and some of our English fragments indicate a still larger size.

The head occupies fully one third of the whole length, and is rather more than a semicircle, and very convex. The inflated glabella occupies more than half the width, taking its measure at the wide front, from which the straight sides converge at an angle of 85°. It is about as wide as its whole length, including the neck-lobe. The glabella is very much rounded in front, and scarcely overhangs the narrow linear margin. Its whole surface is covered with large, coarse tubercles, at equal distances, scarcely more than their diameter apart. There are but slight traces of the upper furrows; but the tumid glabella is strongly separated from and overhangs the linear basal lobe, which has not distinct lateral tubercles, but in the internal cast shows deep pits on either side, above and below the lobe. The neck-segment is strong and broad, wider than the basal lobe. The neck-furrow is continued round the smooth cheek, and separates a broad, strong margin, leaving a subtrigonal space, much of it occupied by the great eye, which varies from half to more than half the length of the cheek, and is placed rather behind the middle of it. The eye, from the depression of the upper eye-lobe, is subhemispheric, not greatly curved, but strongly convex exteriorly, and covered by about fifty-four strong, prominent large lenses,—in rows shortening towards either side; about five in one of the central vertical rows. Externally the lenses are very convex.

Foreign specimens show us that the cornea of this species is very thick, and rises into ridges between the lenses; and it apparently thickens by age internally. At least old specimens (fig. 9) have the substance much thicker, as shown by the great projection of the casts of the supporting cups (fig. 12*), which of course are the spaces occupied by the soft substance which lay underneath the lenses.

In some instances, probably in younger individuals, the projection of the cups is considerably less (fig. 13). In others again, the cups project so little, and the cornea is so thin, that there is little difference of level between the ridges and the cups (fig. 11*a*). I do not think these differences of proportion, nor even the varying numbers of the lenses in different individuals, at all tend to constitute distinct species. Steininger gives us 46 to 50 lenses in one variety, 87 in another, and 130 (probably for the two eyes) in a third. But he has not pointed out any clear distinctions in the species to accompany these differences in the eye. A specimen of *C. bufo*, Green, which is a closely allied form, and may *possibly* be only an extreme variety of our species, has 66 lenses in each eye.

The eye is elevated, the lentiferous surface not sunk in a furrow, but standing prominently out from the cheek, and overhanging its own base (fig. 11). Head-angles rounded. Obscure traces of the facial suture occur below the eye, but practically they are soldered, and the head does not part at the sutures at all.

Thorax (in German and Spanish specimens) with the pleuræ much bent down, so

---

* These cups in the cast of a Trilobite's eye occupy the place of the vitreous body, according to Dr. Burmeister's explanation of the eye-structure. 'Organiz. of Trilobites,' Ray Soc. edit., pl. vi, fig. 4 *d*.

that the axis looks nearly as wide as the pleuræ. It is not in reality nearly so broad. The surface is coarsely granulated, as is that of the tail, when the crust is preserved.

There is a perfect body, with eleven rings, in Mr. Vicary's cabinet, of which a woodcut is here given. The axis is convex, semi-cylindric, tubercular, very slightly nodular on the sides, and little more than half as broad as the pleuræ. These are very convex, and as steeply bent down as in Calymene; the fulcrum is placed at about one third out, and as far from the axis in the hinder as the front rings. The pleural groove is not very deep, and does not reach the long facets. The ends of the pleuræ are recurved and rounded.

Fig. 6.

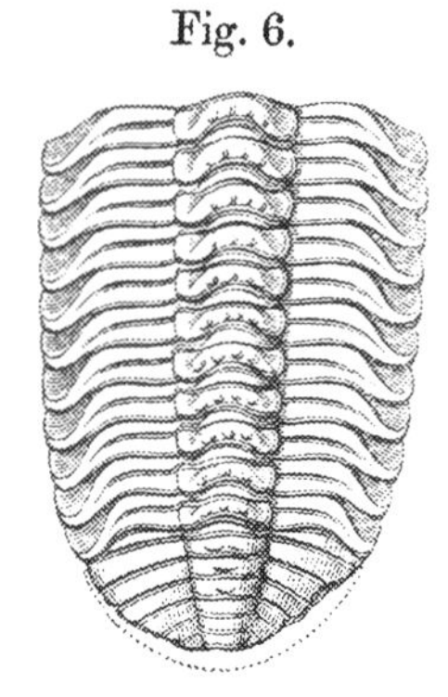

Body and tail of *Phacops latifrons;* Newton Bushell.

Tail in our specimens (figs. 14, 15,) wide, more than semicircular, arched in front, and very convex, with the axis quite elevated, and separated by broad furrows from the convex limb. A few tubercles show on the ribs of the axis. It projects in front, slowly tapers to a rounded end at one sixth from the margin, and is strongly annulated by seven or eight rings (ten in some German specimens) while the convex sides slope quickly down, and have five or six strong ribs, straight, or but little curved, and all but reaching the indistinct margin. There is no flat border. The axis varies in breadth, but is never so much as one third the whole width, usually one fourth. Incurved under margin very convex.

Localities.—LOWER DEVONIAN. Hope and Barton, South Devon; near Liskeard and Totnes, in slates with *Pleurodictyum problematicum.* UPPER DEVONIAN (Petherwin Group)? Newton Bushell. UPPERMOST DEVONIAN (Barnstaple Group). Croyde, Barnstaple, and the neighbourhood; Brushford, Pilton, &c., abundant; Yealm Bridge, north of Launceston (Pattison and Salter).

Foreign localities.—Rhenish Prussia, Belgium, France, Russia, (*P. bufo* takes its place in North America); also the Andes, South America (Mus. Geol. Soc.).

Var. Pl. I, fig. 16.

I figure as a variety the wide pygidium found at Newton Bushell, which differs from the ordinary forms of *P. latifrons,* by the narrower proportion of its more depressed axis to the sides,—little more than one fourth, and in having the six side ribs more direct than usual. It can only be a variety, but the aspect is different to that of the common form.

Locality.—UPPER DEVONIAN.* Newton Bushell.

* I consider the upper quarries of Newton Bushell as the equivalent of the Petherwin beds. This subject requires minute examination, and will be determined chiefly by the labours of local geologists. The upper limestones are clearly not of the same age as the great Plymouth limestone, but nevertheless, they contain a good many of its fossils.

P. (Phacops) Stokesii, *Milne-Edw.* Pl. II, fig. 1—6.

Asaphus macrophthalmus, *Brongniart.* Crustacés Fossiles, pl. i, fig. 5 (*not* fig. 4), 1822.
Calymene macrophthalma, *Buckland.* Bridgw. Treatise, pl. lxiv, fig. 4 (*not* fig. 5), 1836. *Murchison,* Silur. System, pl. xiv, fig. 2, 1837.
— Stokesii, *Milne-Edwards.* Crustacés, 3, 324, 1840.
Portlockia sublævis, *M'Coy.* Sil. Foss. Ireland, pl. iv, fig. 13, 1846.
Phacops Stokesii, *Salter.* Memoirs Geol. Survey, vol. ii, pt. 1, pl. 5, fig. 1, 1848. Quart. Geol. Journal, vol. vii, pl. ix, fig. 2, 1850.
— — *M'Coy.* Synopsis Woodw. Mus., p. 163, 1851.
— — *Salter.* Siluria, 2nd edit., pl. 18, fig. 6, 1859.

P. (*Phacops*) 1—2 *uncias longus, ovatus, granulatus. Glabella parum inflata, frontem impendens, sphærico-trigona, lobis basalibus rotundatis, reliquis omnino* (*etiam intús*) *obscuris. Oculi magni, insuper depressi. Thorax axe angusto. Cauda brevis semicircularis, convexiuscula, immarginata, axi depresso* 5-*annulato, lateribus* 4—5-*sulcatis, sulcis abbreviatis obscuris.*

This is a very common Upper and Middle Silurian fossil, but is never found in Lower Silurian Rocks. Yet it is but imperfectly known, being generally in fragments. And it has been moreover a troublesome one to the palæontologist, owing to the original mistake of Brongniart, in confounding it with the true *macrophthalma.* And, until lately, specimens of the more common *P. Musheni,* our next figured species, were constantly mistaken for it. I believe the above synonyms may all be trusted, and I do not cite the more doubtful ones.

The species is rather a small one, seldom exceeding an inch or an inch and a half in length, and is obtusely ovate and very convex. The head occupies less than a third and more than a fourth of the whole length, and is semicircular, and wider than the thorax; the glabella is an equilateral spherical triangle, overhanging the front and the eyes, and much narrowed behind, where the basal lobes are distinct enough as small tubercles on each side of a linear lobe. This is overhung by the inflated upper lobes, which are confluent, and show only very obscure traces of any of the upper furrows, even on the inner cast, where they are strongest. The median furrows are short and curved, the upper strongly bent midway, as in many species of the subgenus *Phacops.* Surface of glabella covered with a fine granulation.

The cheeks are triangular, with a strong, continuous, marginal furrow. The eyes large, much depressed above, and with the lentiferous surface gently curved, of numerous lenses, and so placed that the forward end nearly touches the glabella, while the base is widely distant, and does not moreover reach the sharp neck-furrow.

The body has a convex axis, narrower than the pleuræ, which have a deep sharp groove for two thirds their length, and are steeply curved down beyond the fulcrum, the latter being placed less than half-way out in the front rings, and at one third behind.

The tail is semicircular, gently convex, but with the axis depressed and flattened; it tapers slowly, has a blunt apex, and is marked by five or six segments. The side-lobes have four or five abbreviated narrow furrows; the upper ones are interlined, but all are inconspicuous.

Our larger figure (fig. 6) is from a doubtful locality in the Wenlock Rocks, but is certainly British. It shows the internal cast of the head sufficiently well. The species grows occasionally to a rather large size, as may be seen by this specimen, and others in the cabinet of Mr. Hollier, of Dudley. But usually the specimens are not larger than our other figures. *P. sublævis* of M'Coy is a very obscure drawing, but the original specimens do not differ from *P. Stokesii*.

Localities and Geol. Range.—LLANDOVERY ROCK, Galway; Ayrshire; Haverfordwest, Pembrokeshire. MAY HILL SANDSTONE, of Tortworth; May Hill; Malvern; Shropshire. WENLOCK ROCKS; Abberley; Malvern; Dudley; Wallsall; North and South Wales; Dingle, West of Ireland; Peebleshire. LUDLOW ROCKS; near Leintwardine, Shropshire; Pentland Hills, Edinburgh.

P. (PHACOPS) NUDUS, n. sp. Pl. VI, figs. 19, 20.

P. parvus, convexus, omnino P. Stokesii simillimus, nisi genis profunde marginatis, oculorum lentibus paucis, caudâ rotundiore, lateribus inflatis, costis distinctioribus.

It is difficult to define the fragments of this species, otherwise than by comparing the parts with the better known and more perfect *P. Stokesii*, from which the species, a really distinct one, differs in the following characters—the general form and the shape of the glabella being extremely like in both cases.

The cheeks are larger, more rounded, and less convex, rather suddenly raised, and with a tumid space between them and the glabella (*d*). The lenses are far less numerous, and have granules in the interspaces (*f*). The tail (fig. 20) is more oblong than a true semicircle, transverse, blunt, and is more depressed. The axis is short, with an obtuse flattened tip, and has six rings; the lateral lobes tumid, with four arched furrows (including the uppermost one), and a very obscure fifth furrow. They do not nearly reach the margin, and are faintly interlined. The tumid sides, being rather strongly divided from the flattened axis, give a peculiar character to the tail, very different from that of *P. Stokesii*.

This new form has unexpectedly turned up in the collections made by the Irish Survey in the wild district of the Dingle Peninsula. In the mountain of Cahirconree, among slates of decidedly Upper Silurian date, occur some limestones, of whose age we are not so clear, but containing the present species, which belongs to an Upper, and not a Lower Silurian group. I have not much hesitation in referring them to the Wenlock, or more probably the May Hill, formation. The *Illænus Barriensis, Spirifer plicatellus,* and other Upper Silurian species, occur with them.

Locality.—WENLOCK or MAY HILL group. Cahirconree Mountain, west side, Dingle. [Mus. Irish Industry, B. 242, 243.]

P. (PHACOPS) MUSHENI, n. sp. Pl. II, figs. 7—12.

P. parvulus, vix ¾-unciæ longus, ovatus, lævis. Glabella oblonga, haud inflata, superne urceolata, postice contracta, lobis basalibus profunde scriptis bituberculatis, reliquis connatis lineisque angustis modo sejunctis; sulci mediani breves, curvi; superiores fracti. Lobus frontalis transversus oblongus. Thorax axe convexo ut pleuris lato, his lentè recurvis fulcro prope axin posito, sulcoque pleurali angusto distincto brevi. Cauda semiovalis convexa immarginata, axe prominulo distincto pauci-annulato, lateribus 3—4-sulcatis abbreviatis angustiori.

A much smaller species than the true *P. Stokesii,* with which it has been very generally confounded. Nevertheless it differs by several important particulars, the chief of which is that the shape of the head is long instead of broad, and the glabella decidedly oblong instead of broad-triangular. The other portions, body and tail,—also differ; the axis of the body is broader, and that of the tail much more prominent, while the tail itself is of a longer shape, less transverse.

The length is seldom more than three fourths of an inch. The head in good specimens is half a broad oval (that of *P. Stokesii* being a semicircle). The glabella is much more than one third the whole width, oblong, only rather broader above than at the eyes, and is urceolate, the sides bulging out between the eyes, and then contracted for the hinder portion. The glabella is not at all inflated, nor does the front overhang in any sense, and the facial suture is not even quite marginal in front.

The eyes are fully half the length of the glabella, and very large and convex, not depressed above, about as deep as they are broad, and containing nearly 130 lenses. These have no visible spaces between them. The eyes are not very prominent, and on a side-view appear sunk, especially towards the front, into the cheek. They vary a little in size and prominence, but not very much so.

The body has a rather convex axis, which is as broad as the sides, or nearly so. The pleuræ are curved down gently at the *fulcrum,* which is at about one third anteriorly and less behind. The pleural groove is narrow, and reaches but little more than half-way along the pleuræ.

The tail is longer than semicircular, very convex, smooth, with a prominent axis (not flattened as in *P. Stokesii*), and reaching fully four fifths the length. It is long, conical ringed above, and smooth towards the bluntish point; the sides are smooth, the three or four faint lateral furrows not reaching half-way over the convex limb, even in the upper portion, and being quite obsolete in the lower. Only the upper ones are interlined. There is no marginal flattened space, nor any concavity; the tail is gently convex to the very edge.

Locality. Wenlock Shale and Limestone.—Malvern, abundant in Wenlock Shale; Dudley. It has not yet been found in other localities.

Subgenus—ACASTE.

PHACOPS (ACASTE) DOWNINGIÆ, *Murch.* Pl. II, figs. 17—36.

CALYMENE MACROPHTHALMA, *Brongniart.* Crust. Foss., pl. i, fig. 4 (*not* fig. 5), 1822.
— — *Buckland.* Bridgwater Treatise, pl. lxiv, fig. 5 (*not* fig. 4), 1836.
— DOWNINGIÆ, *Murchison.* Silurian System, pl. xiv, fig. 3, 1837.
— — *Milne-Edwards.* Crust., iii, 324, 1840.
ASAPHUS SUBCAUDATUS and A. CAWDORI, *Murchison.* Sil. System, pl. vii, figs. 9, 10.
ACASTE DOWNINGIÆ, *Goldfuss.* Syst. Uebersicht der Trilob., Neues Jahrb., 563, 1843.
PHACOPS MACROPHTHALMA, *Burmeister.* Organiz. der Trilob., 139, 140, 1843, ed. 2 (Ray Society), 1846, p. 92.
— DOWNINGIÆ, *Emmerich.* Neues Jahrb., 1845, p. 40, pl. i, fig. 2 (*icon mala*).
— — Translated in Taylor's Scientific Memoirs, 1845, vol. iv, pl. iv, fig. 2.
— — *Salter.* Memoirs Geol. Survey, June, 1848, vol. ii, pt. i, p. 336, pl. v, figs. 2—4; Decade vii (1853), pl. i.
— — *M'Coy.* Synopsis Pal. Foss. Woodw. Mus., 160, 1851.
— — *Murchison.* Siluria, 2nd edit., pl. 18, figs. 2—5, 1859.

P. vix biuncialis, alutaceus, margini frontali capitis angulato. Glabella depressa subparallela, sulcis utrinque tribus distinctis, lobo basali lineari, secundo ovali, superiori transverso, sulco antico ascendente sinuato; lobis omnibus planis, fere ad medium glabellæ, spatio angusto interjecto, extensis; cervice elevato. Oculi modici. Cauda subtrigona, marginata, apice angulato; axi convexo costato, costis 5 *distinctis prædito; lateribus* 5-*costatis, costis duplicatis.*

If I have given a lengthy set of synonyms, they do not represent a moiety of the works in which this very common fossil is noticed. It is one of the really abundant Upper Silurian species, being moreover one of the few trilobites which are common in the Ludlow rocks, as well as in Wenlock strata. No trilobite is more frequent on the Dudley slabs; but it is rare to find it in the underlying shale. I have seen it from many parts of Britain, but do not know that it has ever been described from foreign localities; nevertheless it does occur in the true Upper Ludlow rocks of Nova Scotia, as we learn from the collections brought to the International Exhibition by the Rev. D. Honeyman, in 1862.

An inch and a half long; general form long, ovate, broader in front, the axis following the same lines, and regularly tapering towards the tail. The surface is moderately convex, the axis raised above the sides, not separated by deep furrows except in the head, and more convex in the thorax than in the head or tail.

The head is less than a semicircle, though just twice as long as broad; the general outline rather triangular, from an indentation in the outer margin on each side of the glabella; the front is not produced, but angular. The glabella occupies more than one third the width of the head in front. It has nearly straight, parallel sides, and rises considerably above the cheeks, but is depressed rather than convex, especially the forehead-lobe, which slopes gradually to the narrow marginate front. Neck-lobe strong, broader than the first basal lobes, which are transverse and linear; the middle pair are broader than these, and oval. The lower furrow bends downwards, and reaches the side of the glabella; the middle one is abbreviated, and curves the reverse way. Upper lateral lobe transverse, scarcely triangular, bounded above by a sigmoid furrow, which runs obliquely out above the eye.

All the furrows stretch equally towards the middle of the glabella, leaving but a narrow space between their ends; between the upper pair a short longitudinal depression occurs. The lobes are not swelled between the furrows, but the surface is even, and the furrows shallow (they are, however, sharply defined on the internal cast), the neck and basal furrows strong; the upper ones fainter.

The cheeks are steeply bent down, their outer margin not distinguished by any furrow, and they slope gradually from the eye, without any ridge or groove, beneath the latter; the neck-furrow is continued almost to the angle, which is rounded off, and has a tubercle only in place of a spine. The facial suture cuts the outer margin in a curved line opposite the base of the eye; on the under surface of the head the suture cuts the margin further backward (fig. *b*). In front of the eye it runs along the axal furrow and round the glabella just outside the marginal furrow. It is thus what is called *intramarginal.*

Eyes moderately large, in some specimens (figs. 18, 25, 26, 30) much larger than others; and they rise occasionally to the level of the glabella, but are generally lower. The eye is not very prominent; it is placed half-way up the cheek, near to the upper glabella-lobes, and occupying their length; eye-lobe with a raised outer margin; lentiferous surface broad, with about 155 lenses in the eye, each vertical row containing eight. The cornea is convex over the lenses, and the intermediate flattened spaces are finely granular, the granules forming a rough hexagonal network toward the base of the eye; the lenses are nearly their own diameter apart, but this varies much in different individuals, the space being often much less. (Fig. 36, *d*, *e*.)

On the under side of the head the incurved front portion, or *hypostome*, as in all the genus, is continuous across; it is broad (fig. 36, *b*) and granular, like the upper surface. The labrum is also granulate; it is subquadrate, broadest at base, regularly and strongly convex; a faint concentric furrow runs round its sides and tip, just indicating a narrow margin more flattened than the other parts; there are no lateral furrows, but high up on each side is a small tubercle. The tip is obtusely truncate, with no visible serratures. The labrum is narrower than the glabella, and half its length; but from the position of its base its tip reaches back as far as the middle pair of glabella-furrows.

Thorax considerably longer than the head; the rings not very convex; the axis of nearly equal breadth with the pleuræ. These are traversed by a straight deep groove (fig. 36 *g*), curved rather abruptly down at the fulcrum, which anteriorly occurs at the inner third of their length, and in the posterior ring does not reach further than one fourth. The anterior edge of each pleura is sharpened or facetted to pass under the preceding one, and the posterior edge is thick.

Each pleura is bent forward at its end, which is notched somewhat deeply, and on the *under* side of each, in front of this notch, is placed a tubercle. When the animal was in the act of rolling up, the tubercle served as a buttress to prevent the posterior ring from being pushed too far forward; the tail, too, has similar tubercles on its anterior edge.

The under side, cleared out with great labour and patience by Mr. John Gray, of Hagley, shows two or three other interesting points distinctly.

First. The incurved under portion, which is very narrow in the tail, and not much broader beneath the border of the head. But along the pleuræ a wider strip is turned inwards, which is smooth in this genus, and has a straight inner edge; while it presents on its forward margin the tubercle before mentioned.

The interior ridges of the axis show distinctly along the axal line as short, transverse ridges, more prominent a good deal on their inner margin, which does not even show, except as a slight depression, on the upper side. This broad ridge does not extend into the tail portion.

Var. α, VULGARIS. Pl. II, figs. 17—25.

Figs. 22, 24, show about the ordinary form of the species; with broad glabella, complete glabella-furrows, and pointed front. The tail-furrows are strong, and the apex pointed. The axis of the body, too, is prominent.

Varieties.—Fig. 24 shows some tendency in the narrower glabella towards the variety *constrictus*, figured in the upper part of the plate. It has, however, very large eyes, like the var. *macrops*, but the tail wants the lateral furrows, or rather has them much slighter than usual, and the apex less pointed. The body-axis is narrow and prominent.

Fig. 23, a very large individual, is more convex than usual, and the eyes are depressed so as not to rise nearly to the level of the glabella. The tail is of the ordinary type.

Var. β, MACROPS. Pl. II, figs. 26—29.

The variety *macrops* is distinguished by the very large prominent eyes:—they occupy a large part of the cheek. I counted in one of them 140 lenses. The interspaces are granulated. Fig. 28, which belongs to this marked variety, has a larger head than usual, and the tail has the side-furrows more than usually obsolete. Mr. Ketley's specimen, (fig. 27) has the front glabella-furrows strong, and the pleuræ flattened. The tail is of the usual type, but the side-furrows are obscured.

Var. γ, INFLATUS. Pl. II, figs. 30, 31 (32, 33?).

Fig. 30 shows a remarkable variety. While it retains most of the characters of the ordinary form, it nevertheless puts on a very different aspect. The glabella is greatly swollen, so as to be very convex, instead of flattened, in front. All the furrows are indeed distinct and in their proper situations, but from the inflation of the glabella they appear crowded. The eyes are small. Dr. Grindrod's cabinet and the Museum of Practical Geology are the only collections which I know to contain this variety. Probably fig. 31, a Ledbury specimen, belongs to var. β. It has the upper glabella-furrows all but obsolete.

On the other hand, figs. 32, 33, also from Ledbury, show all the furrows of the head and tail stronger than usual. These are casts of the interior, and the thickening of all the internal ridges is a constant character in Trilobites.

Var. ? δ, SPINOSUS. Woodcut, fig. 7.

Agrees with the ordinary variety α in the glabella, but has short head-spines! Only a single specimen is known; it is in Mr. Edgell's collection. With it, however, occurs a pointed and strongly furrowed tail, which may belong to it, and would mark it as a very distinct form; possibly a species.

FIG. 7.

Phacops Downingiæ, var. δ, *spinosus*. Upper Ludlow rock, Ludlow.

But to pass on to a marked and definite variety, which has not yet received a name, and which would by many be considered a distinct species. I propose, however, only to term it—

Variety or Sub-species[1] ε, CONSTRICTUS.

PHACOPS CONSTRICTUS. Pl. II, figs. 13—16.

Minor, fronte convexiori rotundato haud angulato, oculis magnitudine variis, prope glabellam positis. Cauda rotundata, sulcis obscuris.

[1] I am much disposed to follow the plan adopted by some eminent botanists, and to make a distinction between the occasional variations in form and structure which are usually termed *varieties*, and those more permanent and well-defined groups, which are designated *sub-species*. Regarding these latter, there will always be differences of opinion as to whether they should receive separate specific names, and it seems the most convenient plan to describe them as distinct forms, which may be considered either species or varieties by the student, while their supposed relation to the parent species is indicated by their being grouped under it. We do not yet know what the limits of species and varieties are, and probably there is no real line to be drawn; but all truly distinct forms should receive attention, and, for the purpose of the geologist especially, a marked variety is as useful as a species.

This is smaller than the typical variety, seldom more than one inch long, and has a different aspect, from the rounding of the front and the greater convexity of the forehead-lobe. The glabella has nearly parallel sides, as in the typical form; the front furrow is not stronger than the rest, but is rather more sigmoid. The middle furrows reach the side of the glabella; the neck-segment is prominent and has a distinct tubercle. The eyes are small, with about 130 lenses, and placed very near the glabella.

The axis of the body is convex; the pleuræ more tumid between the grooves than usual. Some young specimens have larger eyes in proportion, and the front furrows obscure.

But these variations with larger or smaller eyes, more distinct or less distinct glabella-furrows, &c., occurring as they do both in the angulated and rounded varieties, oblige me to consider them as of the same species, though the characters above given show that we are dealing with a very distinct variety or *sub-species*.

FIG. 8.

Phacops Downingiæ, var. *cuneatus*, from Llanrwst near Conway.

There is yet another variety, which might be called var. ζ, *cuneatus*. Fig. 8.

Localities and Geol. Range.—The ordinary variety α is found from the May Hill Sandstone to Upper Ludlow Rock. MAY HILL Rocks; Pembrokeshire; Norbury and Bogmine, Shropshire. WOOLHOPE LIMESTONE AND WENLOCK SHALE; Malvern, many localities; Burrington, Shropshire; Usk, Monmouthshire; near Llanrwst, and many places in the Denbighshire grits, N. Wales. Pembrokeshire and Carmarthenshire, S. Wales. WENLOCK LIMESTONE of Dudley, Wallsall; Benthall Edge; Malvern; Abberley, &c. LUDLOW ROCKS, LOWER AND UPPER; Shropshire; Carmarthenshire; Pembrokeshire.

The varieties β, γ, above described, are as yet only known from the Wenlock Rocks; variety γ only from the Wren's Nest and Malvern; but the variety δ is from UPPER LUDLOW strata, Whitcliff, Ludlow (Mr. Edgell's cabinet); and the species or variety ε, *constrictus*, only from the Wenlock shale, of Dudley, Wallsall, and especially Malvern. Var. ζ is from the Denbighshire grits of Llanrwst, near Conway. (Woodw. Mus.)

Foreign distribution.—Nova Scotia, in Upper Ludlow rocks.

P. (ACASTE) APICULATUS, *Salter*. Pl. I, figs. 36—38.

PHACOPS APICULATUS, *Salter*, in Prof. Sedgwick's Synopsis Classific. Pal. Rocks, fasc. 2, Appendix iii, pl. i G, figs. 17—19 (1852).
PORTLOCKIA APICULATA, *M'Coy*. Ibid. (1851), fasc. 1, 162.
PHACOPS APICULATUS, *Salter*. Memoirs Geol. Surv., Decade vii, art. 1, p. 9. (1853.)
— — *id.* Siluria, 2nd ed. (1859), p. 75, Foss. 13, f. 2.

P. (Acaste) omnino P. Downingiæ simillimus, sed capite longiore. Glabella elongata antice convexior, lobis basalibus circumscriptis subtrigonis nec transversis; sulco mediano longiore, supremo distincto. Oculi elongati depressi. Anguli capitis brevissime mucronati. Cauda ad apicem compressa et in apiculum recurvum brevem producta, axi angusto.

Heads and caudal pieces of this small species are not uncommon in the Caradoc rocks of Wales and Shropshire, but are rare in the Llandeilo flags. It is a smaller fossil than *P. Downingiæ*, though much like it, and is readily distinguished by the faint upper lobes of the glabella, and the strongly apiculate or shortly mucronate tail. The entire fossil could not have been above one and a quarter inch long, and is rather depressed than convex.

Head semicircular (smooth?), the front bluntly angulated, or, rather, like a broad Gothic arch. The glabella is long and parallel-sided, but slightly broader in front, and occupies less than half the width of the head. There is scarcely any margin in front of the forehead-lobe, the sides of which, above the eye, slope into the cheek without distinct axal furrows at that point. These are indeed but faint throughout, but are nearly parallel along the two upper lobes, and the basal lobes and neck-segment are scarcely narrower than the upper lobes. The neck-segment is strong and prominent; the basal lobes are transverse-oval, and deeply circumscribed, except on their inner margin. The uppermost furrows straight, oblique, deepest just over the eye, and thence extended nearly to the centre. The middle furrow is very faint, and gently arched upward: it nearly meets the margin, but is so faint that the two upper lobes appear like one.

The eye is small and conical, not much curved, and covers the space of the two upper lobes, leaving a wide space between it and the sharp neck-furrow. There is no marginal furrow to the cheeks on their outer sides, by which character it is easily distinguished from the following species. There is a very short mucro to each of the head-angles.

The tail is broad-triangular, and strikingly recalls that of *P. Downingiæ*. The largest we have is seven lines wide and five long, without the short recurved apiculus. The axis rather narrow, not nearly equal to the limb, regularly conic, ribbed by eight or nine distinct ridges, and the smooth terminal portion is pinched up, as it were, into the short apiculus. The axal furrows are not strong; the sides are convex, declining, and with a broad concave margin, not crossed by the side-furrows, which are five in number, and are deep narrow grooves, interlined pretty strongly on the cast (fig. 37), but less distinctly on the outer surface (fig. 38). Our figures do not express this as well as they ought to do.

Localities.—Upper Llandeilo flag. Treiorwerth, near Llanerchymedd, Anglesea. Caradoc Sandstone. Soudley, Horderly, and many other places in Shropshire; abundant; Cerrig-y-Druidion; Bala Lake; Llangollen; Meifod; Conway River; Pwllheli, &c., in North Wales; Coniston Water and Troutbeck, in Westmoreland.

Phacops (Acaste) mimus, n. sp. Pl. I, fig. 35.

P. (Acaste) minor, capite elongato (angulis obtusis?). Glabella subparallela, haud convexa, lobis superioribus obscuris, basali distincto. Oculi submediani, modici. Genæ lati-marginatæ, sulco antico exarato.

Though at first sight a good deal resembling *P. apiculatus* (p. 28), this little trilobite is found to differ when closely examined. Its glabella is much the same in shape, but rather broader above; the basal lobes not so strongly marked, especially in the cast; the upper furrows less defined, and making a less indentation on the sides; the upper eye-lobe broader and flatter. The cheeks, without being smaller, project less at the (blunt?) angles and more at the sides, and the eye is a little further forward.

But the most striking character, although it seems to be an unimportant one, is the very distinct marginal furrow which subtends the eye closely in front, and leaves a strong, broad, anterior margin to the cheek. It does not, however, quite meet the neck-furrow, which rises towards it and runs nearly to the angle.

This apparently neutral character is nearly the only one on which we can rely; the rest are proportional characters. But the species is a distinct one; and although it is almost too imperfect to name, the little Lower Silurian oasis in Cornwall, from whence it comes, is of so much geological interest, that I venture to distinguish it. As the fossils of that area are wholly dissimilar from those of other portions of the British Silurian rocks, and only comparable with those of the "May" Sandstone of Normandy, it is worth while to note the discrepancy by describing even the more obscure species.

Calymene Arago, De Verneuil, and *Homalonotus Brongniarti*, Deslongschamps, accompany it, and will be described under their appropriate genera. The same set of fossils is found in the remarkable pebble-bed at Budleigh Salterton, in South Devon. See description of the next species.

There is a certain resemblance to *Dalmanites Phillipsi*, Barrande, and even to *Dalm. Hawlei*, Barr., from the Lower Silurian rocks of Bohemia. But *P. mimus* is sufficiently distinct from all.

Localities.—LLANDEILO FLAGS ? of Great Peraver, near St. Austell; in South Cornwall.

P. (ACASTE) INCERTUS, *Deslongschamps*. Pl. I, figs. 27, 28.

ASAPHUS INCERTUS, *Deslongsch.* Trans. Soc. Linn. de Calvados, vol. ii, p. 298, &c., pl. xx, fig. 5, 1825.

PHACOPS (DALMANNIA) INCERTUS, *Rouault.* Bulletin Soc. Géol. Fr., vol. viii, p. 371, 1851.

P. (Acaste) convexus, biuncialis, capite convexo, fronte angulato, caudâ mucronatâ. Glabella antrorsùm parum dilatata, sulcis anticis haud profundis, posticis valde exaratis; mediano et postico arcuatis. Anguli genales brevispinosi. Cauda trigona lata; axi 9-annulato, convexo, in mucronem longum crassum recurvum producto; lateribus 6-sulcatis, sulcis interlineatis.

About two inches long. We have only head and tail, the former convex, with a glabella wider than the cheeks, and subparallel, or rather with the sides gently divergent in front (not abruptly, as in *P. socialis;* the front is bluntly pointed, as in that species).

The glabella-furrows are well marked on the cast, and reach fully two thirds inwards, towards the centre. The upper ones are oblique and sinuous; the middle pair arched forward. The basal pair are by far the strongest, and enclose, in conjunction with the interrupted neck-furrow, a pair of transverse, oblong, nearly circumscribed, and very convex basal lobes, not so large as the middle pair. The contrast between the deep basal furrows and the fainter upper ones is striking, and is expressed by Deslongschamps' term "*postice bituberculato*."[1]

The cheeks are regularly convex, the neck-furrow being strongly marked as far as the angle, which seems to have been armed with shorter spines than in *P. socialis*, Barrande ('Tril. de Bohème,' pl. 26), the species which is most nearly allied to ours. It is not, however, quite certain there were even *short* spines to the head-angles.

The outer side of the cheek is moderately arched, and margined by a fainter furrow than the neck-furrow, but continuous with it up to the angle of the head. The axal furrows are not very strong; and there is no margin in front of the glabella.

The eyes are small, placed centrally on the cheek, not in advance of this position. The facial suture beneath the eye is nearly direct to the outer margin, and vertical in front of the eye.

We have not the body. The tail also resembles that of *P. socialis;* but has a shorter mucro and fewer ribs. The shape is broad-triangular, a good deal wider than long; the axis narrow, conical, and rather convex, and at its apex curved upward and passing into a strong, thick, and greatly recurved mucro, whose exact length we do not know, but which was probably as long as the tail itself. There are six flat side-ribs, somewhat arched, and directed obliquely backwards; they nearly reach the very narrow, flat margin, and are interlined throughout. The sides are tolerably flat, except towards the apex, where they become tumid, and run into the broad base of the thick, recurved spine.

Comparing *P. incertus* with the very nearly allied *P. socialis*, we find the latter with a far more triangular glabella, the furrows of which are more equal; with spinous head-angles, forward eyes, several more rings upon the axis of the tail, which, besides, has more side-ribs and a much more slender spine—the latter less recurved. But the two species are nearly allied, and the group of trilobites, and shells also, which occur in the subjoined locality, are identical with those of the Lower Silurian sandstone of Normandy, where *P. socialis* also occurs, and some other Bohemian forms with them. The subject of the geographical distribution of trilobites is worthy of a separate essay.

Locality. LOWER SILURIAN pebbles, in the "Pebble-bed" of the New Red Sandstone, Budleigh Salterton, South Devon (Mr. Vicary's collection). Also in the "May Sandstone" of Jurques, Normandy (Eudes Deslongschamps).

[1] Deslongschamps' Latin description is short, and not very distinctive. "Clypeo triangulari, angulis brevibus incurvatis......fronte magno convexo, antice acuto, postice bituberculato; genis parvis, oculis lateralibus" (p. 317).

PHACOPS (ACASTE) JAMESII, *Portlock*. Pl. 1, figs. 39—41.

PHACOPS JAMESII, *Portlock*. Geol. Report of Tyrone and Londond., p. 283, pl. 3, fig. 10, 1843.
— — *Salter*. Decades Geol. Survey, No. 7, art. 1, p. 10, 1853.
— — — In Morris's Catalogue, 2nd ed., p. 113, 1854, &c.

P. (Acaste) unciam latus; capite semicirculari, fronte subangulato marginato crasso. Glabella fere plana tuberculata, antice latissima, postice ad dimidium contracta, lateribus rectis: lobo frontali late trigono, oculos impendente; cæteris radiantibus, supremo maximo trigono, medio lineari obliquo haud abbreviato, basali transverso; lobis omnibus fere ad medium glabellæ, spatio angusto interjecto, conniventibus. Genæ declivæ marginatæ, angulis obtusis. Oculi abbreviati valde curvi. Thorax? Cauda rotundata, quam longâ tertiam partem latior, depressa; axi satis magno conico, marginem nullo modo attingente, annulis 8—9; *lateribus sulcatis, sulcis* 6—7 *æqualibus, læviter per totum interlineatis.*

I have seen but four or five specimens of this. But the species is not uncommon in Waterford. It is a marked one, conspicuous for the flatness of the glabella, and the thickened outer border. The cheeks are triangular; the eyes have a peculiar angular upper lobe. The tail is a good deal like that of *Calymene*, and the whole aspect is unusual for *Phacops*. Portlock's figure, which, like all the rest of his plates, were but second-hand office copies of Mr. Dunoyer's beautiful drawings,[1] gives no proper idea of the species, which is named in honour of Lieut.-col. James, of the Ordnance Survey.

P. Jamesii must have been nearly two inches long. It is very gently convex, the head flattened above, the cheeks declining rather steeply. The glabella, covered closely with not very coarse tubercles, occupies much more than half the width of the head, especially in front, where it is very broad, the sides converging behind at about 70°. The forehead-lobe is abruptly wider than the rest, overhanging the eye. It is transverse, subangular in front, with a thickened margin, and behind bounded by the nearly straight upper furrows, which run far towards the middle, and are of equal strength throughout. The other furrows radiate, the middle ones declining towards the base of the eye. The basal ones more direct, nearly parallel to the neck-furrow. The axal furrows are very faint.

The eye is small, strongly curved, set near the glabella, and on a level with it; the upper eye-lobe is pointed, the lentiferous surface imbedded, as it were, in a fold of the cheek, which is here convex, but declines all round, and slopes away to the margin. The cheek itself is triangular, with obtuse outer angles, and is only granular, not tubercular, strongly marginate on the outer side, and with a sharp neck-furrow, which does not meet the marginal furrow.

Tail semicircular, one third wider than broad, depressed (this is not due to pressure)

[1] The original plates were found to be rather too large, and unfortunately had to be cancelled.

Axis rather large, not equal to the limb, conical, rounded at the tip, and reaching less than four fifths of the whole length; annulated throughout by about nine rings; the sides with seven furrows directed obliquely, and reaching nearly to the margin, interlined throughout.

The species is unlike any other, but has perhaps its nearest relations with *P. alifrons*, next described. From that species the shape of the forehead-lobe will distinguish it, while there is only some general resemblance in the shape of the flatter glabella, and especially in the small, greatly curved eye.

Locality. CARADOC ROCKS of Waterford; Tyrone. (Mus. Pract. Geology.)

PHACOPS (ACASTE) ALIFRONS, *Salter*. Pl. I, fig. 31—34.

PHACOPS ALIFRONS, *Salter*, in Appendix to Sedgwick's Synops. Woodw. Foss., fasc. ii, tab. 1 G, figs. 12—14, 1852.
— — *Ibid.* In Decade 7, Geol. Survey, Art. 1, p. 10, 1853; and Morris's Catal., 2nd ed., 1854.

P. (Acaste) capite tuberculato sesquiunciam lato, gibboso, anticè truncato, bis quam longo latiori. Glabella elevata, sed paullum convexa, ad basin angustata, superne dilatata obtusa truncata, lateribus subrectis; lobo frontali brevi transverso limbum crassum impendente, utrâque angulis tumidis cum margine genarum confluentibus; lobis lateralibus tumidis, supremo subtrigono modico, reliquis fere rotundis abbreviatis; genis declivibus marginatis. Oculi curvati. Cauda semicircularis tumida, axi lato convexo 8—9 *annulato, apice obtuso nec marginem attingente; lateribus convexis, costis* 7—8 *simplicibus; margine angusto.*

We are compelled to be minute in the specific characters of these species of Phacops, as they really differ but in proportional characters; yet, if it were allowable to abbreviate, it would certainly be excusable in this case, for in no other species which I know does the outer and upper angle of the glabella run out distinctly into the margin as it does in this fossil. But three or four heads of it are yet known, and two or three caudal shields associated (not in actual contact).

The species was a tolerably large one, the head (in our largest specimens, copied from the figure formerly given by myself in Professor Sedgwick's work) is 1½ inch broad. The entire form may have been 3 inches long. The lesser figure (fig. 33) represents a somewhat smaller, but characteristic specimen. Both are from North Wales, and I do not know it elsewhere.

The head is highly convex, and rather strongly truncated in front, but the glabella, though tumid, is rather depressed on its upper surface. It is very wide in front, the upper lobe overhanging the eye; and being connate with the thickened border of the cheek, seems to be drawn out into it. The lateral lobes are short, the upper largest, but not greatly so, triangular; the upper furrow which bounds it running nearly straight across (not upward, as in *P. Jamesii*); the second, as deep, directed downward and outwards; the rest are short and rounded, the base-lobes narrow, the neck-segment prominent.

Cheeks very convex, separated by deep axal furrows from the glabella, strongly circumscribed in front by the deep marginal furrow, and behind by the neck-furrow; the angles rounded, not spinous? Eyes prominent, small, curved, somewhat sunk, overhung by the angles of the glabella.

Of the tail we have but few specimens; our largest is that figured in Professor Sedgwick's work. It is semicircular and very convex, with a broad, prominent axis reaching nearly the whole length, and blunt at the apex; ringed throughout by eight or nine ribs, and with no smooth terminal portion. The sides are also convex, with about eight simple, radiating ribs, not much curved, and with no intermediate furrows. The ribs abut abruptly against a narrow but distinct margin. The other tail (fig. 34) is less certainly of this species, but is found associated with it. It has a wider margin, and only seven lateral furrows.

Localities.—Caradoc rocks, North Wales; at Capel Garmon, near Llanrwst; also at Penmachno, and Pont-y-Glyn Diffwys, near Corwen; and near Llangynnog, in Montgomeryshire. (Mus. Pract. Geol. and Woodw. Mus.)

Phacops (Acaste) Brongniarti, *Portlock.* Pl. I, figs. 20—25.

♂ ? P. Brongniarti, *Portlock.* Geol. Report, pl. 2, fig. 8 (exclude the references), 1843.
P. Murchisonii. Ibid., fig. 9.
Var. ♀ ? P. Dalmani. Ibid., fig. 7.
P. Brongniarti and P. Dalmani, *Salter*, in Decade 7, Geol. Survey, Art. 1, p. 10, 1853; and in Morris's Catalogue, second edition, pp. 112, 113, 1854.

♂ P. biuncialis, elongatus, granulatus; capite longo trigono, fronte angulato subrecurvo. Glabella ad basin contracta, anticè valde dilatata nec convexa, lobis utrinque tribus radiantibus; lobo antico maximo triangulato, a frontali sulco valido—a medio sulco leviore—sejuncto; lobis infimis minutis hæmisphericis circumscriptis sese remotis. Oculi maximi, a lobo frontali usque ad sulcum cervicalem tracti. Anguli genarum obtusi. Thorax lateribus parallelis, axi convexo angustato, pleuris abruptè deflexis apicibusque rotundatis—fulcro intra medium posito. Cauda trigona, axi longè conico angustissimo ferè ad finem caudæ extenso, 10-*annulato; lateribus* 5-*costatis, costis per totum divisis, nec marginem lævem attingentibus.*

♀ or var. *Dalmani,* figs. 25, 26.

Omnino precedenti simillimus, glabellâ, caudâ, oculis—sed fronte capitis rotundato nec producto.

This is not a difficult species to recognize, once its peculiarities are mastered; but it occurs in a variety of forms, and these are rendered more obscure by the changes the rock has undergone. It is common in the sandy schists of Tyrone, and not unfrequent in North Wales and other places. Colonel Portlock thought it identical with the *P. macrophthalma* of Brongniart and the *P. Downingiæ* of the Wenlock Rocks; and to avoid the confusion of using these controverted names, proposed to term the combined species *P. Brongniarti*, after the author of the 'Crustacés Fossiles.'

But, as I have shown in the Decades of the Survey (*l. c.*), these are really three very distinct species, and we must limit Portlock's name to the species described first by him. His *P. Murchisonii* is only a synonym. But it is with some little doubt I include the *P. Dalmani*, a form with rounded front, but otherwise exactly like the species with which it is associated. As I find Welsh and Irish specimens with an intermediate character, I do not think it can be wrong to unite these two, and I take the pointed form to be the ♂, and the more rounded variety (*P. Dalmani*) to be the ♀ form. The ♂ is the more common of the two.

Two inches long, elongated, and tolerably convex, the head occupying less than one third the whole length; subtriangular, and with the front produced and recurved (in the *P. Brongniarti* ♂; semioval and with a rounded front in the ♀ form, *P. Dalmani*;) the glabella itself is of the same shape in both varieties, tubercular all over, contracted at the base, expanded in front, but not convex, and with radiating lobes. The forehead-lobe is subrhomboidal and transverse, the upper lateral lobe large, triangular, and overhanging all the rest, but not so large as to render them obsolete. The mid-lobe is directed backwards, and is parallel-sided, and about half the length of the front one. The basal lobe is transverse-linear, contracted at the sides, where it forms a tubercle, and the neck-lobe is again broader and thicker and more elevated than the rest. This is less conspicuous in the ♀ form. The cheeks are triangular, and concave outside of the very large eye, which reaches forward beyond all the lateral lobes, and backward to the neck-furrow. It is considerably arched, rather depressed, and has numerous small lenses (about eight in a vertical row, and in all 170 in each eye). The eye-lobe or -lid is strongly furrowed parallel to the lens-bearing surface. The space between the eye-lobe and the glabella is convex, and of an oblong shape. Angles of the cheeks obtuse, not spinous. Thorax with a convex, narrow axis, and with pleuræ which are steeply bent down at about the first third of their length; the fulcrum placed within the middle point; the ends of the pleuræ blunt and arched forward, and their facets large.

The tail is triangular, pointed in the ♂, obtuse in the ♀ form, rather convex; with a long conical axis reaching nearly to the narrow margin, and interlined strongly by intermediate furrows, which project beyond them and quite reach the margin, undulating it most strongly in the ♂ form.

I see no reason for separating the two trilobites quoted above as specific forms. All the characters are more strongly pronounced in the form which I suppose to be the ♂, and which has both the apex of the tail and the front of the head more prominent and produced than in the other variety. The North Welsh specimens are intermediate, the cheeks steeply bent down, the margin of the eye-lobe narrower, and the lenses more numerous; occasionally ten in a row, in the widest part of the eye.

P. macrophthalma of Brongniart, with which the first describer of this species united it, has neither the obsolete basal lobes, nor greatly expanded glabella, nor the large eye of the one we have been describing. And there can now be no question of uniting either of these with the Dudley fossil included by Brongniart (described above as *P. Stokesii*), all subsequent writers having adopted the distinctions.

It is, however, worth while to compare our species with the fossil figured by Dr. Pander, in his 'Beiträge zur Geogn. Russ.,' pl. v, fig. 7, and pl. vi, fig. 9. It has large eyes like ours, but the glabella-lobes differ considerably. And at all events, it is not the *P. macropththalma*, to which he refers it.

Localities. — CARADOC, Tyrone, abundant; Kildare; Wexford; and Waterford, in Ireland. Llanfyllin and Llanwddyn, Montgomeryshire, in North Wales. (Mus. Pract. Geology, &c.)

Sub-genus—CHASMOPS, *M'Coy.*

Though not quite convinced that this is a definable sub-genus or section of Phacops, I adopt it because it conveniently bridges over the space between Acaste proper and the typical species of *Odontochile.* The *P. conophthalmus,* Boeck, is the type of the section, and has to perfection the enlarged upper lobes of the glabella and the expanded forehead-lobe, while the tail is only of moderate size, and has but eight or nine segments.

P. macroura, again, is typical so far as the head is concerned, but has very numerous segments to the tail, a characteristic still further carried out in such species as *P. truncato-caudatus,* which may be considered as linking on the sub-genus *Chasmops* to *Odontochile,* as *P. Jukesii* links it to the ordinary *Acaste.*

Altogether the group has a wide range, and tends to show the unity of the genus *Phacops.*

PHACOPS (CHASMOPS?) JUKESII. Pl. I, figs. 29, 30.

? PHACOPS SCLEROPS, *Dalman.* Palæadæ, t. ii, fig. 1, *g*, icon mala, 1826.
PHACOPS JUKESII, *Salter.* Decades Geol. Surv., No. 7, Art. 1, p. 11, August, 1853.
— *Id.*, in Morris Catal., 2nd ed., 1854.

P. (*Chasmops vel Acaste*) *modicus, unciam latus, granulosus, capite transverso ferè quam longo ter latiore; glabellâ dilatatâ, lobis omnibus distinctis. Glabella subconvexa, frontem impendens, latè triangulata, anticè valdè dilatata, nec cum margine connata, posticè contracta, utrinque triloba; lobo basali transverso lineari, secundo paullo majore rotundato, supremo magno triangulato, frontali maximo transverso oculos imminente, cervicali elevato. Sulci axillares profundi. Genæ latæ marginatæ, transversæ—angulis rotundatis?, sulco cervicali forte exarato, lineâ faciali impressâ.*

It may be doubtful if this species be not included in some of the varieties of *P. sclerops*, Dalman; but it is clearly distinct from his typical form, as we know by specimens in the cabinet of Sir R. I. Murchison. And it well deserves a name, being a most characteristic form, which somewhat tends to connect the ordinary species of Acaste with the more extravagantly lobed species of Chasmops. I scarcely know to which sub-genus it should be more properly referred, and place it first. We have only two specimens.

The head is above an inch wide, and scarcely more than one third its length; convex and granulose, with coarse and fine grains equally all over. The glabella not convex, but greatly dilated in front, and contracted behind to one fourth the width. The forehead-lobe is widely transverse, and overhangs completely the triangular upper lobe, which in its turn is imminent over the mid-lobe, and about twice its size. The middle lobe is, however, wider than the narrow basal lobe, which is not reduced to a mere tubercle on the sides, nor contracted to a much less width than the rather prominent neck-lobe.

The cheek, strongly margined in front, is of a narrow triangular shape, and is deeply divided from the glabella. The eye, much curved, is opposite to and covers the space of the two upper lobes, and is itself overhung by the upper lobe.

The facial suture is deeply marked outside the eye; it nearly bisects the cheek, and runs parallel to the deep neck-furrow.

A young specimen does not differ in any material respects from an older one.

Locality. CARADOC ROCKS of Gelli grin, near Bala, North Wales. (Mus. P. Geology.)

P. (CHASMOPS) MACROURA, *Sjogren*, Pl. IV, figs. 18—23.

ASAPHUS POWISII, *Murchison.* Silurian Syst., pl. 23, fig. 9 (head only), 1837.
DALMANNIA AFFINIS, *Salter.* Memoirs Geol. Survey, vol. ii, pt. 1, pl. v, fig. 5, 1848.
PHACOPS TRUNCATO-CAUDATUS, var. β, AFFINIS, *Id.*, in Decade Geol. Surv., No. 2, Art. 1, p. 7, 1849.

ODONTOCHILE TRUNCATO-CAUDATA, *M'Coy.* Synops. Foss. Woodw. Mus., t. i G, fig. 20 (21 ?) (not *Chasmops Odini,* same plate, figs. 22, 23), 1851.

PHACOPS MACROURA, *Angelin.* Palæont. Suecica, t. vii, figs. 3, 4, 1852.*

P. (*Chasmops*) *ovatus magnus, capite semilunari, vix quam longo bis latiore, tuberculoso, caudâ elongatâ submucronatâ multi-annulatâ. Glabella parùm convexa, anticè valdè dilatata, lobo frontali maximo triangulato oculum imminenti; lobis lateralibus supremis magnis triangulatis quam latis multo longioribus, medianis obsoletis, basalibus contractis, cervicali lato. Genæ convexæ nec gibbæ, declivæ, glabellâ angustiores, angulis in cornua lata longa extensis; oculo majore, sulco circumdato, linea faciali impressâ. Cauda magna trigona convexa, apice acuto, axi lato per* $\frac{4}{5}$ *longitudinis extenso, annulis* 16; *costis lateralibus circa* 16, *ferè rectis, vix interlineatis, sub apice radiantibus.*

All the above names have at various times decorated or disfigured our fossil, which labours under the combined disadvantages of being very common and conspicuous, and seldom at all perfect; and also of belonging to a group of closely allied species which occur in company, and have been hitherto always badly figured. I can hardly hope to have remedied *all* these defects, and only give those synonyms of which I am assured. The name *Phacops conophthalmus,* under which it is generally known in cabinets, really belongs to a species with short head-spines, smaller eyes, wider cheeks, and altogether a more transverse form of head, and a short tail. We must wait for a more complete knowledge of this group of trilobites; and meanwhile our figures represent truly the most conspicuous of these forms, and certainly that figured by Prof. Angelin. The numerous ribs of the tail and the great head-spines readily distinguish it.

The *P. bucculenta,* also figured by Angelin, has a large, many-ribbed tail, but has only a small pair of lateral glabella-lobes instead of the great encroaching pair distinctive of the *P. macroura,* and which have procured for this grotesque-looking species the name among collectors of the 'Cat's-head Trilobite.'

There are other allied species both in Scandinavia and North America; but the group appears to be peculiar to this northern zone, and does not occur in mid-Europe, as M. de Barrande has specially remarked.

General form rather narrow, ovate, blunt but abruptly angulate in front, and shortly pointed behind. The head is about as long as the tail, and not quite twice as broad as long; very convex; semilunar, with a very wide glabella, larger than the cheeks, the sides of which converge behind at about 50°. The cheeks are very convex, deeply margined, and produced behind into exceedingly broad, flat, and rather lengthened parallel spines, at the inner angle of which the strong marginal and still stronger neck-furrow meets.

* The cover of this excellent folio is marked 1851; but in order to preserve priority of date, it is necessary to be exact. It seems to have appeared Jan., 1852.

The lobes of the glabella are as follows:—The forehead or frontal lobe transverse, rhombo-trigonal, depressed in the centre, twice and a half as wide as long, rounded-truncate at the upper angles, and only a very little overhanging the position of the eyes. The great lateral "cat's-ear" lobes are right-angled triangles, with the corners rounded off, their larger angle inwards, and only slightly sinuous above, and they occupy in length full half the side of the glabella, and reach more than one third across it, bounded on all sides by very deep furrows. A narrow, triangular space is left between the bases of these lobes for the lower pairs; the middle lobes are minute, and mere tubercles; the basal lobes more extended, but still narrower than even the base of the upper lateral lobes, which completely overlap them. The neck-furrow is very strong, and equal across; the neck-segment of equal width throughout, but its whole breadth is scarcely one half that of the frontal lobe.

Fig 7.

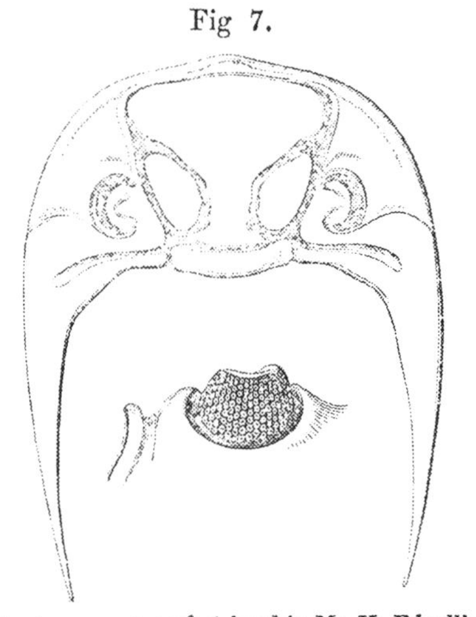

P. macroura, a perfect head in Mr. H. Edgell's collection: the eye magnified.

The axal furrows are very deep, and tolerably straight, the cheeks rising very convex from them, bearing the small, strongly lunate eyes.* These have about 160 lenses rather small, and very little prominent, and the pits containing them have on the interior cast an elevated rim or border. The eyes are placed opposite the centre of the frontal lobe, about half way up the cheek, and their own width from the glabella; a deep channel surrounds its base. All the furrows of the head are deep, the marginal one especially so, and the neck-furrow is much curved to follow the descent of the great head-spines. These latter are broad, flat horns, pressed against the body, and extend back a great way, as shown in our woodcut.

The surface of the head—the glabella especially—is largely tubercular; and in fig. 20 a curious double row of deep puncta runs half way down the forehead-lobe from the emarginate front. But this is perhaps not repeated in other specimens.

The facial suture is not strictly marginal in front, but a little above the margin. Outside the eye it takes a forward curve, and here it lies in an impressed groove. The labrum (fig. 21) has a very broad, arched base, and a blunt-pointed end. The body of the labrum is strongly separated from the wings, and its sides are parallel for a short distance, then converge to the rounded obtuse point. There are two deep concentric furrows; the outer one quite continuous, very strong, and remote from the apex. Above these the organ is convex and roughly tubercular; beyond them it is flat. There is a deep perforation (*a*) in the broad basal wing, which indicates the place of the ascending process of attachment.†

* I think Prof. Angelin's figures are not correct as to the eyes. The cornea must have been thin.

† This varies in shape in various species. The form above indicated is a rare one.

The tail, from very perfect specimens (figs. 19, 22, 23) is subtriangular, with rounded sides and apex; pointed, highly convex; the sides regularly so, without any flattened border. The axis is broad, and very gently convex; much more than half of the width of the broad limb, regularly conical, and reaching four fifths the length of the tail. It is annulated to the very tip by sixteen sharp rings. The sides are deeply grooved by sixteen* nearly direct strong furrows, faintly interlined throughout, which gradually radiate downwards, till from the tip of the axis they become longitudinal. Three or four longitudinal ridges run from the tip to the end of the tail, thus completing the fan-like arrangement.

The incurved portion of the tail, seen in the small specimen (fig. 23), is narrow and granulated, as the whole tail appears to have been.

Locality.—CARADOC SANDSTONES of Shropshire; plentiful in many places. (Cabinets of Messrs. Lightbody and H. W. Edgell; and in the Ludlow Museum, and Mus. P. Geology).

P. (CHASMOPS) CONOPHTHALMUS, *Boeck?* Pl. IV, figs. 24, 25, and Pl. VII, fig. 25.

TRILOB. CONICOPHTHALMUS, *Boeck.* Gæa Norvegica, i, 4, 1838.
PHACOPS — *Emmrich.* Dissert., 21, 1845.
CALYMENE ODINI, *Eichwald.* Sil. Syst. Esthland.
— — *De Verneuil.* Geol. Russia, t. xxvii, fig. 8.
PHACOPS SCLEROPS, *Burmeister.* Org. Trilob., ed. 2, tab. iv, fig. 5, 1846.
— CONOPHTHALMUS. Ibid., p. 91.
— FELINUS, *Salter* (MS.). Ibid., Appendix of Species, p. 125.
— CONOPHTHALMUS, *Angelin.* Palæont. Suecica, t. vii, figs. 5, 6, 1852.
— — *Salter.* Decades Geol. Survey, vii, Art. 1, p. 7 (exclude other synonyms than those above), 1853.
CHASMOPS ODINI, *M'Coy.* Synopsis Woodw. Mus., t. i G, figs. 22, 23, 1851.

P. (Chasmops) ovatus, capite valdè transverso, quam longo ferè ter latiore, granuloso convexo [caudâ brevi, 8—9 costatâ]. Glabella convexa, anticè valdè dilatata, lobo frontali maximo triangulato, oculum parvum imminente, lobis lateralibus supremis magnis rotundato-trigonis, medianis obsoletis, basalibus linearibus, cervicali lato. Genæ convexæ ferè gibbæ, marginatæ, angulis in cornua brevia extensis, oculo brevi, lineâ faciali impressâ. Cauda [associata] lata, punctata, vix marginata, axi conico 9—10 annulato; costis lateralibus 8 arcuatis, duplicatis.

The head of this species differs at a glance from the one above described, in being so much more transverse, only half as long as it is broad, and rather semicircular than semilunar. Though fig. 25 (copied from my own original figure in the 'Cambridge Synopsis') is not quite perfect—it is only the internal cast—there is enough to show that it agrees better with Angelin's *P. conophthalmus* than the other species; and the tail

* Angelin's figure has 19 axal, and 18 side ribs.

figured with it in the same work, will agree pretty well with Burmeister's figure of *P. conophthalmus*, in his later and improved edition;* much better than with Angelin's figure, which surely represents the tail as too pointed, and he describes it with twelve lateral ribs to the tail. We are still at sea about this species, but must accept Angelin's figure as the type, as he has access to all the described Swedish species. I only describe our own, and believe it will by and by constitute a new species.

The head is transverse, semicircular but obtuse in front, and more than twice as wide as long; very convex; granular, not tubercular; and with the wide glabella occupying much more than one third of the head. It is nearly a right-angled triangle, with the corners rounded off. The forehead-lobe is widely transverse, and considerably overhangs the eyes. The upper lateral lobes are large, but not expanded forwards, only behind, and rounded-trigonal; their transverse diameter fully equal to, if not greater than the fore and aft measurement. The mid-lobes are quite obsolete, and the basal one transverse-linear, and smaller than the neck-lobe. The cheeks are equilateral triangles, very convex, and bear centrally the small eye, which is subtended by a much deeper furrow than in *P. macroura*. The neck-furrow is strong, and continuous with the marginal furrow, but the margin is in no part very broad. Head-angles short-spinous.

The tail (which occurs in the same locality, and probably belongs to the same) is transverse, "its length nine lines, against a breadth of fourteen lines; the axis convex, of ten segments, and about two thirds the width of the side-lobes in front; abruptly narrowed after the fifth segment. There are about eight lateral segments" (M'Coy). The ribs are much arched, and reach quite to the margin; strongly interlined throughout. The surface of the tail is strongly punctate.

This obtuse, short caudal portion is very like that figured by Burmeister in the Ray edition of his work. It is more like the tail-piece of a large *Calymene* than a *Phacops*. Not having access to the Cambridge specimens at present, I had thought of leaving this and some other doubtful species for a supplement, but prefer to present it, with other fragments, in our Plate VI, and so complete what we at present know of this obscure species. Collectors will do well to pay special attention to this fossil, as it is a characteristic northern form.

Locality.—CARADOC SLATES of Llansaintffraid, Glyn Ceiriog, south of Llangollen, North Wales. Also at Acton Scott, in Shropshire (Mr. H. W. Edgell).

I omit other localities, but one or the other of the above two species occur throughout North Wales, at Llanfyllin, Llanwddyn, and Meifod, Montgomeryshire; Pwllheli and Bettws-y-coed, Caernarvonshire. In South Wales, at Llandeilo. In Westmoreland, at Coniston, Troutbeck, and Rother Bridge. In Clare and Kildare, Ireland.

* Ray edition, 1846. He describes the tail as having nine lateral ribs—the number present in a specimen from Christiania in the Mus. Pract. Geology.

P. (Chasmops) amphora, *Salter*. Pl. IV, fig. 16.

Phacops (Dalmania) amphora, *Salter*. Decade Geol. Survey, vii. Art. 1, p. 12. 1853.

P. (*Chasmops*) *caudâ magnâ biunciali elongatâ, convexissimâ, fere semicylindricâ. sulcis axalibus fere obsoletis. Axis latus nec eminens, marginem angustum inflexum haud attingens, in annulos* 16 *subplanos divisus, apice obtuso. Latera valde curvata deflexa, costis* 14—15 *planis, sulcis acutis separantibus, costâ quâque lineâ medianâ lævi elevatâ sub cortice tenui impressâ; apice obtuso* (*emarginato?*).

This species must be closely allied to the preceding. It differs only in proportions, but markedly so, being greatly more convex in the only portion we know—the tail; it is almost semicylindrical in shape, and the axis much less distinct from the sides than in *P. macroura*. It is also, but somewhat more remotely, allied to *P. truncato-caudatus*.

The large tail, which must have been fully two inches long, is but a little more than this in breadth. It is more than half a cylinder, the curvature being remarkably regular and even, the axis scarcely projecting beyond the general convexity, and scarcely divided from the sides by any depth of axal furrow. The axis is broadly conical, and below rounded off towards its apex; it reaches fully five-sixths the whole length of the tail; and is made up of sixteen rather flat rings. The four or five front ones show a false articular surface, distinctly; the real articular surface is very large and distinct.

The side-lobes are in reality half as wide again as the axis, but on a front view do not appear so, owing to the great convexity; the surface is curved steeply down, and is scored by fifteen deep sharp furrows, separating sixteen prominent flat ribs, much wider than the furrows. Each rib has an interrupted chain-like dividing line down its middle. The same occurs, somewhat exaggerated, beneath the crust, on the cast. But the perforations here are so arranged as to leave a crest-like ridge. The dividing lines extend nearly to the end of the ribs, which cease rather suddenly near the steep, almost vertical or inflexed, marginal rim.

The shell or crust of this species is remarkably thin. Except some species of *Homalonotus*, I do not know any kind of trilobite which has so cylindrical a form.

Locality. Caradoc limestone of Grug, N. of Llandeilo, Carmarthenshire. Collected and presented to the Mus. Pract. Geology, by Mr. Williams, of Llandeilo.

P. (Chasmops) truncato-caudatus, *Portlock*. Pl. IV, figs. 13—15.

Phacops truncato-caudatus, *Portlock*. Geol. Rep. Tyrone, l. c., pl. ii, figs. 1—4, 1843; *Paradoxides Bucephali*, ibid. (labrum only), pl. i, fig. 8.
— — *Salter*. In *Morris*, Catal., 2d ed., 1854.
— — (Dalmania) *Id*. Decade II, Geol. Survey, Art. 1, p. 7, 1849; (exclude the synonyms of *D. affinis*, for which see *P. macroura*).

P. (*Chasmops*) *granulatus, capite antice truncato, oculis maximis subdepressis, pleuris obtusis, fulcro ab axe valde remoto. Cauda axe* 18-*annulato angusto, costis lateralibus* 14—16, *apice emarginato.*

An obtusely ovate form, strongly truncate in front, and obtusely pointed behind. The head widely transverse, fully two and a half times as wide as long, and with the glabella so expanded along the front as to occupy more than half the width of the head, and overhang the middle of the large eyes. Thence the glabella abruptly narrows to half this width, sloping inwards to the base of the middle lobe; but from this point it is again parallel-sided to the base (in *P. macroura* the shape is triangular, here it is turbinate).

The upper glabella-lobes are quite overhung by the frontal lobe, and are not greatly larger than the succeeding lobes. They are right-angled triangles, with the obtuse right angle outwards, not inwards as in the preceding species. The space between them is fully equal to the width of the lobes themselves; their length is two thirds that of the great eye. The middle lobes are rather smaller, narrower, and somewhat linear, but tumid at the sides—(and not overhung and rendered obsolete as in *P. macroura*). The basal lobes are still narrower, more linear, and less tumid; the neck-lobe of the same shape as the basal lobes. The whole glabella, as well as the cheeks, and all the rest of the head is thickly covered with large granulose tubercles.

The cheeks are not large, and are chiefly occupied by the semilunar eyes, which reach from the front furrows nearly to the neck-furrow. They are much curved, the eye-lobe strongly margined, with an angular fold along the middle. Cheek-spines large.

Axis of the body-rings gradually widening behind; rather convex, especially along the middle part, which is wholly covered with granules. The sides of the axis are nodose on every joint. The axal furrows strong, the pleuræ not very much wider than the axis, flat as far as the remote fulcrum, which is fully two thirds out. Pleural groove deep, very oblique, and leaving a strong node on the proximal or inner end of the pleura, against the axis. This is a conspicuous character. Tubercles, larger and smaller ones, are visible over the whole surface. Tail broadly half-ovate, convex, but regularly so, except the concave narrow margin. The axis not very prominent nor strongly separated from the sides, tapering, narrow, with straight sides, composed of eighteen rings nodular (as the axis of the body, but on a smaller scale). Sides of tail depressed, with fourteen or sixteen flattened radiating ribs, with a somewhat sigmoid curve; the ribs separated by sharp furrows which run almost to the margin. The ribs are granular throughout; there is an imperfect double row of granules on each flattened rib—the space between the rows representing the interlining furrow, which is not present in this species.

The apex of the tail is, as in most of the allied forms, emarginate and a little raised; a ridge runs to it from the angular termination of the axis, as in kindred species.

The *labrum* or *epistoma*, fig. 15, has the wide arched base characteristic of the subgenus. It is gently convex and somewhat angular in form (but subparallel at the base) with a shallow notch for the ascending process. The concentric pair of furrows are deep;

the outer pair continuous as a marginal furrow, the inner very oblique. The apex is angular, but blunt.

Localities. Caradoc of Desertcreat, &c., Tyrone. (Mus. Pract. Geology).

Phacops (Chasmops) Bailyi, n. sp. Pl. VII, figs. 21—24.

P. (Chasmops) satis magnus, capite subangulato; glabellâ dilatatâ, lobis omnibus distinctis; caudâque (associatâ) rotundatâ multiannulatâ. Glabella longa elevata tuberculosa lobo antico rhomboideo transverso, lobis lateralibus inæqualibus, supremo magno rotundato trigono, secundo paullum abbreviato, basali majori, cervicali expanso. Oculi elevati. Cauda semi-ovalis, apice rotundato obtuso; axe longo conico 10—11-*annulato; costis lateralibus* 11 *planis, arcuatis, lineisque intermediis distinctis ad apicem unâ cum primariis connatis.*

It is scarcely doubtful that this is a *Chasmops* rather than *Odontochile*. It certainly has many affinities with *P. truncato-caudatus*, to which Mr. Baily referred it in the Irish collection. He has judiciously preserved every fragment, and in this way we have been enabled to reconstruct the species.*

The head is rather elongate, the glabella somewhat depressed and widely clavate, broad above the wide forehead, and actually overhanging the eye, and narrowed behind; all the lobes distinct and somewhat radiated. The forehead-lobe is transverse and rhomboidal, separated by an arched furrow from the rounded, triangular, upper lateral lobe, and this by an oblique downward furrow from the much smaller mid-lobe, which is triangular in the reverse direction. The basal lobes are large, and tumid at the sides; the cervical lobe is wide. The eyes prominent, and placed rather forward, on a somewhat sharply elevated cheek. The margin of the cheek narrow, and not strongly defined. The facial suture runs in a broad ogive some distance within the front margin, which is slightly angulated.

Of the tail we have one external and three interior casts, which show it to have been a large semioval plate, broader than long, but rather abruptly narrowed behind, and with a rounded obtuse end. The axis is narrow and conical (but broader than in *P. truncato-caudatus*), reaches five-sixths the length of the tail, and has ten or eleven rings at least. The axal furrows are deep, and the sides are convex, and scored by eleven narrow sharp furrows; the ribs between which are flat and interlined all along, much in the same way as in *P. macroura*. The upper ribs are duplicated very strongly. The margin is very narrow.

It is very like *P. macroura*, above figured, and differs from *P. truncato-caudatus*, fig. 13, both in the shape and proportions of the glabella-lobes, and in the much narrower segments of the tail, and it does not need comparison with other British species.

Locality.—Caradoc slates of Tramore, Co. Waterford. (Mus. Irish Industry, Nos. B 647, 650, 652.)

* Many species are lost from want of care in the collocation of fragments of specimens. Very often, indeed, in slate-rocks, the species must be truly reconstructed from the *disjecta membra*, and with judgment this may generally be safely done.

Section.—Odontochile.

If there were some doubts of the limits of the subgenus Chasmops, there can be none at all of the general character of that which Emmrich called *Dalmania*, and which name was rightly changed to *Odontochile* by Corda, *Dalmania* having been in previous employ.

Large expanded forms, with the glabella wider in front, and with the lobes nearly equal or only gradually increasing forwards; a depressed habit, and a caudal shield composed of many segments (technically more than eleven), give a combination of characters easily enough recognisable among the large Trilobites, which chiefly occur in Upper Silurian rocks; a few began in the Lower Silurian, a few extend into the Devonian; but the bulk of the subgenus are to be found in Wenlock and Ludlow rocks, and the genus is worldwide.

We have in this section the largest and most conspicuous of the *Phacopidæ*. And while the section *Acaste* is chiefly Lower Silurian, and *Chasmops* wholly so,—the large flattened species which form this group, with their numerous tail-segments, are almost all Upper Silurian. They commence, so far as I know, in the May Hill Sandstone. The most renowned of them all is the *P. caudatus*, and it is at the same time the most typical.

I commence with those species in which the caudal portion is least extravagantly developed:—

Phacops (Odontochile) obtusicaudatus, *Salter*. Pl. I, figs. 42—45.

Phacops obtusicaudatus, *Salter*, in Sedgwick and M'Coy's Synopsis Foss. of the Woodw. Mus., pl. i G, figs. 15, 16.
— — *Id.* Mem. Geol. Survey, Dec. II, pl. i, p, 7, note.

P. (Odont.) uncialis, capite elongato semielliptico tuberculato; caudâ brevi. Glabella longa, lobo antico magno rotundato, lateralibus subæquis, radiatis, sulcis longis profundis. Oculi elongati, antici. Genæ angustæ scrobiculatæ, spinis modicis. Cauda trigona, apice angulato, axe rotundato longo 12-*annulato; lateribus recti-sulcatis, sulcis* 10, *omnibus duplicatis; margine angustissimo.*

This is known only in one locality; but it is a common fossil there, and is accompanied by but very few other fossils. It is a gregarious species, occurring in crowds upon the faces of the "bated" or cleaved rock in Coldwell quarry, Westmoreland.

The head is half elliptical, frequently an inch and a half long, the width then being about two inches; the general form that of a pointed arch; the glabella narrowed, long-oblong, not much broader in front than behind, the forehead-lobe as long as the rest of the glabella (including the prominent neck-segment), and almost *round*. The side-lobes are very strong, and divided by deep, radiating furrows, deepest and strongest inwards, and reaching far towards the centre. The upper lobe trigonal, not greatly larger than the mid-lobes; the basal lobe narrow, almost linear externally (and clavate in the cast). The whole glabella covered by coarse tubercles and fine granules intermixed.

The eyes are large, reniform, elongate, compressed laterally instead of longitudinally, and placed in advance of the centre of the cheek, half their own length from the neck-furrow, and touching the glabella in front. The cheeks curve steeply down, are scrobiculate, not tubercular, margined nearly all round by the narrow furrow, and provided with head-spines half as long as the glabella.

The facial suture curves largely forward outside the eye, and cuts the margin about half-way up the cheek.

We have no body-rings. The tail is broadly triangular, pointed below, but not produced into any mucro, and with a very narrow equal margin. The conical and very convex axis nearly reaches the tip, annulated all the way down, and has about twelve sharp rings. The sides are strongly radiated; nine or ten strongly duplicated ribs running on each side to the narrow margin. The intermediate furrows are straight at their origin and terminations, and at the latter point they are so deep as to make the margin pretty strongly undulate.

Locality.—CARADOC SLATE of Coldwell, Westmoreland; in beds over the Coniston limestone. [Brit. Mus. Mus. P. Geology. Woodwardian Museum, &c.]

PHACOPS (ODONTOCHILE) MUCRONATUS, *Brongn.?* Pl. IV, figs. 11, 12.

[ENTOMOLITHUS, *Linnæus.* Iter Vestrogoth. p. 88, *cum fig.* 1747.
ENTOMOSTRACITES CAUDATUS, *Wahlenberg.* Nova Acta Reg. Soc. Upsal, tom. viii, p. 28, t. ii, fig. 3, 1821.
ASAPHUS MUCRONATUS, *Brong.* Crust., p. 24, t. iii, fig. 9, 1822.
— — *Dalman.* Palæadæ, p. 236, t. ii. fig. 3, *a*, *b*, 1826.
— — *Hisinger.* Lethæa Suecica, p. 13, 1837.
PHACOPS — *Emmrich.* Dissert. (1839); Neues Jahrbuch, 1845.
— — *Angelin.* Palæont. Suecica, tab. viii, fig. 1, 2, 1852.]
— — *Salter.* Decade VII, Art. 1, p. 12, 1853.

P. triuncialis et supra. Glabella convexa, antice parum dilatata, utrinque lobis tribus subæqualibus transversis, sulcis longis satisque profundis sese separatis. Cauda triangularis acuta; axe convexo limbo plano haud æquali, in annulos 9 *et appendicem diviso* [*appendice in apicem caudæ brevimucronatum percurrente?*]; *lateribus costis* 8 *planis, sulcis* 7 *angustis acutis valde curvatis et cum tot lineis intermediis profundioribus ad apices confusis; margine angusto subundulato.*

From the number of authors who have described the Swedish species, it must be a common one, and yet there is no good figure extant. Angelin's, the latest, is worse than Dalman's, which appears to be the best, for Wahlenberg's figure is so very unlike the fossil that, were it not for his description and statement of locality, it would be impossible to recognise it. Brongniart has only copied Wahlenberg, who appears to have been chiefly attentive to the course of the facial suture, and allowed his artist to take his own course.

To make evident the distinctions between our British fossil (which is certainly very nearly allied to the Scandinavian one), and the true *P. mucronatus*, I subjoin a figure from one of Sir R. I. Murchison's specimens.

FIG. 10.

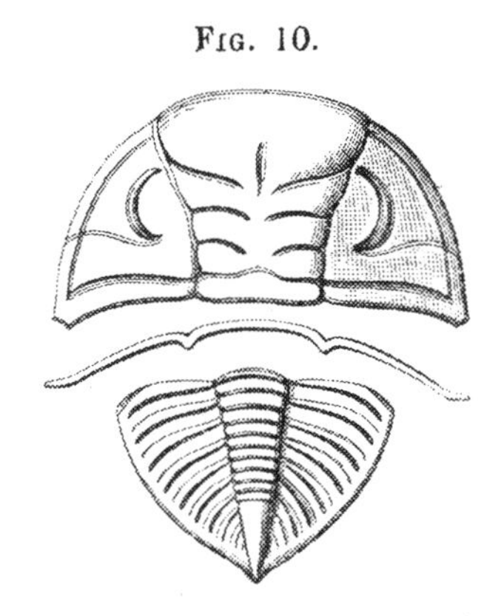

Swedish specimen of *P. mucronatus*. Lower Silurian; Ålleberg Mountain, Vestrogothia.*

It will be seen readily that the head we here figure has less regular and equal lobes than the British one, and this is a character of high importance in the family. And the tail, while it has a great resemblance, has not the strongly curved and deeply impressed ends to the side furrows, which give so conspicuous a character to our fossil. I am justified, therefore, in placing a ? against the same: indeed, had it not been already referred in the Decades to *P. mucronatus*, I would have preferred keeping it distinct under the name *P. appendiculatus*. For the present I leave it in doubt.

Description. Our specimen of the head only shows a broad glabella with three equal pairs of lobes, which are divided by nearly straight somewhat radiating furrows: the two lower pair of lobes are linear and direct; the upper somewhat clavate, but not triangular, and separated by a slightly sinuous furrow from the wide transverse forehead-lobe. All the lobes are tumid; and a convex median ridge, of equal breadth throughout, divides the lobes of one side from those of the other. A deep central pit in the cast occurs upon the forehead-lobe, as in many species.

Of the triangular tail we have more complete evidence, though even of this only a fragment.

The axis is broad and convex, of nine strong rings, and a terminal portion or appendix, of which we do not know the length. While the axis is convex, the sides are very flat, have the forward edge strongly curved, and are deeply grooved by seven arched duplicated furrows. The backward curve of these near the margin is considerable, and the secondary furrows are as strong or stronger than the primary, and coalesce with them at the tips in deep grooves, so as to undulate the very narrow margin considerably.

* "*Loc. Nat.* In stratis argillaceis *E*, Vestrogothiæ, in montibus Alleberg et Mössberg, alibique." Angelin, 'Palæont. Suecica,' part 1, 1852.

Locality.—CARADOC SLATES of Pen-y-Rhiw, west of Bala, over the volcanic ash-bed. (Mus. Pract. Geology, figs. 11, 12.) The other locality given in the Decade above quoted is erroneous; the specimen referred to is a species of *Ogygia*, which will be figured further on.

PHACOPS (ODONTOCHILE) IMBRICATULUS, *Angelin?* Pl. IV, fig. 10.

[PHACOPS IMBRICATULUS, *Angelin*, Palæont. Suec. t. vii, fig. 5, 1852. ?]

This I do not wish yet to name. It is a new British species, but scarcely perfect enough to assure me it is not the same as one of Green's published casts.

The pygidium is seven-eighths of an inch long, excluding any mucro it may have possessed, and the width about one and a half inch. It is rather flat, the centre raised in a low pyramidal form, and the sides sloping outwards and gently curved down near the very narrow, regular, smooth margin. We have two specimens, in both of which the axis is not one fourth the entire width at tip; it is regularly conical to within a very short distance from the margin, annulated by about twelve or thirteen flattened rings, and has an appendice beyond, as far as the prominent tip; the rings of the axis are angulated forwards. The axal furrows slight but distinct, the pleural furrows on the side-lobes widely patent, and nearly straight, except a slight angular backward bend near their abrupt ends. There are ten lateral ribs, all strongly duplicate to their very tips; the intermediate furrows not keeping quite separate, but combined at the ends, and nearly as strong as the principal ones. This strong duplication gives a very striking character. The incurved margin is regular and narrow all the way round.

It is very like *P. mucronatus* from Sweden; but perfect specimens of that species (see woodcut, supra p. 47) show that species to have had a narrower tail with wider margin, and the head is flattened, with arched lower glabella-lobes. It must be distinct. But the *P. imbricatulus*, on the other hand, is quite like ours; it has ten dichotomous side-ribs, and a very distinct margin. The axis has ten distinct, and five or six terminal joints. The tail is mucronate. Altogether it is very like our species.

Locality.—MAY HILL SANDSTONE (UPPER LLANDOVERY), of Nash Scar, Presteign, collected by Mr. J. E. Davis, of the Middle Temple. (Mus. Pract. Geology.)

Ph. (Odontochile) caudatus, *Brünnich.* Pl. III, figs. 4—18, and Pl. IV, figs 1—5, including varieties.

Trilobus caudatus, *Brünnich.* Kjobenh. Sellsk. Skrifter, Nye Samml., vol. i, p. 392, 1781.
— — *Schlotheim.* Nachträge, 35, 11, 1823.
Asaphus — *Brongniart.* Crust. Foss., t. ii, fig. 4 (not iii, fig. 9), 1822.
— — *Dalman.* Palæadæ, t. ii, fig. 4, 1826.
— — *Green.* Monogr. Tril. N. Amer., cast 17, 1832.
— — *Buckland.* Bridgw. Treat., pl. xlv, figs. 9—11, 1836.
— — *Murchison.* Sil. System, pl. vii, fig. 8 *a*; and *A. tuberculato-caudatus*, ibid., fig. 8 *b*, 1837.
— — *Milne-Edwards.* Crust., vol. iii, p. 308; and *A. tuberculato-caudatus*, 1840.
— myops, *König.* Icones Sectiles Foss., fig. 53, 1825 (our Pl. III, fig. 13, is from the original specimen in the British Museum).
Phacops caudatus, *Emmerich.* Dissert., 1839.
— — *Goldfuss.* Neues Jahrb., 1843.
— — *Burmeister.* Org. Tril., 2nd edit., t. iv, fig. 9, 1846.
— — *Salter.* Decade ii, Geol. Surv., 1849 (exclude syn. *P. longicaudatus*), pl. i, figs. 1—12, and fig. 15.
Dalmania caudata, *Emmerich.* Neues Jahrbuch, 1845.
— — *Salter.* Memoirs Geol. Surv., vol. ii, pt. 1, 1848.
Odontochile caudata, *M'Coy.* Synopsis Pal. Foss. Woodw. Mus., p. 160, 1851; Pl. III, figs. 4—17.
Var. α, vulgaris. Most of the above synonyms, Pl. III, figs. 4—17.
Var. β, tuberculato-caudatus, *Murchison.* Pl. IV, fig. 1.
Var. γ, nexilis, *Salter.* Pl. IV. figs. 3—5.
Var. δ, aculeatus, *Salter.* Pl. III, fig. 18.

P. (*Odontochile*) *magnus,* 3—4*-pollicaris; ovatus, sub-convexus, capite angusto, fronte plus minusve angulato. Oculi magni, pyramidati. Pleuræ sulcatæ* (*fulcro ad dimidium posito*) *posticæ recurvæ. Cauda apice brevi-mucronato, axi* 11—12*-annulato; costis lateralibus* 7, *rarius* 6, *arcuatis, duplicatis, ad marginem lævem abruptè terminatis, recurvis.*

Four of the varieties are here figured, and I have no doubt there are others. Those selected illustrate the very close connection that exists between this fossil and the so-called species *P. longicaudatus.*

In the Decade of the Survey above quoted I ventured to unite these two species, and have still little doubt that the series of figures here given will convince naturalists that all are derived from one stock. But it will not be so easy to convince paleontologists; and as there really are definite characters on which we can rely, it is thought best to follow the general practice, and describe them separately.

The cabinets of Dr. Grindrod, of Malvern, Mr. E. Hollier, Mr. Tennant, Mr. Mushen, and the British Museum, have furnished the chief materials.

Description.—Four or five inches long; ovate; length as to width varying from 10 to 6, to $8\frac{1}{2}$ to 6; gently convex; the thorax longer than either head or tail; the latter about equal to the head, excluding the produced mucro; axis of the body much more than half the width of the sides.

Head semicircular, exclusive of the slightly produced front or the long stout posterior horns, which are broad at the base and reach the seventh or eighth thoracic ring. Glabella coarsely tubercular, widened above, more than one third the width of head; the forehead-lobe a transverse rhomb, separated from the upper side-lobes by deep, broad, oblique furrows, which nearly meet; the lateral lobes and neck-lobe all nearly equal in width, and together equalling the forehead-lobe; the first and second lobes run straight across, and leave but a narrow ridge down the middle; strong axal furrows separate the glabella from the triangular cheeks, on the surface of which the large pyramidal eyes occupy a variable extent; a lunate depression surrounds their outer edge, and this is sometimes extended over more than half the width of the cheek, sometimes barely half, giving a considerable difference in appearance. The length of the eye, which is sharply curved, is uniformly from the first basal furrow to the middle of the upper one; eyelid with a deep concentric groove; lentiferous surface considerably broader forwards; lenses about 240, 8 or 10 in a vertical row. In several perfectly preserved specimens the cornea is present, and it is distinctly convex over each lens, the intermediate surface being ornamented with tubercles and granules, so as to leave no doubt of its being the true exterior; it is, too, continuous with the other portions of the preserved shell. The cornea does not rise, as in some other species, into elevated ridges between the lenses, but is flat in some specimens, in others it sinks down between them. Occasionally a lens (or two) is undeveloped and very small, among other perfect ones. The facial suture exactly circumscribes the glabella in front, falls perpendicularly to the eye, and arches outwards and upwards from its lower angle to the outer margin of the head, on the lower face of which it extends along the margin further back than on the upper surface (Pl. IV, fig. 4). The cheek-pieces are united in one (as in all the genus) across the front, and there is no extra piece (or rostral shield) inserted between them above the labrum.

The latter is parabolic, obtusely pointed, the sides a little waved, and the base of attachment extended; a transverse furrow within the lip is connected with a marginal longitudinal one on each side; a pair of lateral strong indentations indicate a second furrow above.

Round the head a strong furrow separates a broad margin, but ceases at the base of the spines, and there nearly meets the equally strong neck-furrow, which is arched down at its end. The margin in front of the head is more or less produced into an obtuse point.

Thorax with the central lobe not separated by any strong furrow from the pleuræ, which in English specimens are half as wide again as the axis; the axis is a little fusiform, not broader than the base of the glabella, and either has the rings smooth, tubercled at their outer edges, or with two tubercles on the central prominent part. All these variations occur on the same thorax. The front pleuræ are flat for the first half, then gently curved down, and a little backwards, at the fulcrum; they are divided by a moderately strong curved furrow for most of the length, and their edge sharpened anteriorly (the flat surface not crossing the furrow, as in Acaste), for rolling or bending. The extremities are truncate, and pointed at the hinder angle; posterior pleuræ with the fulcrum at less than half, much more curved back and strongly pointed, overlapping the upper corners of the tail. Whole surface of thorax finely granular.

Tail with the sides meeting below at nearly a right angle, and produced more or less to a sharp point beyond this; axis not much raised, and gently tapering, ending a little abruptly at some distance from the point; it has eleven or twelve ribs, which are occasionally tubercular, like the thorax, sometimes much so; the sides have six, seven, or sometimes eight narrow ribs, strongly arched down at their ends, where they abut suddenly on the smooth margin; each is sharply defined behind, and duplicated on its forward edge by a fine rib along its whole length. The number of ribs appears to be as great in the young trilobites as in the adult form.

The above description, extracted from the Decade, will serve for the general account of the species. I distinguish the following varieties more particularly.

VAR. *α*, VULGARIS. Pl. III, figs. 4—17.

Form narrow-ovate. Head broad, rounded, or slightly angulated only in front. Glabella as broad as the cheeks. Tail truly triangular, simply pointed, scarcely at all acuminate.

There is no essential difference between any of the specimens of this common variety. All have the rounded or arched front margin (not at all produced) to the head; the glabella at its widest part much more than one third the whole width of the head. The length of the head as near as possible half the width of the same; the axis broad; the tail shorter than the width, and simply pointed, with a very short mucro, or none at all.

They vary a little among themselves in the following particulars. Some have the glabella more strongly marked out, and more decidedly narrowed behind. Some have the glabella-furrows a little deeper. The eyes are larger in some (as in the varieties of *P. Downingiæ*, before described), and more prominent; but I have elsewhere referred these differences to sex. The male had probably more prominent eyes.

Some specimens have the characteristic tubercles on the sides of the axis, and the central ones on the tail much stronger than others. We have figured these tubercular varieties in Pl. III, figs. 15—17. As these seem to be chiefly characteristic in the young state, and are much more obscure in the full-grown animal, I will here notice them particularly as they occur in the young and older states.

Young specimens. Pl. III, fig. 15; Pl. IV, fig. 2 (see also Decade 2, Geol. Surv., pl. i, fig. 7).

These differ in several particulars from the adult. The head-spines are smaller in proportion, and the head-margin very narrow. The tail, too, is much smaller proportionally, and more pointed, even mucronate, the mucro being in some cases nearly as long as the tail.* The sides of the tail have fewer ribs, and the axis is not so strongly marked out as in the adult.

In the very neat and characteristic young specimen, Pl. III, fig. 15, while the general character of var. α is preserved, it is in some respects like the var. γ, hereafter described.

Surface-ornaments and Tubercles.—The tubercles of the surface are larger in proportion in the young specimens. On the axis they are particularly strong. In one of Mr. Mushen's specimens (supposed to be a distinct species by De Koninck, and called by him *P. Mushenianus*, MS.), the second, fifth, and eighth thorax-rings are ornamented by a pair of tubercles, the fourth and seventh rings of the tail-axis, and the first and fourth lateral ribs. Our figured specimen (Pl. III, fig. 15) has the same ornaments, but they are too small to be shown in the figure. This central double row of tubercles is very conspicuous, and must surely have some special meaning. On all the young specimens a distinct central space is marked out by these pairs. On all, the forehead-lobe of the glabella has one pair wide apart and one pair more approximate. Each of the hinder lobes of the glabella has a pair (in one of Mr. Mushen's fossils even the front head-margin has a pair).

Then, as before stated, the second, fifth, and eighth thorax-rings show them—the fourth, seventh, eighth, and twelfth of the tail, &c. And so we get a median line of double tubercles, which are only more conspicuous in the young because in these the tubercles of the general ornamented surface are not so large; and consequently these, which do not much increase in size as the animal grows, lose their importance, so far as the exterior is concerned.†

FIGS. 11, 12.

Symmetrical cutaneous glands in the crust of P. caudatus.

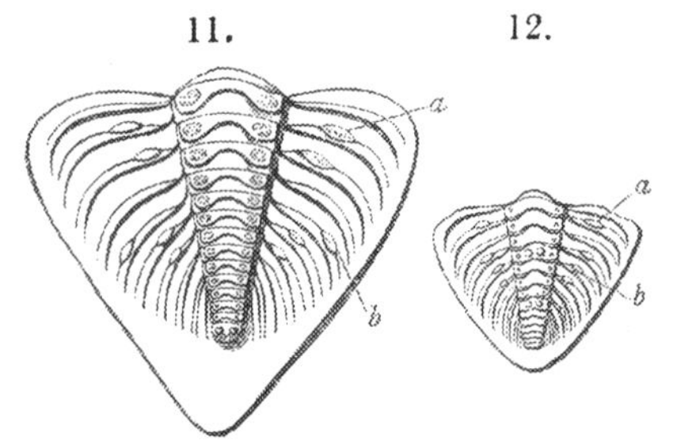

11. Interior cast from Ledbury.
12. Exterior of Malvern specimen. Both in Mus. Pract. Geology.

But that these tubercles are of real significance in the structure of the animal is evident enough, if we examine the fine interior cast figured in Pl. III, fig. 12. I have reproduced this specimen in the woodcut (fig. 11), and also the exterior of the tubercular variety (fig. 12), which is the same specimen

* This character is in ordinary cases eventually lost, the mucro not growing in proportion to the rest; but occasionally it is preserved, as in var. δ, *aculeatus* (Pl. III, fig. 18).

† M. Barrande has specially noted the same fact, with regard to the *Sao hirsuta* of the Bohemian basin.

as that given in Pl. III, fig. 17. It is from Malvern, as is also fig. 16, with the same prominent ornament.

Here, in this full-grown and very large specimen, it is shown that not only are the tubercles more than superficial ornaments, for they are distinctly marked out by shallow oval pits within the crust, but that they are actually far more numerous within, a pair occurring in fig. 11 on every joint of the axis. It will be observed that they occur in the adult in precisely the same relative position as in the young, viz., externally at intervals on the axis, and on the second, third, fifth, sixth, and seventh lateral ribs, or rather on the intermediate duplicating furrows of these ribs.

Their position on the side ribs is remarkable. The first pair *a*, occur on the second rib about half way out; the next pair closer in to the axis. In like manner those on the fifth segment *b*, are far out, and those on the two succeeding ribs gradually close in. This regularity of position should be taken into account, as it indicates that there is some structural importance in them.*

Var. β, TUBERCULATO-CAUDATUS. Pl. IV, fig. 1.

PHACOPS TUBERCULATO-CAUDATUS. *Murchison*, Sil. System, pl. vii, fig. 8 *b*, 1837.
— CAUDATUS. Siluria, 2nd ed., pl. xviii, fig. 1, 1839.

Form broad-ovate. Head decidedly angulated in front. Glabella not so wide as the cheeks. Tail triangular, pointed, and a little produced.

Comparing this with the ordinary variety, we find that the principal differences reside in the generally broader form, the length being to the breadth about as 6 to 4, or rather longer. This excludes the extreme mucro, which is broken off in our largest specimen.

Besides this, the margin of the head is broader, and more sharply distinguished from the head by a deep marginal furrow. The glabella is somewhat narrower and shorter, with few large tubercles. The eyes quite as large and conspicuous as in the larger-eyed forms of the ordinary variety, and which I suppose to be the male (♂), and the front is more angular; it is produced, but not nearly to the extent of the anterior mucro in the next species, *P. longicaudatus*.

The thorax has remarkably large tubercles on the axis, and a central pair on the second ring, and the pleuræ on this specimen are tumid at their origin.

The axis of the tail is short and rather flat, and the side lobes have seven or eight very prominent ribs; on the fourth, fifth, and sixth of which the remarkable gland-like tubercles show strongly. The upper one on the first rib is not so prominent. The tail is strongly margined, and the mucro (broken off) must from its shape have been short and rather suddenly acuminate.

* They are occasionally even more numerous than here indicated, occurring as tubercles on several of the lateral segments of the body-rings, as well as on those of the tail (specimen in Mr. Edgell's collection), see also Pl. III, fig. 23.

These are all the characters that appear to me worthy of note in this fine variety. I think the large specimen figured by Murchison must belong to it; the tubercles are arranged in the same fashion. But, the front being broken off, it is not quite certain. Nor do I know where to find that specimen.

Locality.—WENLOCK LIMESTONE, Dudley? (Murchison); LOWER LUDLOW, Ledbury Tunnel (Dr. Grindrod's cabinet); Dudley Tunnel (Mr. Edgell's collection).

Var. γ, NEXILIS. Pl. IV, figs. 3—5.

P. CAUDATUS, *Salter*, Decade Geol. Surv., ii, pl. i, figs. 10, 11, only.

Form broad-ovate. Length to width as eight and a half to six. Head arched, obtusely pointed in front, and more than twice as wide as long. Glabella very narrow, not nearly one third the width of the head. Tail broad-acuminate, not produced, its axis very narrow.

This neat variety was figured in Decade 2 of the Geol. Survey, but does not seem to have been specially noticed by any author. It is not very common, except at one locality, where it is in profusion. At Vinnal Hill, Ludlow, the fine mudstones of the Lower Ludlow Rock are full of this trilobite.

It is remarkable for the narrow axis to the head, a good deal resembling in this particular the next species, but differing entirely in the extremely short general form, which is rounder than even the common variety, and with scarcely an angular, much less a mucronate, front. The forehead-lobe is large in proportion to the others; the upper lateral lobe not much larger than the rest; and the tail, abruptly acuminate, and very broad for the species, has a very narrow axis, like the next species. The side-lobes are marked by the usual seven curved ribs, strongly duplicate.

FIG. 13.

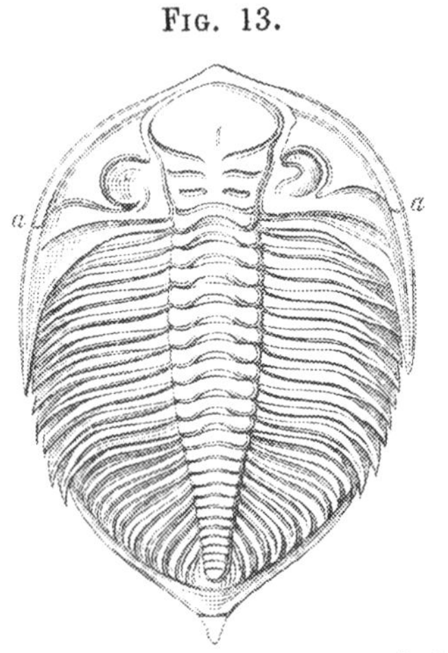

P. caudatus, var. γ, *nexilis*, Lower Ludlow, near Ludlow. Mr. W. Edgell's cabinet.

Our Pl. IV, fig. 4, is the same specimen as that figured in the Decade 2, pl. i, and shows the true shape of the first or front ring, bearing the eyes in this subgenus. The narrow riband-like band in front of the glabella is all on the upper surface, not, as in the previous subgenera, marginal, or beneath the margin.

This last variety brings us very near to the characters of the *P. longicaudatus*, but has still the abrupt, not the mucronated, front. And while the axis is very narrow, the general shape is rounder, instead of longer, than the ordinary forms.

Var. δ, ACULEATUS. Pl. III, fig. 18.

I have before referred to this rather uncommon form (p. 52, note).

Localities.—[It has been quoted from Lower Silurian rocks, by myself (the Caradoc strata of Robeston Wathen and Narberth, in Pembrokeshire, for instance); but I do not know that it really descends below the MAY HILL SANDSTONE.]

MAY HILL SANDSTONE (or Upper Llandovery, Murchison), Dingle, Ireland; Barr, Walsall. WOOLHOPE LIMESTONE and SHALE, Malvern; Presteign; Woolhope; Walsall, and from the DENBIGHSHIRE GRITS of N. Wales. WENLOCK LIMESTONE and SHALE; the Malverns; Shropshire; North and South Wales, everywhere; Westmoreland; Dingle, &c. LOWER LUDLOW and AYMESTRY LIMESTONE of Shropshire and Herefordshire; of Dudley Tunnel; of Malvern. UPPER LUDLOW, Ludlow; rare in this formation (R. Lightbody).

The above are localities in general for the common variety. Var. β is found in WENLOCK LIMESTONE (Murchison), and our figured specimen from LOWER LUDLOW, Ledbury tunnel, in Dr. Grindrod's cabinet. Var. γ is only known as yet in LOWER LUDLOW rocks, and specially at Vinnal Hill, Ludlow (Mus. Pract. Geol.); also at Newton, in the same neighbourhood (Mr. Edgell's cabinet). Var. δ is from Dudley.

PHAC. (ODONTOCHILE) LONGICAUDATUS, *Murch.* Pl. III, figs. 19—28.

TRILOBITE WITH CAUDAL PROCESS, *Parkinson.* Organic Remains, vol. ii, t. xvii, fig. 17, 1811.
ASAPHUS LONGICAUDATUS, *Murchison.* Silur. Syst., pl. xiv, figs. 11—14, 1837.
— — *Milne-Edwards.* Crustacés, iii, 308, 1840.
PHACOPS LONGICAUDATUS, *Emmerich, Goldfuss, &c.* Loc. cit., 1845, 1846.
— MUCRONATUS, *Burmeister.* Org. Tril., t. iv, fig. 9 (exclude synonyms), 1843, and ed. 2 (Ray ed., 1846).
— CAUDATUS, β LONGICAUDATUS, *Salter.* Decades Geol. Survey, ii, pl. i, figs. 13, 14.
Var. α, ARMIGER. Pl. III, figs. 19—21 (*P. longicaudatus* of authors).
Var. β, GRINDRODIANUS. Pl. III, figs. 22—28, new variety.

P. (*Odontoch.*) *medius,* 2—3-*pollicaris, ovatus elongatus subconvexus, capite lato, fronte mucronato. Oculi maximi. Pleuræ sulcatæ recurvæ, fulcro ad dimidium. Cauda trigona longimucronata; axe* 15—16-*annulato, ad apicem haud eminente; costis lateralibus* 7 *arcuatis, duplicatis, ad marginem lævem abruptè terminatis, recurvis.*

Distinguished easily at a glance from the more common *P. caudatus* by the greatly elongate tail, but much less distinct when closely examined, since the general habit and even the minute structure is similar to that of the Wenlock species.

In all the specimens I have seen (and there are few perfect ones extant of the long-tailed species) the following differences are observable:—1st. The axis is invariably narrower, not one third the width of the head, always quite as much so as in the *P. caudatus,* var. *nexilis,* which differs from the other varieties in this particular, and yet has *less* trace of a mucronate front than any of them. 2nd. The abrupt mucro (not mere angulation) to the front. 3rd. The smaller tail, with more arched furrows. 4th. The great terminal mucro, always equal to the length of the tail itself, and often half as long again.

There is evidently, however, a nearer connection between the var. β. (*Grindrodianus*) from Malvern, and the var. γ of the last species. But while in this variety β, which comes nearest to *P. caudatus*, the mucrones of the head and tail are shortened, the narrow axis still easily distinguishes it from the varieties of that species; and the variety γ, *nexilis* of the *P. caudatus*, which has really a narrow axis, has nearly lost all trace of the processes, and has scarcely a mucro at all. It is not difficult, therefore, to distinguish the two species.

Description.—Full-grown specimens must have been fully six inches long, including the tail. The form is elongate, and less convex than in *P. caudatus*. In both varieties the head is less than twice as wide as its length, but is still transverse in shape, its glabella scarce one third the whole width. The glabella at its front part is parallel-sided. Forehead-lobe transverse-oval, well separated from the side-lobes, which are tolerably equal,—the upper one not greatly largest. Eyes large, pyramidal, and placed very near the glabella, so as almost to touch it in front, and strongly curved,* reaching below to the neck-furrow, and often overhanging it.

The axis of the body and tail narrow in proportion to the sides, sometimes not one fourth the width in the body-rings, and barely one fifth in the tail. The surface of the body-rings is tubercular. Hinder pleuræ much bent back. Fulcrum at about one half. Pleural groove shallow, ending in a point beyond the fulcrum. Tail small in proportion, with few (5—7) side-furrows, much curved and strongly interlined. Mucro abrupt, linear, as long or longer than the rest of the tail. The gland-like tubercles sometimes present, but often inconspicuous throughout.

The above characters are common to the two following varieties.

Var. *a*, ARMIGER. Pl. II, figs. 19—21.

FIG. 14.

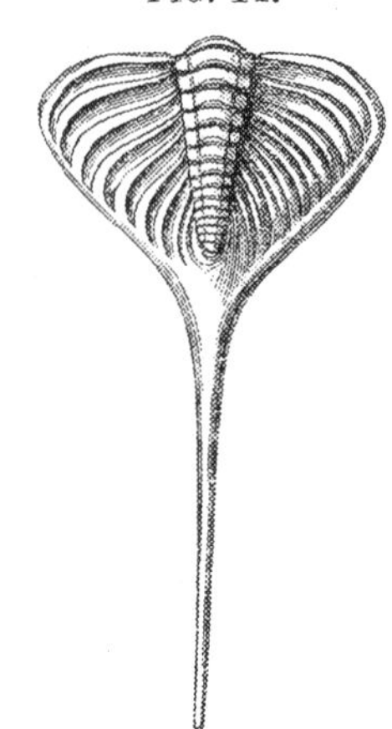

Head, including the strong front mucro, much more than half as long as wide. Glabella not one third the width of the head. Marginal furrows strong. Head-spines as long as the cheek. Mucro of tail much larger than the rest of the tail; side-furrows 7.

I figure a remarkable specimen from Mr. Edgell's cabinet, in which the spine is considerably more than twice the length of the axis, and is expanded at the tip, more than is shown in our woodcut.

One specimen of Dr. Grindrod's shows clearly the passage from this form to the next variety, *P. Grindrodianus*, having the mucro quite as long as usual, but having only five lateral ribs, and tubercles on the axis and sides. Usually *P. longicaudatus*, var. *armiger*, shows seven lateral ribs, and has no distinct tubercles.

* Much more so than in *P. caudatus*.

Localities.—Var. *a*, WOOLHOPE LIMESTONE, Presteign; WENLOCK SHALE of Cheney Longville and Burrington, Shropshire (abundant); Usk, Monmouthshire; also Carmarthenshire, Brecon; and Marloes Bay, Pembrokeshire, more rarely. It is quoted also from WENLOCK LIMESTONE, Malvern, in the Survey lists.

Var. *β*, GRINDRODIANUS. Pl. III, figs. 2—28.

Head, including the short front mucro, somewhat more than half as long as wide. Glabella, at its widest, not more than one third the width of the head. Forehead-lobe prominent. Tubercles on head, body, and tail, conspicuous. Tail with few (5—6) side-furrows, and about ten axal rings; the spine not more than equal to the rest of the tail.

This variety, of which we fortunately can examine a fine suite of specimens, differs from the typical form in points which do not affect the specific characters given above. It is only less extravagantly armed,—the front, rear, and side-spines being all shorter. The side-spines reach to the seventh segment. The surface shows the tubercles characteristic of the young,—in the adult state; both on the forehead-lobes and the lower lobes of the glabella, on the axis of the body,[1] and on the fourth and eighth segment of the axis of the tail. The gland-like tubercles on the fourth side-pleuræ are not always conspicuous.

The mucro is shorter, seldom equalling the tail itself in length. The side-ribs are five, strongly duplicate to the very end, and the margin very narrow.

Locality.—Var. *β*, WENLOCK SHALE. It is known only at Malvern Tunnel and Dudley.

PHACOPS WEAVERI, *Salter*. Pl. III, figs. 1—3; Pl. IV, figs. 6—9.

PHACOPS WEAVERI, *Salter*. Decades Geol. Surv., ii, art. 1, p. 7, pl. i, fig. 16.
— — *Id.* Morris's Catal., 2nd, ed., p. 114, 1854.

P. lævis, capite quam in P. caud. *nisi trigono lobisque glabellæ tumidioribus. Cauda triangularis, fere æquilatera, multicostata, convexa, apice acuto brevimucronato; axi* 13—16-*annulato, costis lateralibus* 10—12 *simplicibus, vix curvis, ad marginem angustum abruptè terminatis.*

We know but little yet of this fossil. It is fragmentary as yet; and though I believe the specimens here figured all belong to one species, I would prefer to keep the name for the fossil of the May Hill Sandstone.

FIG. 15.

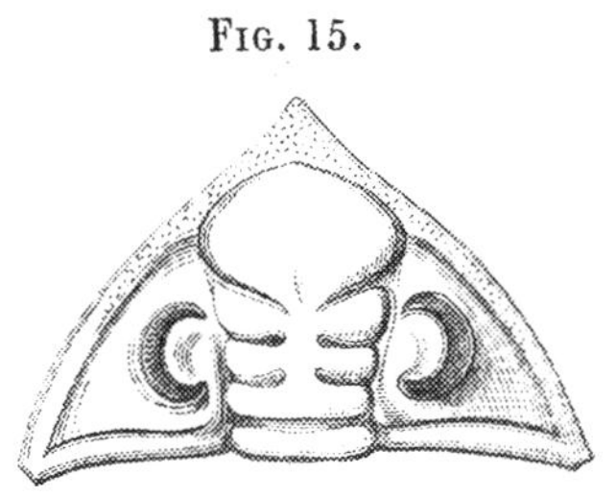

Head of *P. Weaveri?* Canada.[2]

The head is very imperfect; what there is of it strongly resembles *P. caudatus*, except that the lobes are a little more tumid; the forehead-lobe is more strongly marked

[1] It seems that when the tubercles exist on the sides of the axis, they are not present towards the centre, and *vice versâ*. But it is not every segment which has the tubercles at all.

[2] I have good reason to believe this Canadian species, brought by Sir W. Logan from Ile Percè,

8

off, and somewhat overhanging the side-lobes. The eye does not reach the neck-furrow.

The tail is broad-triangular, wider than long; the sides a little convex; the apex short-mucronate; the axis narrow, conical, ribbed by about sixteen rings; the sides very convex, with nine to ten arched simple ribs, scarcely at all interlined; the margin narrow, smooth.

Locality.—May Hill Sandstone, "Long's Quarry, Damory Bridge, Charfield, and Woodford Green, Tortworth;" (Decade 2, Geol. Surv., p. 7).

Figs. 8, 9, show a variety with ribs slightly duplicated on the interior cast, but with a short mucro. It is from Marloes Bay, Pembrokeshire (Mus. Pract. Geology).

Larger form of the species.

Pl. III, figs. 2, 3.

The large caudal shields here figured clearly belong to a species distinct from *P. caudatus*. They are one and a half inch long, and probably reached two inches in width when perfect. The general form agrees well with that above described, and it has the numerous, slightly curved ribs, scarcely duplicated at all, and with the interstices deeply concave. They are very little bent at the extremities. Our fig. 2 is better in this respect than fig. 3. The margin is equal and narrow all round.

P. imbricatulus of Angelin has ten dichotomous side-ribs, and a very distinct margin; else it is not unlike our species.

Locality.—Ludlow Rocks? Horseshoe Farm, Tortworth. It is possible there is a boss of May Hill Sandstone at this place.

Section—Cryphæus (p. 15).

This subgenus alone would decide against the folly of classifying Trilobites by such a character as the pattern of the tail. All the species known in Europe or North America are allied to *Odontochile*, and follow naturally that subgenus. But in the southern hemisphere we have compact forms allied to *Acaste*, which have the tail spinose.

Gaspè, Lower Canada, to be identical with ours. The tail is a little more mucronate. The head agrees nearly with our fragments, and will at all events give collectors a good idea of the probable shape. It is remarkably triangular, even for this subgenus.

PHAC. (CRYPHÆUS) PUNCTATUS, *Steininger?* Pl. I, figs. 17—19.

OLENUS PUNCTATUS, *Steininger*. Mém. de Soc. Géol. Fr., i, p. 356, vol. i. pl. 22, 1833.
— — *Emmerich*. Dissertation, p. 55, 1839.
ASAPHUS ARACHNOIDES, *Goldfuss*. Leonhard und Bronn's Neues Jahrbuch, p. 561, tab. v, fig. 3, 1843.
CALYMENE — *Höninghaus*, pamphlet, Crefeld, 1835.
PLEURACANTHUS ARACHNOIDES, *Milne-Edwards*, Crust., iii, p. 329, 1840.
— PUNCTATUS, *Roemer*. Rheinisch Uebergangs., p. 82, 1844.
PHACOPS ARACHNOIDES, *Burmeister*, Org. Trilob., 1st ed., 1843; 2d ed., p. 96, 1846.
— LACINIATUS, *Salter*, in Morris's Catalogue, 2nd ed., p. 113, 1854 (not of *Roemer*).

Phacops (*Cryphæus*) "*scuto capitis in medio marginis antici acuto, angulis posticis valde productis; limbo scuti caudalis decies spinosa.*"—Burm., l. c.

The only notices I can find of this as a British species are by myself—first, in the Decades of the Survey, No. 2, under the description of *P. caudatus*, p. 8, where it is referred to *P. arachnoides* of Goldfuss; and afterwards in Morris's Catalogue, 2nd ed., 1854, where it is referred, I think wrongly, to *P. laciniatus*, Roem. The fragments known were so very obscure, that enough could only be seen to make it certain we had one of the species of this remarkable group, in which the tail-margin has undergone its fullest expansion.

In order to give the English collector a notion of what he may expect to find, I have drawn a perfect specimen from Germany, nearly following Burmeister. And, after comparing with the best specimens at my command, I think the first reference was right, and that the British species is the well-known, but still rarely perfect, *Olenus punctatus* of Steininger, a name which should have been retained for this fossil, as Steininger's figure is quite recognizable.

FIG. 16.

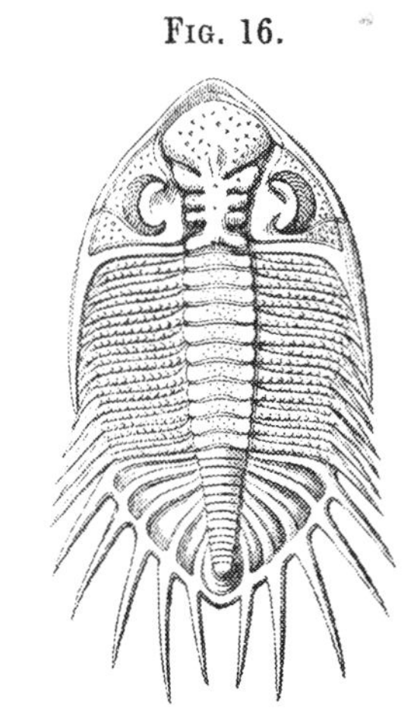

Phacops (*Cryphæus*) *punctatus*, Stein., from Burmeister's work.

I purposely avoid describing the British fragments, as the inspection of the plate will show that they only indicate a species with narrow, long glabella, with tolerably equal lobes, a granulated surface, a pointed front to the head (not produced, however, as in the foreign specimens), and longer head-spines than in the German specimens. The eyes appear to be not quite so large.

The tail shows ten spines, which are rather shorter, less convergent, and more radiating than Eifel specimens warrant.[1] But they are narrow, not broad, fin-like plates, as in *P. laciniatus*, and longer, too, in proportion than in that species; the glabella, too, is nar-

[1] Goldfuss's figure is more faithful than Prof. Burmeister's.

rower and more pointed, and the lobes less crowded. It need not be compared with *P. stellifer*, Burm., as that species has eleven broadish spines.

We must wait a little for more perfect information.[1] It is certain that a very beautiful Trilobite is yet to be found in Cornwall and South Devon, and collectors should look carefully for it.

Locality.—Middle and Lower Devonian. Torquay, South Devon; Slates of Liskeard, Cornwall (Coll. Mr. Pengelly and Mus. P. Geology).

Family—CHEIRURIDÆ.

A group whose limits are not yet fully defined, if, as I suppose, it should include all the forms cited below. The Cheiruridæ proper have really more of the characters of the Phacopidæ than appears at first sight, for these loosely built Trilobites have the eyes reticulate externally, though on a smaller scale than *Phacops*, and the facial suture ends upon the exterior, not the posterior, margin. The normal number of body-rings is eleven. With these points of resemblance, important as they are, the analogy ceases, for the short, oblique furrow and free ends of the nodular pleuræ give a very marked aspect; and the tail is of few joints, and the ends of their pleuræ are also free.

The genera thus included are *Cheirurus* proper and its subgenera, and *Sphærexochus*, with *Amphion* and *Placoparia*.[2] *Staurocephalus* and *Deiphon* must follow these.

But the other group, *Encrinurus*, *Cybele*, *Cromus*, *Zethus*, &c., though they agree in the marked position of the facial suture, have so many differences of habit that I am doubtful if they should not form a distinct family. The long prominent eyes may or may not be granulated; we have no evidence of it. But there is a strong resemblance to the family of the Acidaspidæ, which follows next, and I shall reserve the description of both of these groups till all the available material is collected.

CHEIRURUS, *Beyrich*, 1845.

Head strongly trilobed; glabella with three lateral lobes, the basal ones circumscribed; eyes facetted minutely; facial suture ending on the external margin [under side of head with a transverse rostral shield, *Barrande*]. Cheeks generally scrobiculate. Labrum inflated, oblong, truncate, with a pair of furrows and small lateral auricles.[3] Thorax of ten, eleven, or twelve joints, variously sculptured in the different subgenera.

[1] Goldfuss's introductory remark to his 'Essay on Trilobites' is still very applicable to the group, if not quite germane to the subject. I beg to refer the curious reader to it ('Neues Jahrb.,' 1843, p. 537).

[2] There is also a blind Devonian genus of this group, from Africa, in Lower Devonian,—*Typhloniscus*.

[3] See Plate VI, figs. 12, 13. We want a term, and I propose the above for these lateral portions of the labrum, very distinct in *Cheirurus*, *Lichas*, &c.

Subgenera.

Crotalocephalus, Cheirurus, Eccoptochile, Actinopeltis; and most probably *Sphærexochus*, ought to be added. It is simply convenient to keep this last as a distinct genus.

Subgenus—CROTALOCEPHALUS, *Salter*, 1853.

Glabella oblong, broadest in front, with furrows continuous across. Body-rings 11, nodular.

Devonian. Upper Silurian.

Subgenus—CHEIRURUS, *Auctorum.*

Glabella broadest in front, with furrows all distinct, but not meeting across. Body-rings 11, nodular.

Upper and Lower Silurian.

Subgenus—ECCOPTOCHILE, *Corda*, 1847.

Glabella ovoid, not much inflated, with long furrows. Body-rings 12, grooved imperfectly, not nodular.

Upper Cambrian? Lower Silurian.

Subgenus—ACTINOPELTIS, *Corda*, 1847.

Glabella ovoid, often much inflated, and with the anterior furrows nearly obsolete. Body-rings 10, grooved imperfectly, subcylindrical, not nodular.

Many of these lead directly to the genus *Sphærexochus*, which has only the characters of *Actinopeltis* in an exaggerated form. Its labrum is like that of *Actinopeltis*, and its pleuræ are subcylindrical, and the body-rings 11.

Subgenus—CROTALOCEPHALUS.

CH. (CROTALOCEPHALUS) ARTICULATUS, *Münst.?* Pl. VI, figs. 7, 8.

CALYMENE ARTICULATA, *Münster.* Beiträge, vol. iii, pl. v, fig. 7, 1840.
— STERNBERGII, *Phillips* (not *Münster* nor *Boeck*). Pal. Foss., fig. 247, 1841.
CALYMENE — *Burmeister.* Organ. Trilob., 2nd ed., p. 116, 1846.
CHEIRURUS ARTICULATUS, *Salter.* Decade Geol. Surv. 7, pl. i, art. 1, p. 10, 1853.
— — *Id.* Morris's Catal., 2nd ed., p. 103, 1854.

C. capite $1\frac{1}{2}$ *unciam lato, longo, convexissimo, articulato. Glabella subcylindrica, anticè latior, lobo frontali semicirculari, supremo lineari, secundo subtriangulari, basalibus trigonis subplanis longis, ferè medium glabellæ attingentibus, cervicali triangulato lato. Genæ declives convexæ, profundè scriptæ et marginatæ, spinis divergentibus brevibus. Oculi haud ad glabellam appositi, ad lobum medianum retracti. Junior cum senioribus formâ convenit.*

When described for the Decades of the Survey, we possessed but one or two imperfect heads. Now we have excellent specimens from Mr. Pengelly's cabinet. These show the true shape of the glabella and the fixed—but not the free—cheeks. Lest there should be a mistake, I have only given the characters from the British specimens, which differ from Münster's figure in having a shorter glabella, with the terminal lobe especially not so long, (Münster's figure has it almost circular, which is not likely), and narrower cheeks; and that author also describes his species as having *great* wings, or free cheeks, like those of his *C. propinqua.* If his figures were quite trustworthy,[1] our species must be distinct, and I should, in that case, propose the term *C. Pengellyi* for this fine Devonian fossil.

Our species has the head eleven lines long, by one and a half inch broad, the glabella occupying the whole length, and regularly tapering from a breadth of eight lines in front to six at the base. The forehead-lobe is semicircular, and occupies just as much length as the remaining lobes, excluding the broad, subtriangular neck-segment. The middle and upper lobes are separated by complete, deep, transverse furrows, and are of equal width, the upper pair forming a transverse linear band, the middle pair also transverse, but rudely triangular where they abut on the truly triangular basal lobes. These reach far across, and are separated by a depressed space barely equal to half the width of one of the lobes. The basal lobes are less tumid than the rest, the upper lobe more so than the middle one, and the forehead-lobe more tumid than any. These proportions are maintained in young specimens (see fig. 8). The surface of the glabella is coarsely tubercular.

The cheeks, separated by deep axal furrows, are convex, rather steeply declining, their full width equal to that of the glabella at its base, and the position of the eye-line is such that the free cheek is not above one third the length of the head (in Münster's figure it is much more), with a strong thickened margin and marginal furrow, within which the cheek is deeply punctured, and appears tubercular; the head-spines diverge, and are short and narrow.

The profile-view (fig. 7, *b*) gives a good idea of the convexity of the head.

[1] Count Münster gives us good reason to distrust his accuracy, for he figures the lower glabella-furrows as continuous across, instead of bending down to join the triangular basal lobes. This is impossible. He has represented his so-called *C. Sternbergii*, and *C. propinqua*, which are probably but one species, as having these broad wings, but there is an appearance of error about all his figures.

The specimens figured are destitute of body and tail; perhaps our Devonian friends can supply better examples.

Locality.—MIDDLE DEVONIAN LIMESTONE of Lammaton, Torquay (Mr. Pengelly's and Mr. Vicary's cabinets, Mus. Pract. Geology).

CHEIRURUS BIMUCRONATUS, *Murch.* Pl. V, figs. 1—5 ; Pl. VI, figs. 9—18.

PARADOXIDES BIMUCRONATUS, *Murchison.* Sil. System, pl. xiv, figs. 8, 9, 1837.
— — *Milne-Edwards.* Crust., iii, p. 343, 1840.
CALYMENE SPECIOSA, *Hisinger.* Leth. Suecica, 2nd suppl., t. xxxix, fig. 2, 1840 (not of *Dalman*).
ARGES BIMUCRONATUS, *Goldfuss.* Neues Jahrbuch., p. 544, 1843.
CHEIRURUS — *Beyrich.* Ueber einige Böhmische Tril., pp. 18, 19, 1845.
— INSIGNIS, *Id.* Ibid., p. 12, fig. I, 1.
— — *Barrande.* Notice Préliminaire, p. 49. 1846.
— — *Id.* Syst. Sil. de Bohême, pl. xli, figs. 1—13 (optimè), 1852.
— SPECIOSUS, *Salter.* Memoirs Geol. Surv., vol. ii, pt. 1, pl. vii, figs. 4, 5, 6.
CERAURUS WILLIAMSII, *M'Coy.* Ann. and Mag. Nat. Hist., p. 408, 1849.
— — *Id.* Pal. Foss. Woodw. Mus., pl. i, F, fig. 13, 1851.
CHEIRURUS BIMUCRONATUS, *Salter.* Decade 7, pl. ii, 1853.

C. (*Cheirurus*) *grandis,* 8—10- *uncialis. Glabella supernè latior, sulco frontali et oculari obliquis propè medium glabellæ terminatis; lobis inferis trigonis* (*ætate rotundioribus*) *sejunctis. Genæ glabellâ angustiores, oculis medianis, spinis posticis parallelis. Thorax pleuris trituberculatis. Cauda parvula, utrinque pleuris tribus subæqualibus ad basin brevisulcatis, apicibus robustis in ætate arcuatis, sæpe mucrone centrali brevi.*

Few Trilobites are more common than this in the border-counties of England and Wales, and few are so seldom found in a perfect state. We owe the beautiful specimens here figured to the care and diligence of the Dudley collectors, who, following the good example long ago set by Mr. John Gray, are patiently developing the shale specimens which used to be neglected. It is found that the shale matrix preserves the fossils best. Mr. Gray is still at work, and is the master of this art. Our good friend, Mr. C. Ketley, of Smethwick, is hardly less skilful, and we are much indebted to both these gentlemen. Mr. Allport, of Birmingham, Mr. E. Hollier, jun., of Dudley, and others, have lent us choice materials. We now know the whole of this species, which, when first described, was one of the most obscure of all our Trilobites. It has a wide range, from Caradoc to Ludlow Rocks inclusive. It is found in North Europe, but not in America; and it is an excellent type of the genus *Cheirurus.* Neither unduly inflated in the head, nor greatly depressed, as in some of the Lower Silurian species, it is a typical form, with glabella-furrows running partly and not completely across, and with an oblong shape of glabella, the aberrant species having it oval, and some even globular, as we shall see further on.

It was first described as a *Paradoxides* by Murchison, who only knew the caudal portion; the head was figured about the same time by Hisinger, under the name *C. speciosa* of Dalman, an identification which there is now every reason to doubt, as the synonymy has been cleared up by the labours of Angelin, to whom we should be still more indebted were his descriptions less brief. As I have endeavoured in the Decades of the Survey[1] to show the history of the species, I need not go over all that ground again, and I retain the name *C. bimucronatus*, as already well known and adopted by several authors; indeed, it appears to have been the earliest.

Var. *α*, BIMUCRONATUS. Pl. VI. figs. 15, 16, 17. *Cauda mucronibus posticis angustis, centrali nullo.* Decade 7; Geol. Survey, pl. ii, figs. 4, 5, 6.

Var. *β*, CENTRALIS. Pl. VI, figs. 9, 10, 11, 12, 13, 14, 18. *Cauda mucronibus subæqualibus, centrali brevi distincto.* Mem. Geol. Surv., l. c., fig. 7; Decade 7; pl. ii, fig. 16.

The following is the description given in the Decade, with some modifications, and the omission of details.

One of the larger Trilobites; it measures occasionally nine (not twelve) inches, but probably not more. The more perfect specimens found at Dudley are not above two or three inches long, those from Malvern are larger, and those from the Caradoc limestones of South Wales and Ireland the largest of all. Length to breadth as 3 to 2; the head occupies fully one third the length, and is a little broader than the body. General form moderately convex and oblong, but narrowed suddenly towards the posterior end; the sides of the thorax and tail deeply serrated by the projecting ends of the segments. The animal is sometimes found half coiled up; the pointed ends of the pleuræ closing together, and overlapping each other.

Head rather more than a semicircle, the obtuse front projecting; glabella gently convex, equal in breadth at the base to the cheeks, above considerably broader, marked with three strong furrows on each side, besides the neck-furrow, the lowest being directed obliquely downwards, and joining the neck-furrow before reaching the middle; it thus encloses a spherical triangle as a basal lobe. In older specimens this lobe is somewhat squarer, and the furrow more curved. The other furrows curve but little downward, and are variable in length, but usually extend more than one third across the glabella on each side. The furrows on the glabella, as well as the axal furrows, are sharp, but not broad or deep exteriorly, although they are so on casts of the inner surface. Forehead-lobe of

[1] Decade 7, 'Geol. Surv.,' plate ii.

moderate size, half as long as the entire glabella, and on the sides overhanging the other lobes; in front it is somewhat produced, and occupies all the margin. The glabella is neither gibbous nor depressed, a line taken from the front edge to the neck-furrow presenting a regular and gentle convexity. Cheeks subtriangular, not so wide as long, with a broadish margin distinctly separated by a furrow, which meets the strong, straight neck-furrow at the posterior angles; these angles are spinous, the spine short and directed backwards. The eye is placed more than half way up the cheek, and not close to the glabella; it is opposite the middle furrow, and is rather small, supported by a raised rim below; the eyelid is narrow and indented, the lentiferous surface very convex, supine, and covered with minute, closely set, convex facets, with no spaces between them. In some specimens each facet has a minute pit upon it, but this is due to wear; the lenses being regularly convex when perfect. These are figured in the Survey Decades. Above the eye the facial suture takes a vertical course, and cuts the margin exactly where the axal furrow ends on it; below the eye it turns directly outwards to the smooth border, which it cuts considerably in advance of the posterior angle, and in an oblique direction, so that it reaches further back on the lower side than on the upper.

We do not know the course of the suture in front; it is probably direct across, beneath the front margin, and Barrande describes a rostral shield on the under surface. The surface of the glabella is sparsely covered with small granules (fig. 9), the cheeks are largely scrobiculate, and the wings or free cheeks have their border smooth, and only scabrous on its outer edge; they are sometimes dilated a little in advance of the facial suture. Labrum (Pl. VI, figs. 12, 13) large, ovate, oblong, very convex; its length generally one fourth more than the width, but in appearance often more; broadest near the strongly arched base of insertion, from which the central convexity rises immediately, and reaches nearly to the tip. A rather deep furrow surrounds the central portion, and separates it clearly from the more or less tumid margin. The furrow becomes deepest near the rounded shoulder, which we have called the 'auricle,' followed by a deep notch, above which the ascending processes (*a*, *a*) take their origin. The apex of the labrum is truncate, the corners angular, or even mucronate. Besides the distinct sulcus, which separates the border all round, there is a short oblique furrow higher up on each side. The whole surface of the labrum is closely scabrous (Pl. V, fig. 5); the convex portion has, besides, scattered, larger granules. The organ is hollow when viewed from the inner side (Pl. VI, fig. 13), and the structure there observable is such as has been described by Barrande.[1] There are two ascending pro-

[1] M. Barrande, 'Neues Jahrbuch.,' 1847, p. 389, has given a full description of the 'hypostome' of Cheirurus. He describes the ascending processes *a*, *a* (Flügel), as bent upwards at right angles to the surface of the organ, and uniting with the upper crust along the line of the dorsal or axal furrow, with a broad base of attachment, reaching from the upper to the middle glabella-furrow. In Phacops it has nearly the same position. He also describes a second organ, of the same size and shape, but less convex in all its parts, lying immediately behind the hypostome, between it and the upper crust of the head. This organ he calls "epistoma;" and he has seen it both in *Cheir. insignis* and a species of *Phacops*. It has

cesses (*a*, *a*) rising from the ends of the basal or front margin, and directed obliquely backwards; and on the sides (*b*, *b*) are two inflated, broadly triangular portions characteristic of the genus. These triangular curved plates give the appearance of thickness on viewing the organ from the side (fig. 13*c*); but the general surface on the inner side is concave, answering to the great convexity of the outer side.

Thorax much longer than the head, but narrower, and for most part of it parallel-sided, of eleven gently convex rings, which are very minutely scabrous; the axis is narrower than the glabella, of nearly equal width all the way down, but scarcely so wide as the pleuræ. These are linear, and directed straight outwards for two thirds their length, then curved a little backwards, and tapering to a sharp point. The fulcrum, placed at about one third, is of singular structure; a small, semi-oval piece (fig. 14, *a*) is attached to the posterior edge of each pleura; and against this piece abuts a similar tubercle (*b*), placed on the front edge of each; and the two pieces, forming together a narrow oval tubercle, are insulated by a deep sulcus from the body of the pleura, which is also constricted and furrowed across at this point so as to have the outer and pointed portion (*c*) quite distinctly separated from the small inner one. The latter is very strongly divided into two tumid lobes by a short oblique sulcus, and just beyond the constriction the outer portion rises into a stout boss, giving the tri-tuberculate form characteristic of the genus. The line of the fulcral-points is parallel to the axis for its whole length, and the constriction beneath them, though not very marked on the upper crust (fig. 14), produces a longitudinal ridge on the under surface and a corresponding strong furrow in casts (fig. 14*d*).

Tail, in all our specimens, very much narrower than the body, with three strong spinous lateral lobes on each side directed backwards, the outer ones a little divergent and longest; all extend equally backwards; the tail is therefore truncate, but exclusive of the spines, it is broad-triangular, following somewhat the shape of the axis; it is marked on each side by four short, deep puncta or furrows, which do not run to the margin, even in young individuals. The axis is convex and short-conical, of three distinct ribs, and a small terminal piece, the last very obscurely indicated; there is, between the lowest spines in the ordinary Wenlock forms, sometimes a blunt, sometimes an acute mucro, and occasionally none at all.

Variations.—The following have been observed:—In a Dudley specimen the front or forehead-lobe occupies much more than half the length of the glabella, the side lobes being, therefore, more crowded. In another Dudley specimen a large tubercle occurs in the middle of the forehead-lobe. In some individuals the glabella widens more above; in others it is nearly parallel-sided, and the lateral furrows vary in length. The head-spines occasionally reach the third thorax-segment. The margin of the cheek in one specimen is notched at the facial suture. The axis of the thorax is sometimes, though rarely,

never yet occurred to our observation, nor apparently to others. Yet it seems not unlikely that two plates, an upper and under lip, should form the channel to the mouth.

as wide as the pleuræ. The most important variations occur in the tail. In figs. 10, 15, we have represented the spines as all directed backwards, and the two central ones closely approximate; they are so in the large Ledbury specimen figured in the 'Silurian System,' where, too, they are shorter than the outer spines. In others they are a little space apart. In a Lower Silurian specimen we have seen a small tubercle appear between, and in our var. β, figs. 9, 18, a decided, though short, mucro protrudes. Lastly, as a monstrous variety from the Silurian rocks of Kildare (we have reason to think it of the same species), we have one with a wider interval and a bifid mucro. In old specimens, as well as in var. β, the spines diverge much more than in most of those here figured. Perhaps some of these variations are due to sex. It is observable in some specimens (fig. 9) that a double row of tubercles, like those observed in Phacops (p. 52), occurs down the axis of the thorax.

Affinities.—Among a host of kindred species in this prolific genus, the only Bohemian fossil with which it is really necessary to compare this is the *C. insignis*, Beyrich. Barrande's figures leave nothing to desire, and I confess I know not how to separate the two. *C. insignis* has a somewhat different habit, and the central tail-spine is more prominently developed. But except this, and having a less overhanging glabella, I do not know how to distinguish the Bohemian form, nor can my friend Mr. H. Woodward see any differences of value.

With the *C. speciosus*, as figured by Hisinger, I believe ours to be identical. *Ch.* (*Calymene*) *ornatus* of Dalman has the head very like, as we learn from Angelin's figure; but the upper glabella-lobe of that species is not nearly so long as ours, nor is it wide enough above. The British species is variable enough; but it is, for all that, distinct from those above quoted.

Localities.—CARADOC to LUDLOW ROCKS; CARADOC, South Wales, near Haverfordwest, abundant; North Wales; Kildare, Ireland. LLANDOVERY ROCKS of Goleugoed, near Llandovery, South Wales; Mullock, Ayrshire; Galway, &c. MAY HILL SANDSTONE, Norbury; (Pl. V, fig. 3). Also in the purple Shales, Onny River, Shropshire. WOOLHOPE LIMESTONE, Presteign; Malvern. WENLOCK LIMESTONE and SHALE, everywhere. Wenlock strata of South Wales, and West of Ireland. AYMESTRY LIMESTONE, Downton Castle, Ludlow. Also in Sweden and Bohemia (Upper Silurian).

CH. (ACTINOPELTIS) JUVENIS, *Salter*. Pl. V, figs. 9—12.

CHEIRURUS JUVENIS, *Salter*. Memoirs Geol. Survey, vol. ii, pt. 1, pl. vii, figs. 1—3, (exclude 3, *b*, which belongs to the next species), 1853.
— CLAVIFRONS, *Id.* Ibid., Errata, p. viii.
— — *Id.*, in Appendix Synops. Palæoz. Fossils of Woodw. Mus., t. 1 F, fig. 11; t. 1 G, fig. 9, 1851.
CERAURUS CLAVIFRONS, *M'Coy*. Ibid., p. 154 (but not t. 1 F, fig. 12, for which see *Sphærexochus boops*).

This rather common fossil has been bandied about, in search of a godfather, for some time, and, having found a resting-place in the verdict of Prof. M'Coy and myself, must be again disturbed; for there seems little doubt that, whatever Dalman's *C. clavifrons* may be, it is not a species with a narrow front to the glabella. Prof. Angelin's figures, quoted under the next species, give us a clue through the uncertainty which has hung over this species; but unfortunately his scanty description does not enable us to clear it up. Prof. Sars and Dr. Boeck described first one and then a different Trilobite under this name; and neither was the true species of Dalman, who, as if to justify all the confusion that has since resulted, placed two distinct species together in the Stockholm Museum under the name *C. clavifrons*, one of which has the glabella narrow in front and parallel-sided, and could not have suggested the term (Angelin, 'Pal. Suec.,' t. xxxviii, fig. 9); to this one Prof. Angelin has, however, applied the name; the other, with a subclavate glabella and minute eyes (pl. xxxviii, fig. 10, which does deserve the name *C. clavifrons*, and to which Prof. Angelin attaches that synonym and the note as to Dalman's authority), he by some strange fatality calls a new species, *Cyrtometopus affinis!*

We shall never have done with the confusion of these forms, unless we restrict Dalman's name to the species called *C. affinis* by Angelin. In that case I believe our next species will have to forego its baptismal privileges. But pending the settlement of this (and the difficulty is increased tenfold in a genus like Cheirurus, full of closely related and very similar forms), I shall revert to my old name, *Ch. juvenis*, which sufficiently expresses the smooth contour of the head. It is certainly distinct from all the Swedish forms, though resembling all which have figured under the above name; and the synonyms given above may, I think, be trusted; all the specimens have been seen by myself.

C. juvenis is a large trilobite for this section of the genus, and must have measured three inches in length. It has not yet occurred perfect; but the glabella is rather common, both in Wales and Ireland. It is an ovate and very convex mass, equally attenuate at each end, gibbous along the median line, and regularly convex along a line taken from front to back, not projecting near the base, and scarcely overhanging the front. The furrows are three on each side, the basal one strong, and completely circumscribing the oval basal lobe; the middle and upper ones gently curved, and equally remote from each other and the basal lobe. The front furrows are placed very forward.

The cheeks are steeply curved down, but not abruptly so; the neck-furrow strong, the surface scrobiculate, the eye placed behind the middle furrow, and therefore about half way up the cheek. We have not the free cheeks, but they probably showed a narrower margin than in the following species. The glabella is covered equally with granules and tubercles. Head-spines short.

Tail with very unequal lobes, the outer ones by far the largest, and produced into stout, slightly divergent spines, with a broad subquadrate base, marked by a short, deep, pleural groove. The two inner pairs are rapidly smaller, the last quite minute. But these vary in length and obtuseness. In our figured specimen they are not longer than the broad, long, conical axis of four joints, which intrudes its last joint between the short terminal mucrones, and which is, at its upper portion, as broad as its pleuræ, which have a wide, straight base of articulation. The articular portion of the axis is also large.

Fig. 9 represents the outer granulated surface of the glabella. The furrows are not so deep as in the cast (fig. 10); but the cheeks are equally scrobiculate, and more strongly so than in our next species.

Comparing with other species, we find, first, that the original *C. clavifrons* of Dalman is, according to Angelin's figure, a species with a narrow glabella, somewhat gibbous at the base, and with very broad and deep-fixed cheeks, with strong head-spines. The basal lobes of the glabella are prominent and circular. The *C. affinis* of Angelin, which was placed by Dalman with this, and seems by its shape to have been the actual species intended by him, has a subclavate glabella, and also very forward, minute eyes. The glabella of *C. tumidus*, Ang., a good deal resembles the large corresponding part in our species; but it differs essentially in not having the basal furrows complete, and the base is very broad. *C. gibbus*, Ang., does not need a very close comparison, while the several species arranged under *Sphærexochus* by Angelin must wait for more perfect materials to decide their collocation.

Localities.—Caradoc Slates and Limestones of Wales, Westmoreland, and Ireland. In North Wales,—west of Bala Lake, Rhiwlas, &c.; Corwen; Cerrig-y-Druidion, &c.; Llanfyllin and other places in Montgomeryshire. In South Wales,—Sholes Hook, Haverfordwest; abundant. Coniston, Westmoreland (M'Coy); Wexford and Chair of Kildare (Mus. P. Geology, Woodw. Mus., &c.).

Ch. (Actinopeltis) octolobatus, *M'Coy*. Pl. V, figs. 13, 14.

? Calymene clavifrons, *Dalm.* Vetensk. Akad. Handl., p. 75, 1826.
? — — *Lovén.* Ib., p. 63, 1845.
Sphærexochus clavifrons, *Salter.* Memoirs Geol. Survey, vol. ii, pt. 1, pl. 7, fig. 3, 1849.
Ceraurus octolobatus, *M'Coy.* Synopsis Foss. Woodw. Mus., pl. 1G, fig. 10, 1851.
Cheirurus — *Salter.* Decades Geol. Surv. 7, art. 2, p. 11, 1853.
Cyrtometopus affinis, *Angelin.* Palæont. Suecica, t. xxxix, fig. 10 (1855, not 1854), (and quoting *C. clavifrons* of Dalman from the originals in the Swedish Museum!)

Ch. (Actinopeltis) minor, ovatus; capite gibbo; glabellâ oblongâ nec ovatâ; genis declivibus; caudâ 8-lobatâ, brevi. Glabella convexissima tuberculata granulosa, anticè paullulum dilatata, posticè sulcis completis lobos rotundos circumdantibus; sulcis anticis longis. Genæ declives, externæ parvulæ, ferè perpendiculares, scrobiculatæ; oculis minutis. Pleuræ spinosæ. Cauda transversa bis quam longâ latior, subplana, axe haud eminente; pleuris utrinque 4 æqualibus ovatis planis, apicibus omnibus ovatis retrorsis, nec radiatis.

A pretty species, which I was glad to recognize as an old friend, when Mr. Henry Woodward called my attention to it in the British Museum. *Ch. octolobatus* had long been described from the tail only; but it was not suspected that it belonged to the section Actinopeltis, and was so nearly allied to the *Ch. clavifrons* as to make its separation a matter of critical distinction. If it be the *Cyrt. affinis* of Angelin (he has only figured the head in a side-view, and described it as smooth), our British name must give way to the prior one, for it seems to be really the long-contested *C. clavifrons* of Dalman.

It is a small species, the three caudal shields known not indicating a fossil much larger than one and a half inch long, of which the gibbous head occupies rather more than a third, and is a little broader than the body, with steep, almost vertical cheeks, which, measured in their own full width, are equally wide with the glabella, but do not appear so in a front-view. The glabella is oval-oblong, blunt in front, and rather broadest there, where it overhangs a narrow, very distinct margin. A line taken from front to back is regularly convex, and the glabella is much elevated, both above the front margin and the narrow neck-segment. The glabella-furrows are a basal pair, which completely surround the rotund-oval basal lobes, only somewhat fainter where they join the neck-furrow; and two upper pairs, rather long, placed at equal distances, the middle furrow being about the length of the basal lobe apart from it, and as remote from the front one. The front furrow comes thus very forward, and on a line with the deep marginal furrow of the cheek, opposite the notch between the facial suture and this margin. The cheek is rudely triangular, strongly margined all round (spinose, probably, at the head-angles), and with the very small eye placed on the side of the declivity, and a good way in advance of the middle of the cheek, opposite the ocular furrow. The free cheek is small and triangular, bisected by the deep marginal furrow. The cheeks are scrobiculate all over, the glabella covered with fine granules, and scattered larger tubercles, very equally.

Body-segments eleven, with a broad axis, equal to the pleuræ, and only abruptly tapering in the two or three last segments. The pleuræ semicylindrical, the posterior portion being so broad and convex as to occupy nearly the whole width of each,[1] the fulcrum placed rather far out, and the pleuræ then bent downward and a little backward, facetted distinctly, and ending in a point. The hinder ones are most bent back, and follow the curve of the tail-border.

Tail transverse-oblong, nearly rectangular, the segments ending regularly behind, along a nearly straight line. There are four on each side, the front ones bent at right angles, and all nearly equal, with bluntish ovate tips. No space between the two terminal ones. The axis short, of two segments and a very small terminal portion.

[1] There is no central line in this, to indicate a pleural groove, as in Eccoptochile (and some species of Actinopeltis). It would appear that this character is not even of subgeneric value in some cases.

Comparing the side-view fig. 13, *a*, with the side-view fig. 11 of the nearly allied species, *C. juvenis*, we at once see that it is distinct, and at the same time how very closely these species of Actinopeltis approach each other. The glabella of *C. juvenis* is oval, not oblong, and the free cheek less triangular, and a good deal larger, judging from the much smaller fixed cheek left behind. The eyes, too, are placed far more backward, and the neck-segment is more prominent. And if I am right, as I believe, in referring the tail (fig. 12) to that species, there is, of course, no need to compare them minutely.

Cyrt. affinis of Angelin is very like ours; I think it can only be a synonym.

Localities.—Greenish CARADOC Sandstone of Pinwhapple, Ayrshire (Brit. Mus.); Rhiwlas, Bala, and Cerrig-y-Druidion, Denbighshire (Mus. P. Geology).

CH. (CHEIRURUS) GELASINOSUS, *Portlock.* Pl. V, figs. 6—8.

AMPHION GELASINOSUS, *Portlock.* Geol. Rep., t. iii, fig. 4, 1843.
ARGES PLANOSPINOSUS, *Id.* Ibid., t. v, fig. 9.
CHEIRURUS GELASINOSUS, *Beyrich.* Böhm. Tril., i, p. 19, 1845.
— PLANISPINOSUS, *Bronn.* Ind. Pal., 1848.
— GELASINOSUS, *Salter.* Quart. Geol. Journal, vol. vii, pl. viii, fig. 1, 1851. Decade 7 Geol. Survey, art. 2, p. 11, 1853. Siluria, 2nd ed., p. 538, 1859.

Ch. (Cheirurus) depressus, 3-uncialis. Caput transversum, glabellâ rectangulari, sulcis brevibus transversis, lobo frontali brevi, basalibus oblongis transversis vix circumscriptis, uno ab altero spatio æquali sejunctis; genis latis, marginibus depressis, spinis brevibus. Cauda (hîc haud dubiè referta) lata, segmentis utrinque tribus latis, ad basin longe adnatis, acuminatis; primo in appendicem longam producto, secundum longè excedente, hoc tertium brevem superante; axe 4-annulato, articulo ultimo minimo angusto, nec apicem caudæ profundè emarginatum attingente.

The specimens figured by General Portlock[1] no doubt belong to one and the same species; and the form has been recognized by both Beyrich, Barrande, and myself, as a *Cheirurus*. It is rather a remarkable one, for the extreme width and shortness of the head; but the tail is of the ordinary form for the genus.

We have specimens of the head nearly two inches wide; the whole length may have been three and a half inches, of which the short, wide head measures only ten lines. The glabella is not so wide as the cheeks, and is rectangular-oblong, with the upper angles slightly rounded, and the sides strictly parallel. The furrows equal in length, rather long, oblique, and but little curved downward, the basal ones complete, and enclosing triangular oblong lobes, separated from each other by a space equal to their own diameter.

[1] While these pages are printing, we have to record the loss, at an advanced age, of this distinguished man. His work in various departments will endure long, for it was done well.

The neck-furrow strong. The forehead-lobe is small, generally equal in breadth to the rest, but no wider; the other lobes equal in length and width. The axal furrows are deep and strong; in them, opposite the front furrows, are deep pits on each side in the cast, which indicate the place for the processes, which join the corresponding projections of the labrum. The cheeks are convex, wide-triangular, with a broad margin, and a narrow, deep neck-furrow, continuous, as in all the genus, with the outer marginal furrows. The free cheek is small; the eye small, and placed midway on the most convex part of the cheek; the facial suture curving strongly inwards above the eye, and outwards and upwards behind it. Spines short? and rather thick, directed backward. The whole cheek is closely pitted, and the glabella and neck-segment sparsely covered with coarse tubercles.

The tail (*Arges planispinosus*, Portl.) is very flat, and moderately transverse, with a broad triangular axis of three rings, and a minute terminal appendix, which does not quite reach the notched apex. The upper side-lobes or pleuræ are largest and longest, broad-lanceolate, with a short pleural groove, and directed outwards; the second pair ovato-lanceolate, and diverging but little; the terminal pair of the same shape, but parallel, and with a deep notch between them. All the lobes are adnate, but strongly separated by furrows, and extend backward to about the same distance. Surface unknown.

Localities.—CARADOC. Tyrone (figs. 7, 8); Ayrshire (fig. 6), in the Craig Head Limestone (Mr. J. C. Moore), occurring with Heliolites, Petraia, and Pleurorhynchus. (All three figured specimens are in the Mus. Pract. Geology.)

CH. (CHEIRURUS) CANCRURUS, *Salter*. Pl. V, fig. 15 (and 16?).

CHEIRURUS GELASINOSUS, *M'Coy*. Synops. Sil. Foss. Ireland, p. 44 (not of *Portlock*). 1846.
— CANCRURUS, *Salter*. Decade Geol. Surv., No. 7, Art. 2, p. 11, 1853.
— — *Id.* Siluria, 2nd ed., p. 538, 1859.

Ch. (*Cheirurus*) *satis magnus, caudâ lineas* 20 *latâ, transversâ, apice abruptè truncato præmorso; axe lato, annulis quatuor subæqualibus, tertio à quarto punctis binis remotis solùm separato; lateribus spinis quatuor longis sub-parallelis, ad basin adnatis, transversis, apicibus lente decurvatis; basalibus utriusque lateris longo intervallo remotis.*

A most remarkable species, in which the four lateral lobes of the tail start horizontally from the broad axis, instead of gradually converging beneath it, and leave its apex bare; the breadth of this space being increased by the outward direction of the spines themselves, which begin to curve downwards only when they have attained half their length. The appearance of the perfect tail is just like that of a crab; *præmorsus* might have been an appropriate specific name. *C. obtusatus*, a Bohemian species, somewhat resembles this; but the spines are radiating, not parallel. None of Angelin's *Lower Silurian* Swedish species are very like this; *C. clavifrons*, Dalm., and *C. Sarsii*, Angelin, have some relation to it; but the *C. conformis*, from the Wenlock of Gottland, is apparently a near ally, and but for the swelled terminal lobe and wider cheeks, might have been identified with ours (see his pl. xxi, fig. 3.)

The species is named from the tail only. It is an inch and three quarters broad, and quite transverse, consisting of both axis and pleuræ of four segments, which are less connected together than usual, even in this genus; and the pleuræ are less metamorphosed than in any that I know. The axis is very broad, and regularly conical; broadly truncate, reaching to the end of the tail. There are four side-lobes or pleuræ, the first two of which are equal in length, and the others but little shorter. All are directed straight outwards for a third of the length, and then curve gently backward, leaving the whole truncate blunt tip of the axis exposed.

C. cancrurus? Head, fig. 16.

The head, which is here provisionally associated with the species, may belong to quite a different one; for it is much elongated, the length equal to two thirds the breadth. The glabella, including the very prominent neck-segment, is not far short of being twice as long as broad. The basal lobes complete and spherical-triangular, the upper furrows short and somewhat curved. The forehead-lobe is as long as all the rest, exclusive of the neck-lobe. The surface finely granular. The eye placed far back. The fixed cheeks narrow and scrobiculate. The margin strongly marked, but narrow. Spines?

Locality. CARADOC LIMESTONE of Chair of Kildare, Ireland (Mus. Pract. Geol. The head in the cabinet of Sir R. Griffith, Dublin).

Section.—ECCOPTOCHILE. *See* p. 61.

CH. (ECCOPTOCHILE) SEDGWICKI. Pl. V, fig. 17.

CRYPHÆUS SEDGWICKI, *M'Coy*. Ann. Nat. Hist., 2d ser., vol. iv, p. 406, 1849.
ECCOPTOCHILE — *Id.* Synops. Pal. Foss. Woodw. Mus., p. 155, pl. i, F, fig. 14, 1851.
CHEIRURUS? — *Salter*. Morris's Catal., 2nd ed., p. 107, 1854.
— — *Id.* Siluria, 2nd ed., p. 538, 1859.

Ch. (*Eccoptoch.*) *triuncialis, rectangularis, subplanus, axe angusto, caudâ latissimâ. Caput* (*imperfectum*) *semicirculare?, glabellâ genis angustiore, subclavatâ, sulcis decurvis, basalibus ferè completis; oculis anticis, genis latis profundè scriptis. Thorax segmentis* 12, *axe convexo, angusto, latitudine pleuræ dimidium haud efficiente; pleuris latis planis punctato-sulcatis, apicibus acutis lentè decurvis, secundo et tertio sese conniventibus. Cauda lata, axe brevi turbinato* 4-*annulato, annulo terminali trigono apicem emarginatum attingente; pleuris* 6, *primum rectis (duobus primis semisulcatis) dein arcuatis, apicibus retrorsis foliaceis ovatis subæqualibus.*

I cannot do better than extract Prof. M'Coy's excellent description from the work above quoted, and I think he is fully justified in placing this Lower Silurian form in the subgenus Eccoptochile. It is, indeed, a near ally of *Cheirurus claviger*, Beyrich, from which, however, its rectangular, not ovate, shape, and broad transverse flattened tail easily distinguish it.

"Cephalic shield nearly semicircular; glabella slightly clavate, smooth, with three segmental furrows on each side; the posterior pair longest, turning backwards and inwards nearly to the neck-furrow, enclosing a triangular space on each side longer than wide, the width rather less than that of the undivided portion of the glabella beneath their bases, the two anterior pair of furrows shorter; cheeks broad, gently convex, closely and coarsely pitted; neck-segment thick, smooth; eye-line extending with a slight curve to the margin in front of the angles; axal lobe very convex, narrow, slightly tapering, nearly parallel-sided, smooth, of twelve segments; three similar rings belong to the pygidium, a fourth terminal one being obtusely trigonal; the side-lobes are flattened, and more than double the width of the axal lobe; pleuræ nearly straight, narrow, and, for the greater part of their length, flattened, and having a broad, nearly mesial, pleural sulcus, deeply punctured like the cheeks, dividing each into two parts; the posterior largest, and forming a thick, smooth rounded ridge in the distal third of its length, bent down and a little backwards, swelling to a thick, narrow ridge in the middle, the sides and the extremity expanding into a broad, thin, foliaceous appendage; the pygidium terminates in six broad, ovate, leaf-like, semi-membranous flaps. Length of thorax and pygidium two inches two lines, width two inches three lines; width of axal lobe six lines.

"This magnificent Trilobite can only be confounded with the *Eccoptochile clavigera*, Beyrich, from which it is distinguished by the much greater width of the lateral lobes of the thorax, and the thin, flat, leaf-like appendages of the pygidium, which in that species resemble thick, pear-shaped clubs."

Locality. LLANDEILO FLAGS, two miles north of Builth, Radnorshire (Woodwardian Mus.). Also Abereiddy Bay, Cardiganshire (Mr. H. Hicks).

CH. (ECCOPTOCHILE) FREDERICI, *n. sp.* Pl. V, figs. 18—21.

CHEIRURUS FREDERICI, *Salter*, in Mem. Geol. Surv., vol. 5, (ined.), pl. viii, fig. 1—3.

Ch. (Eccoptochile) triuncialis et ultra, capite transverso, longispinoso. Glabella oblonga parallela seu anticè angustior, granulosa, sulcis distinctis abbreviatis obliquis quasi radiatis, lobo antico parvo trigono. Genæ scrobiculatæ nec scriptæ, longispinosæ. Oculi omnino antici, parvi. Thorax axe modico, ferè ut pleuras lato, his convexis rectis, profundè sulcatis, apicibus recurvis 4, longispinosis. Cauda major, axe 4-(vel.- 7 ?)annulato, longo, limbo sulcato et in spinas 6 retrorsùm flexas subparallelas productis.

This Trilobite, found about five years since by Messrs. Homfray and Ash, in the slates of Portmadoc, is a member of a group more characteristic of higher beds. There cannot be much doubt of the genus, and though the long spinous hinder pleuræ are a new character for this section of *Cheirurus*, it can only be compared with such species as the *Ch.* (*Eccopt.*) *claviger*, before mentioned, and the *E. Sedgwicki*. From both, the spines—and the deeply grooved, not punctate, pleuræ—distinguish it easily; I am not quite sure if fig. 21 belongs to this or to a distinct species; it is much larger, and has more joints in the tail.

Occasionally as much as three and a half inches long, and rather wide (all our specimens are much compressed). The head is wide, less than semicircular; the outer cheeks, occupying the greater part of the head, are margined all round by a continuous furrow, and produced into long head-spines; the border smooth. The eye is very far forward, as in *E. Sedgwicki*, and the facial suture so forward as to cut the outer margin much in advance of the middle of the head, separating a very small free cheek.

The glabella is not nearly so wide as the cheeks; it is parallel-sided, apparently not much longer than broad, and has the furrows very distinctly marked. These are three on each side, and rather deep, all straight and inclined a little upwards, the front ones especially, so as to have a radiate appearance. They reach more than one third across the glabella, and thus leave but a narrow space down the centre. The front lobe, marked out by the two converging front furrows, is a wide triangle, and does not occupy more than a third of the length of the glabella, the surface of which is granulated.

Thorax apparently of only eleven flattened rings, of which the axis is not quite so wide as the pleuræ; these last are convex and deeply grooved along the middle, almost to the ends; each is produced into a long, sharp spine, bent backward in all the segments; but in the four or five hinder ones the spines are fully equal to the length of the pleuræ themselves.

The tail, which can hardly be distinguished from the thorax, has in the more perfect specimen (fig. 18), only four joints and a terminal piece to the axis. In fig. 21 there are six or seven rings. The smaller specimen has the tail-spines more lateral. In the larger one they reach more toward the end of the tail; but part of this difference may be due to pressure.

C. Eryx, a species described by Mr. Billings from the Quebec limestones,* has a larger number of tail-spines; but it is a closely allied form.

Locality. Upper Tremadoc slates; Garth Hill, and Penclogwyn, Portmadoc (D.

* In the 'Canadian Naturalist and Geologist,' vol. v, p. 322.

Homfray, Esq.). Mr. Frederic Ash, formerly of Portmadoc, contributed many specimens of Tremadoc fossils to the Museum of Pract. Geology; and the species is named after him, to record his labours in the district.

SPHÆREXOCHUS, *Beyrich.*

Eyes facetted minutely; head very convex, almost globular; the cheeks not scrobiculate; facial suture ending on the external margin, near the angles, in front continuous and sub-marginal; glabella large and nearly spherical, with three furrows on each side, the two upper very obscure, the lower strong and curved down to the neck-furrow; thorax of eleven joints, without any furrows; tail of three segments, free at their ends; labrum subtrigonal, with a marginal furrow, but without lateral furrows. No rostral shield.

As the labrum of the last described section of Cheirurus, viz., *Actinopeltis*, has a very similar labrum, it might be difficult to say in what, except in proportion, and in the extreme convexity of all the parts, this very abnormal Trilobite differs from that subgenus. However, as Sphærexochus is widely recognised, and as, moreover, there appears to be no rostral shield (that of Cheirurus is well established by Barrande), it is convenient to adopt the name.

SPHÆREXOCHUS MIRUS, *Beyrich.* Pl. VII, figs. 1—6.

CALYMENE CLAVIFRONS, *Hisinger.* Leth. Suec., 2nd Supp., t. xxxvii, fig. 1, 1840.
SPHÆREXOCHUS MIRUS, *Beyrich.* Ueber einige Böhm. Tril., p. 21, 1845.
— — *Id.* —— Zweite Stück, t. i, fig. 8, 1846.
— CALVUS, *M'Coy.* Syn. Sil. Foss. Ireland, pl. 4, fig. 10, 1846.
— MIRUS, *Corda.* Prodom. einer Monog. Böhm. Tril., fig. 72, 1847.
— — *Barrande.* Syst. Sil. de Bohême, vol. i, pl. xlii, figs. 11, 18, 1853.
— — *Salter.* Decades Geol. Surv. No. 7, pl. iii, 1853.
— — *Id.* Morris's Catal. 2nd ed., p. 115, 1854.

A few years ago this was one of our rare British Trilobites. Now, the accident of a tunnel being driven through the rich formations of the Malvern Hills has rendered it one of the most frequent ornaments of the cabinet. Messrs. Gray and Fletcher obtained the species first from the Dudley Limestone. Sir R. Griffith found it in the Caradoc Limestone of Kildare, and Prof. M'Coy published it almost immediately after Beyrich's description had appeared in 1846. The Geological Survey collected it in abundance from the same locality in Ireland; and now our friends, Dr. Grindrod, of Malvern; Messrs. Hollier and Ketley, of Dudley; and Messrs. Allport, of Birmingham, have contributed excellent materials, which enable us to complete the account.

Larger specimens have lately occurred; and the finest I know, in Dr. Grindrod's collection, is nearly two inches long. Of this the head occupies more than a third of the whole length, and the glabella is very large, occupying, as seen from above, four fifths of the width, and quite overhanging the narrow front margin. It is, excluding the neck-segment, nearly a true hemisphere, and has a pair of large orbicular lobes at the base, deeply circumscribed, and further apart from each other than their own diameter. The furrow that bounds each of these lobes is broad, sharp, and equal in depth all round, leaving no communication with the body of the glabella (figs. 4, 5, 6). Above these lobes, on each side, are two faint impressed lines, which represent the upper furrows (see fig. 5, *a*); of these the one next to the round basal lobe is placed at a less distance from it than the diameter of that lobe, at about the point of the head's greatest width, and the upper one at an equal distance in advance of it towards the front. The cheeks are small in comparison with the glabella, and hang vertically from its sides (fig. 5, *cc*), like a pair of lappets from a cap or helmet; they are oblong, and have a thickened margin. The small convex eye is placed very near the glabella, and below the middle of the head; the facial suture runs outwards from it, and reaches the exterior margin, which it cuts obliquely, a little in front of the posterior angle, as is fig. 6; in front of the eye it continues parallel to the glabella, and runs along the edge of the narrow front margin, leaving the free cheeks connected beneath by a narrow band (fig. 5, *b*). The free cheek is hatchet-shaped; and the small eye (fig. 5, *c*) occupies the inner corner, supported on a fold of the crust, which truncates, or even indents it below. The eye is thus pushed up into a supine position; it is short, oblong, and very convex. The lenses are numerous, larger in size than the granulations of the general surface, and placed near together, less than half their diameter apart. In this specimen we have not the outer surface sufficiently perfect to enable us to say whether the cornea is raised into facets (as Barrande thinks) or not; from the inferior surface (fig. 5, *d*) the lenses have fallen out, leaving pits which indicate their size. The posterior corners of the head are rounded off, and they bear, instead of a spine, only a small tubercle (fig. 6), which is placed far inwards.

The labrum has not yet been found in England, but it is figured in M. de Barrande's plates, and we reproduce it from a Bohemian specimen (fig. 5, *e*). It is trapezoidal or inverted-pyramidal, half an inch wide by four lines long, straight at the base, where it is much broader than it is long, and the apex is truncate and slightly emarginate. A broad, shallow furrow runs round the end and sides, leaving only a small central convexity of the same shape as the labrum. This convexity is not indented by any lateral furrows. The auricles (see p. 60) are brought to a level with the base, and thus no notch is left between them and the ascending processes, as in those of forms of *Cheirurus* which most resemble it (see Barrande's plates, xl, fig. 30, and xlii, fig. 19).

The surface of the head is covered by a fine, close granulation (fig. 6), which occupies also the free cheeks or wings; it is, therefore, one of the subgeneric distinctions from *Cheirurus*, in which the cheeks are always pitted or scrobiculate.

Thorax parallel-sided, scarcely tapering backwards, of eleven thick rounded rings; the axis as wide as the sides, and of equal breadth throughout, very convex; each joint much raised and rounded (see fig. 4). Pleuræ horizontal as far as the fulcrum (fig. 6, *d*), and then abruptly deflexed, and from this point the pleura tapers outwards to a conical blunt point, which, at the extreme tip, is a little bent forwards. The fulcrum is placed at rather less than half way from the axis; but in the last segment it approaches much nearer,—to about one third (fig. 6, *f*). Its place is indicated by a protuberance, both on the forward and hinder edge of each segment (fig. 6, *d* and *f*); but these swellings are not isolated tubercles, as in *Cheirurus*, nor are there any oblique or longitndinal furrows on the pleuræ, as in that genus, to break up the uniform convex surface of the segment.

Tail about semicircular, truncate; the axis conical, its base of two depressed close-set rings, its apex of one long triangular joint, which is separated from the second joint by a deep depression; from thence it is flattened, or even depressed, for some distance, but suddenly rises to an obtuse and elevated tip (fig. 6, *g*). When seen endwise the tail presents a bent appearance. The sides are composed of three obtuse convex lobes, which scarcely project on the margin; the upper one follows the bend of the hindermost pleura, the second is less curved, the third parallel to the axis; all are deflected, so that an end-view of the tail gives an angular outline, very distinctly seen in the Decade figure above quoted.

The entire surface of the thorax (fig. 10) and tail, like that of the head, is covered with a fine granulation, the grains of equal size throughout.

Variations.—Our Dudley specimens have the tail somewhat shorter and wider, and the terminal joint of the axis therefore shorter, than those from Bohemia. Irish specimens (figs. 14, 15) are more like the foreign ones in this respect. The space between the lower glabella-lobes is least in these Irish specimens, though some of them have it considerably wider than the diameter of the lobes; in a Wexford specimen the space is proportionally as wide as in those from Dudley, which often have the lobes as far apart as in Bohemian examples.

The species we have to compare this with are many. These are:—the *S. scabridus* of Angelin, which differs but little, and may be identical; the *S. angustifrons*, id., which has flat, expanded lobes to the tail; *S. deflexus*, *S. granulatus*, *S. conformis*, and *S. Wegelini*, of the same author, have incomplete basal lobes; one figured, but not named, by Dr. Beyrich in his second paper (1846), which has the lobes of the tail lengthened out, and the terminal joint of the axis short.* An eighth is that figured in the lower part of our plate (figs. 27, 28), possibly a *Cheirurus* of the *Actinopeltis* group; but having such marked basal lobes, I regard it as a *Sphærexochus* for the present. The last I shall notice is from the Lower Silurian Rocks of Thibet, and has been figured from Col. Strachey's

* This ought to receive a name. The genus is too scanty to render it inconvenient. *S. Beyrichii* would do very well for it.

work.* The projecting bullate lobes of the glabella, in the Indian fossil, give the aspect of staring eyes, and suggested the term *S. idiotes*. Most of the above-mentioned are apparently, but rare species. The really cosmopolitan fossil is the *S. mirus*, which has been found in Bohemia, Sweden, Britain, and North America.

Localities. — CARADOC ROCKS; Chair of Kildare, Co. Kildare; Carrickadaggan, Co. Wexford; Biggar, Lanarkshire (specimens all in Mus. P. Geology). WOOLHOPE LIMESTONE; Malvern? WENLOCK LIMESTONE and SHALE; Dudley, Walsall, Malvern, abundant.

Abroad it is found in Lower Silurian strata of Dalecarlia (Hisinger), Upper Silurian of Bohemia (Barrande), and in Ohio, North America. From the latter locality I have seen specimens in Sir C. Lyell's collection, and M. de Verneuil also quotes it from thence.

SPHÆREXOCHUS? BOOPS, *n. sp.* Pl. VI, figs. 27, 28.

CHEIRURUS CLAVIFRONS, *M'Coy*. Synopsis Pal. Foss. Woodw. Mus., pl. i, F, fig. 12, 1851 (not the other figures, nor of *Dalman*, nor *Angelin*, nor *Sars* and *Boeck*).

I must give this a name, both because of the paucity of British forms of this genus, and to call further attention to it. I had long named it as a distinct species in the Jermyn Street Museum; and lately I have found, by the correspondence (and a careful drawing) of the talented Mr. Harry Seeley, of Cambridge, that M'Coy's figure quoted above represents a more complete specimen than the one in Jermyn Street (fig. 27). M'Coy's specimen is correctly represented in fig. 28. There is some doubt of the genus, for it may belong to the section *Actinopeltis* of the group *Cheirurus*, but the general character is much that of *Sphærexochus*, and we do not yet know the limits of these two subgenera, for such they assuredly are.

I shall content myself with an English description, and only point out its characteristics. It is a larger form than *S. mirus*. The glabella is of an oblong-ovoid shape, very convex, almost gibbous, but not so greatly so as is that of the *Cyrtometopus gibbus*, Angelin, a very near ally.

The basal lobes occupy nearly one half of the length of the glabella; they are wider than long, and somewhat oblique,—their shape compressed-sphæroidal, not truly round, and on the inner side they are connected with the body of the glabella by a depressed neck, about half as wide as the width of the lobe itself. The pair of lobes stand apart about as far as their shorter diameter. The neck-furrow is strong, and much arched forwards,

* The work is yet unpublished. But the plates and descriptions have long been printed. The Trilobites (nine species) are figured in Plate I. Among them is a new type allied to *Cheirurus*, but without eye or facial suture (*Prosopiscus*, Salter).

and is very prominent and elevated, but not broad; neither of our specimens show its full width.

The strong convexity of the great eye-like lobes suggests some such name as the above.

Locality.—CARADOC SLATES of Sholes Hook, Haverfordwest (fig. 27, Mus. P. Geology), Applethwaite Common, Westmoreland (fig. 28, Woodwardian Mus.).

AMPHION, *Pander*, 1830.*

Head short, transverse, without spinose angles. Glabella nearly rectangular, with three pairs of furrows, the front pair generally approximate on the anterior margin, and always enclosing a very small forehead-lobe, often narrower than the glabella. Eyes small [reticulate, *Angelin*]; facial suture behind the eyes ending on the exterior margin a little in advance of the rounded angles; labrum entire, pointed, convex, margined all round; body with fifteen to eighteen rings, the pleuræ without grooves; tail of few segments, the axis short, the pleuræ with free terminations [includes *Pliomera* of Angelin, 1852].

Amphion is just one of those genera which show the near connection between the *Cheiruridæ* and *Calymene*, to which it is allied by the elongate form, numerous body-segments, glabella not widened in front, and facial suture ending nearly in the angles. On the other hand, it is like *Zethus* and *Cybele* by the grooveless pleuræ and free segments of the tail, which extend beyond the margin as in those genera, and close behind the shortened tail-axis; also by the entire labrum.

From both groups the peculiar shape of the glabella and the numerous unfurrowed pleuræ distinguish it. The typical species is the *Asaphus Fischeri* of Eichwald, which was called *Calymene polytoma* by Dalman. Pander established the genus *Amphion*, but again altered the specific name. In *A. Fischeri* there are eighteen body-rings.

AMPHION PSEUDO-ARTICULATUS, *Portlock*. Pl. VI, figs. 29, 30.

AMPHION PSEUDO-ARTICULATUS, *Portlock*. Geol. Rep. Londond. Tyrone, pl. iii, fig. 5, 1840.
— — *Salter*, in Morris's Catal., 2nd edit. p. 99, 1854.
— — *Id.* Siluria, 2nd ed., p. 537, 1859.

A. triuncialis, longus, axe glabellâque convexis, fronte haud contracto. Caput semicirculare, margine angusto. Glabella subparallela, convexa, lobis brevibus, subradiatis, antico quam glabellâ lato, transverso, lineari. Oculi retrorsi, prope glabellam positi. Thorax segmentis 17, *axe convexo. Cauda expansa, pleuris subclavatis* 5, *duobus terminalibus post axin abbreviatum subplanum omnino connatis.*

* 'Beiträge zur Geognos. des Russich. Reiches,' Petersburg, 4to.

It was a proof of the late General Portlock's scientific acumen to have decided this to be an *Amphion*, a genus, be it remembered, never recognised in Britain before his work appeared; for assuredly the characters of the head are unlike those of the typical species, and yet there is no manner of doubt that we must admit it, and enlarge the generic character so to do. The forehead-lobe is much wider in this than in the other species, as our woodcut will show. (Fig. 17.)

The species probably grew three inches long, as indicated by fig. 30. The head is transverse, semicircular, or nearly so; the glabella, which is moderately convex, occupying fully one third, and regularly, but very slightly, tapering backwards to the small neck-segment. The furrows are three on each side, very short, and somewhat radiating, enclosing a linear basal lobe, a clavate middle one opposite the eye, and a subrectangular upper lobe, between which and the very short, wide, and transverse forehead-lobe is only a short, straight furrow; but beneath this it is continued as a faint curved depression (not shown in our figure), so as to follow the direction of the middle or ocular furrow. The forehead-lobe (and this is unusual for the genus) is as wide as the rest, and is a narrow linear segment. It has no central furrow, such as exists in *A. Fischeri*, nor any crenulate border in front; indeed, the front margin must have been very narrow, as our figure (woodcut 17) indicates.

Fig. 17.

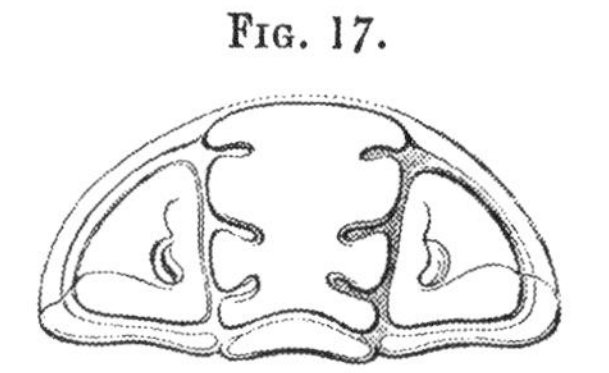

The cheeks are triangular and gently convex, strongly and rather deeply marginal, the margin being not so broad externally as in *A. Fischeri*, and the cheek consequently more triangular and less oblong. The eye is placed far inwards for the genus, and only one third up the cheek. We have not the actual eye. It was small, but is not preserved in our specimen. The facial suture curves largely out and upward beyond the eye, and cuts the obtuse head-angles.

As the glabella is more convex than usual in the genus, so is the axis of the thorax which follows it, and which is not so wide as the sides, but wider in proportion than other species. The pleuræ are each convex, especially within (in the cast), and have the fulcral point at one third, whence they bend backward and curve downward. No pleural groove shows on the exterior surface, but within the crust the furrow is visible on the anterior edge of the segment.[1] The extremities are curved, and apparently rounded, but this last may be deceptive.

The tail is remarkable, and in our largest specimen (fig. 30) shows well the characters which separate it from *A. Fischeri*. Its parabolic flattened axis, divided into five rings, reaches barely more than half the length, divided by only faint axal furrows from the five side pleuræ, of which four are well distinct on each side, and have a sublinear but some-

[1] This is usual in all genera with ungrooved pleuræ, and hence the distinction of those groups which have "plevre à sillon," from those which have "plevre à bourrelet" is an artificial one. Cheirurus and Sphærexochus show the intermediate character.

what clavate form, the truncate ends being oblique. The two terminal ones close so completely behind the axis that the suture is soldered, and they appear as a single rectangular plate, with a terminal notch only. The whole tail is gently and regularly convex, and the appearance thus given is very peculiar.

Locality.—CARADOC SLATE of Tramore, County Waterford (Mus. P. Geol.).

AMPHION BENEVOLENS, n. sp. Pl. VI, fig. 31.

A. minor, capite (solùm cognoto) lentè convexo, latimarginato, margine frontali incrassato nec crenulato. Glabella lentè convexa anticè latior, lobis longis transversis, antico abbreviato triangulari, vix plus quam dimidium frontis efficiente; sulco centrali nullo. Oculi retrorsi, á glabellá paullum remoti. Reliqua absunt.

Much more nearly like the Russian species than the one above described. *A. benevolens*, named in honour of Mr. Nevins, of Waterford, differs from that species in its proportions, and in the presence of a plain, thickened, anterior margin instead of the crenulate border visible in the Scandinavian fossil. It is about the same size. We have only the glabella and a part of the cheeks, which show the eye to have been also very much nearer the glabella than in the species just quoted.

Head seven lines long and about fourteen wide, semicircular, a little pointed in front, gently and regularly convex, the glabella being just as long as broad above, and tapering slowly behind—the axal furrows quite straight. A thick margin runs round the front, quite free from corrugations, and with a small tubercle in the centre, the division between this margin and the glabella being feeble for the extent of the forehead-lobe, which occupies rather more than half the whole width of the glabella in front, and has a pair of very oblique, faint, straight furrows to bound it. Outside this the marginal furrow is as deep (in the cast, which is all we possess) as the abrupt axal furrows of the head. The middle and basal furrows are long, reaching more than one third across the glabella; the middle one straight at first, then gently decurved, the lower one quite straight, and all of them deepest at their inner termination. The neck-furrow rises considerably toward the middle, so as to make the basal lobes cuneate, but neither the neck-segment nor any of the lobes are tumid. All partake of the regular and gentle convexity of the head.

The eye is placed opposite the median lobe; it is small, but elevated, and surrounded by a rather deep furrow, and it is only about twice as far from the glabella as from the neck-furrow (in *A. Fischeri* it is three or four times as remote).

Locality.—CARADOC SLATE of Newtown, Waterford, in company with *Phacops Jamesii*, described at p. 32.

A. PAUPER, n. sp. Pl. VI, fig. 32.

Omnino precedenti simillimus, nisi margine antico angustiore, lobis glabellæ longioribus radiatis, postico sinuato, mediano valdè obliquo recto, antico obsoleto. Glabella lentè convexiuscula, lateribus paullum arcuatis. Oculi subremoti.

One specimen only has been preserved of this neat species, which is truly distinct. It is about the same size as the preceding, from which, at a glance, you may distinguish it by the long glabella-furrows, and when closely examined it is found that there is one pair absent, viz., the obsolete anterior ones. A faint marking only indicates their proper position.

Glabella, including the narrow front border, as long as broad, regularly and gently convex, arched slightly in front; the sides not straight, but curved outwards; the upper angles rectangular. Anterior border narrow, not thickened. Anterior furrows quite obsolete. Median furrows starting from the upper angle or a little below it, straight, oblique, and reaching far towards the centre. Lower furrows situated opposite the eye, and reaching nearly as far as the upper ones. The middle lobes are thus subcuneate, the basal lobes broad-linear. The neck-furrow is distinct, but shallow; the neck-segment linear, but not so wide as the basal lobes; none of the glabella-furrows are thickened at their terminations. Axal furrows not deep. Position of eye doubtful, but probably further forward than in *A. benevolens*. The neck-furrow on the cheek is strong.

Compared with the preceding species, *A. pauper* differs in nearly every part. Instead of a thick front margin, it has a narrow one; the anterior furrows are obsolete, a very curious character, and peculiar to this species. The median ones oblique and longer than the basal furrows, which are sinuous instead of straight. Lastly, the head is less convex, and all the furrows—neck-furrow, axal-furrow, and glabella-furrows—less strong. The outline of the glabella is barrel-shaped, not rectilinear.

Locality.—One specimen only is known, from the CARADOC ROCKS of Tramore, where it occurs with the preceding (Mus. Irish Industry, B. 643).

Genus—STAUROCEPHALUS, *Barrande*, 1846.

Head cruciform, with long clavate glabella, greatly swelled in front into a hemispheric lobe;—the base narrow, cylindric, with two pairs of lateral furrows. Cheeks convex, with pedunculate eyes, and serrate edges. Facial suture ending on the external margin. Body-rings ten, without pleural grooves, pointed. Tail of few segments, the apices of the pleuræ free.—BARRANDE.

STAUROCEPHALUS MURCHISONI, *Barrande*. Pl. VII, figs. 13—20.

STAUROCEPHALUS	MURCHISONI.	*Barrande*.	Prodrom. Sil. Syst. Bohême, p. 53, 1846.
—	—	„	Syst. Sil. Bohême, pl. xliii, fig. 28—32 1852.
—	—	*M'Coy*.	Synopsis Woodw. Mus., pl. i F, fig. 15, 1855.
—	—	*Salter*.	Morris's Catal., 2nd. ed. p. 115, 1854.
—	—	*Id*.	Siluria, 2nd ed. p. 540, 1859.
—	—	*Id*.	Decade 11. Geol. Surv., pl. 5, fig. 1-5, 1865.

S. ovatus, tuberculosus, oculis remotiusculis, margine genarum spinoso. Cauda quadrata, pleuris sex, omnibus æqualibus retrorsis parallelis, haud divaricatis.

One of the most curious, if not one of the most conspicuous, of our British species. The globular head, or rather glabella, set on its narrow stalk-like base; the gibbous cheeks, projecting eyes, serrate border, and spiny comb-like pleuræ and tail—combine to give a most unusual and extravagant appearance to the fossil. It is seldom found perfect; but the skill of the Dudley naturalists has long been exercised on it, and specimens are now to be found in several cabinets. Mr. Hollier's and Mr. Ketley's very fine specimens are the principal ones figured. Our fig. 18 is from the Museum of Practical Geology: it was formerly in Mr. E. Davis's collection, and is from Presteign.

About an inch long, of which the gibbous head occupies more than two fifths. This is longer than broad, roughly triangular in general outline, but not truly so. It appears rather four-lobed, or like the heraldic "fleur-de-lis," the truly globular front occupying more than half the length of the glabella, and being at least three times as wide as its semicylindrical base, from which it is abruptly cut off by a transverse furrow; the base is marked by two distinct lateral lobes, besides the neck-furrow. The cheeks reach forward about half way up this globular portion; and the central part is unusually raised, so as nearly to be on a level with the glabella. The cylindrical eyes are on the most convex part of the cheeks, and are directed outwards, scarcely forwards. The margin is distinct in front of the head, has a very narrow prominent ridge, and furnished on each side with about fourteen truncate spines; the cheek-spine is directed backwards and but slightly outwards, abrupt at its origin, and not reaching beyond the two or three first body-rings. The facial suture cuts the outer border in a direct line from the base of the eye.

All the prominent parts of the head are covered with larger and smaller tubercles; they only fail on the deeper furrows and the truly vertical outer half of the cheeks. They are conspicuous on the border and even on the cheek-spines.

The body and tail united are slightly longer than the head, the body of ten rings many times longer than the short square tail, and the axis about one fourth the whole width and highly convex, especially in front. There are no axal furrows to separate the gibbous axis from the horizontal portion of the pleuræ; and these soon curve downward, and are abrupt and steep on the sides.

The pleuræ are semicylindrical, the front portion, separated by the pleural groove, being very narrow in this and allied genera, placed on the forward margin, and scarcely visible.[1] The apices curve much backward, and in the hinder pleuræ again a little upward, and are produced into strong spines beyond the ovate facetted portion. And all along these pleuræ, and over the axis, tubercles are placed at equal distances, except that the central prominent tubercle fails on the alternate rings of the axis, and the intervening ones, especially the ninth, are stronger than any other tubercles, and remind us of the spines on *Encrinurus,* a genus not yet described in these pages.

The tail is nearly square, concave rather than flat,—the short conical axis of four rings not easily separable from the sides, which are composed of three flat, broad, spinous pleuræ, directed backwards and quite parallel, so as to give a comb-like appearance. A few tubercles are scattered on the surface.

Locality and Geological Position.—Caradoc Rocks, Rhiwlas, near Bala, N. Wales (figs. 19, 20). Woolhope Limestone and Shale, Corton, Presteign (fig. 18). Wenlock Limestone, Dudley and Malvern.

S. globiceps, *Portlock.* Pl. VII, fig. 21.

Ceraurus globiceps, *Portlock.* Geol. Rep. Tyrone, p. 257, tab. i, fig. 7, 1843.
Staurocephalus globiceps, *Salter.* Morris's Catal., 2nd ed. p. 115, 1854.
— — *Id.* Decade 11. Geol. Survey, pl. v, fig. 6, 1865.

S. ovatus, granosus, caudâ utrinque elongatâ, spinis divergentibus. Glabella stipite brevi vix lobato. Oculi approximati. Spinæ genales et pleurales diffusæ. Cauda brevis, pleuris primariis longè extensis, latis; reliquis—?

[1] Yet I doubt the propriety of making this character so important in classification as Barrande has done. The pleural groove is always present in one form or another. In this case it is anterior, in Cheirurus it is very short and oblique.

A much smaller species than the preceding, and distinct from it by abundant characters of shape and habit. The divergent spines of head, thorax, and tail, enable us at once to recognise it; and of the latter the remarkable extended first pair of pleuræ (the rest of the tail is lost) show a near connection with the *S. unicus*, next described.

Only two good specimens, 10 lines long, are known. The head is equal to the thorax in length, and longer than the caudal portion. It has a very large globular front, longer than the square stipes, and granular all over. This stalk or base seems to be only furrowed beneath. The cheeks are granular, gibbous, with a prominent eye on the front edge near the glabella; and directed forward, not outward, with a broad plain margin, and widely divergent spines.

Fig. 18.

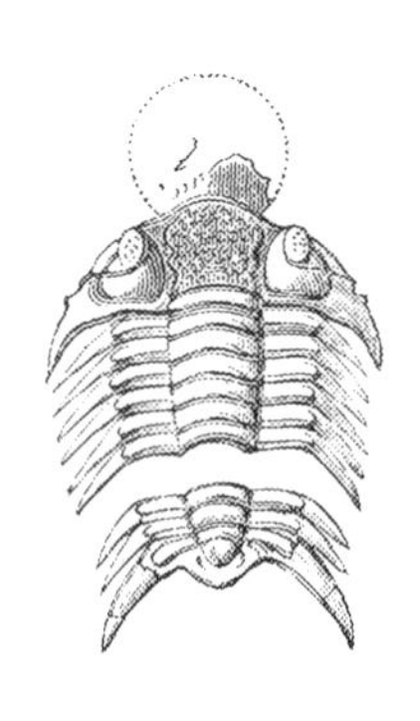

S. globiceps, Portlock, from Ayrshire. Mus. P. Geology.

A very finely preserved specimen, fig. 18, lately acquired by the Geological Survey, shows the body and tail-spines very well, and the free cheek furnished with short spines.

The axis of the thorax is cylindrical, and as wide as the stalk of the glabella. The pleuræ flat as far as the fulcrum, which is less remote than the width of the axis; strongly tuberculate at this point, and thence with patent (not recurved) spines, as long as the portion within the fulcra. The thorax tapers backward rather rapidly to the tail, which has a short three-ribbed axis; and the upper pair of its pleuræ are very much expanded widely divergent, and more arched than in our figure, which also represents the thoracic pleuræ as less curved than they really are. The hinder portion of the tail is absent on our specimen.

Locality. — Caradoc Rocks of Desertcreat, Tyrone; also Ayrshire,—a solitary specimen, figured in the woodcut, (Mus. Pract. Geology).

A third form, very abnormal in its characters and of large size, was named in MS. *S. Maclareni* by Prof. Wyville Thomson, after the veteran Scotch geologist in whose company it was found. It is, however, Prof. Thomson's previously described *Acidaspis unica.* As he has mislaid his own full description, I may supply the following notes from his specimens, and others presented to the Museum of Practical Geology by himself.

S.? unicus, *Wyv. Thomson.* Pl. VII, figs. 22—24.

Acidaspis unica, *Thomson.* Quart. Geol. Journ., vol. xiii, pl. vi, fig. 13; 1857.
Staurocephalus? unicus, *Salter.* Decade 11; Geol. Surv., Art. 5, 1865.

S. $1\frac{1}{4}$*-uncialis, oblongus, sparsè granulosus, glabellâ gibbâ eminentissimâ, corpore plano, caudâ expansâ transversâ. Caput latum, glabellâ clavatâ elevatâ frontem longè impendente, â genis punctatis distinctissimâ; margine* [illegible]*isso utrinque bispinoso. Pleuræ subplanæ,* *sulcatæ, rectæ, apicibus abruptè recurvis. Cauda lata brevis, axe appendiculato, pleuris primariis latissimis spatulatis, margine postico truncato.*

In the absence of a figure sufficiently complete (for the one quoted above is very defective) it is necessary to give a rather full diagnosis of this remarkable form, which tends to show the passage of the Cheirurid into the Acidaspid family. Indeed, if Prof. Thomson be correct in figuring 12 segments to the body, the species is abnormal for either Acidaspis or Staurocephalus. The shape of the head shows clearly enough that it is to Staurocephalus, or else to one of the sections of Cheirurus, that this bizarre fossil must be referred. *Cheirurus* often has 12 segments, *Acidaspis* 9 or 10, *Staurocephalus* only 10. The grooved pleuræ are unlike Staurocephalus, but like the section Eccoptochile among the genus Cheirurus. But no Cheirurus has so clavate a glabella, though a tendency towards it is exhibited in some species, and *Sphærocoryphe* of Angelin is very near to our fossil. There is an evident relation, too, with *Lichas* in this form, both in the shape of the tail and the character of the pleuræ. But the external position of the facial suture—far up the cheek—easily distinguishes it from that genus. I do not further describe it, as it has already been fully noticed, though not figured, in Decade 11 of the Geological Survey, just published.

Locality.—Caradoc Schists, at the base of the " Orthoceratite and Graptolite Flags," Penwhapple Glen, Ayrshire (Wyv. Thomson).

Staurocephalus. Sp. Pl. VII, fig. 25.

An imperfect fossil, but distinct from *S. unicus.* It has tuberculate pleuræ.

Locality.—Caradoc. Ayrshire (Prof. Thomson's cabinet).

Deiphon, Barrande. 1850.

Deiphon, the most abnormal in appearance of the whole Cheirurid family, is not found, when closely examined, to depart very much from the ordinary type. Except in the extreme reduction of the cheeks, and inflation of the glabella, the head might well pass for that of a *Cheirurus* of the section *Actinopeltis;* while the body and tail, extravagant as they seem at first, have all the usual characters of the family. We have now perfect specimens, and can improve Barrande's description.

Form somewhat circular; very loosely built, and produced into spines on the margin. The head composed of a globular glabella without furrows, long-spinous fixed cheeks, and minute free cheeks,—the facial suture ending on the exterior margin. The eye prominent, not stalked. Hypostome narrow, granular, without any rostral shield. Labrum hexagonal, with a truncate end, small lateral auricles, and an unfurrowed hemispherical centre; the base is narrow and arched. Thorax of ten joints with a very convex axis, and linear ungrooved pleuræ with free curved spinous ends. The tail is short, with a minute axis of four joints, and is quite truncate below; its two upper pleuræ developed into great divergent spines—the rest obsolete, more so even than in *Staurocephalus globiceps.*

DEIPHON FORBESI, *Barrande.* Pl. VII, figs. 1—12.

DEIPHON FORBESII, *Barrande*, in Haidinger's Berichte, p. 6, 1850.
— — *Salter.* Morris' Catal., 2nd. ed., p. 106, 1854.
— — „ Siluria, 2nd ed., pp. 539, 262, 1859; and quoted in many general works.

D. uncialis, latus, granulosus. Glabella sphærico-quadrata, oculos impendens, latitudine fere genis æqualis, granulis creberrimis, magnis minoribusque mixtis. Lobus cervicalis angustus. Spinæ a base glabellæ orientes; in juniore curvæ, tenuiores; in ætate crassæ, sinuosæ, retroflexæ, haud patentes. Thorax pleuris antrorsúm curvis, ad basin contractis, apicibus recurvis; anticis posticisque brevioribus; ultimá brevissimá, et sæpissime caudam adhærente, néc connatá. Cauda lata, ad basin truncata, axe 4-annulato stellato depresso, pleuris solúm duobus, crassis, patentibus, apicibus recurvis.

I have been diffuse in the diagnosis, for the reason that I suspect more than one species is confounded under the name of this most odd-looking, rare, and precious Trilobite. Not that fragments are uncommon, they are frequent in the Dudley slabs; but perfect specimens are of the rarest occurrence; and we are fortunate in being able to present naturalists with the complete form.

The specimens figured in our plate show that the species must have grown fully an inch long; and as the breadth is greater than the length, it must have measured one and a quarter inch from tip to tip of the curved spines. The glabella forms the greater part of the head, is hemispherical, rather quadrate in age, but truly hemispheric in the junior stages, or rather subspherical, for it forms at least three quarters of a sphere, and in front it is only somewhat less convex beneath than on the dorsal surface. In very old specimens, fig. 6, the large glabella overhangs the eyes, which are placed close to it on the forward edge of the narrow spine-like cheeks. These arch upward more in young specimens, so as to be like the figure given by Barrande; but are more horizontal at their origin in the adult. Hence they curve outward and backward (most backward in the older state) and are longer than the width of the glabella itself: they are also covered with a more imbricate granulation; see fig. 9. The eyes are supported by folds of the crust, which occur both on the fixed cheek behind the eye and on the small triangular free cheek, which last projects a little at the facial suture beyond the spine, and is more finely granular than the spine itself.

The eyes are (according to Barrande) coarsely granular;—they appear to be more finely so in ours, which, from several differences of proportion in all the parts, I suspect to be a different species. If so, I shall call it after the distinguished palæontologist who discovered the genus. It may be only a local variety.

The thorax, now for the first time figured, has a narrow convex axis, of ten rings, tapering slowly backward, and strongly distinguished from the gently curved pleuræ. These arch a little upward, and at the recurved pointed apices strongly downward, giving a singular aspect, as of the prickles of a *Geum* or a teazel-head. The pleuræ are semi-cylindric, contracted at the base, most convex along the median line, and flatter towards the tips. They are quite separate from each other, not touching, so far as I can see, along the fulcral edge; nor is there any distinct fulcrum, only a little expansion on the forward edge. We have omitted an enlarged view of the thorax-ring on the plate, and supply it here.

Deiphon Forbesii. Enlarged view from above, not endwise, of seventh thorax-ring.

The front pleuræ are shorter than those which follow them, to accommodate the former to the shape of the head-spines; the hinder pleuræ, in like manner from the seventh, begin to shorten; and the last is so short and slender, and so frequently attached to the tail (not connate with it, however, as Barrande supposed and figured it), as to look like a part of that organ.

Tail—a pair of widely divergent broad-pointed prongs, strongly curved back at the tips; and more divergent in the adult (figs. 11, 12) than in the younger state (figs. 1, 2, 3). The end is quite truncate, the base of attachment narrow; and the axis, of three or four rings arranged in a stellate fashion, is depressed below the whole of the general surface.

Locality.—Wenlock Limestone of Dudley and Wallsall; Wenlock Shale of Malvern Tunnel. I do not know it elsewhere in Britain.

Foreign distribution.—Etage E. Bohemia. The genus occurs in Regio D—E. in Sweden (Angelin).

I do not enter upon the description of those doubtful members of the Cheirurid group—the genera *Cybele, Encrinurus,* &c., for the reasons before assigned, viz., the imperfection of the materials, and the doubt I still feel whether these genera should not form a distinct family group. I shall also omit the family *Acidaspidæ,* the materials for which are more complete, but the knowledge we yet have of them is not so. It is better to leave these, and begin upon those groups of which we have abundant materials. I shall therefore follow on with the families *Calymenidæ* and *Asaphidæ.*

CALYMENIDÆ, *Brongniart.*

The group so named must for the present be restricted to two genera, both of which are the most common of all Trilobites, and grow to a large size. If we admit more than *Calymene* and *Homalonotus* into the family, we shall have a heterogeneous group, some of which have the facial suture ending in the posterior angles, and others on the posterior margin—some with more and others with less than thirteen rings to the body. But if these latter are considered to form a distinct family—the Conocephalidæ, the two genera above named constitute a most natural group. *Calymene* has the trilobation of the body complete. *Homalonotus* is singularly deficient in this respect. Some of the species, indeed, of this last genus are sufficiently trilobed to connect them with *Calymene*, others are so slightly lobed that the term Trilobite is a misnomer as applied to them. But in the essential characters of thirteen body-rings, and a notched labrum attached to a distinct rostral shield, both agree; and the facial suture in each ends exactly on the angle, a position midway, it will be observed, between that in the two groups already described, and that which it assumes in all the 'Primordial' genera of the Conocephalidæ, which have it posterior. Restricted thus, we have a numerous set of species, belonging to two natural genera, included in this group. They have a thick granulated crust, often ornamented and even spinous superficially, but without produced spines or angles to any part of the margin.

CALYMENE, *Brongniart.*

One of the most graceful and compact of all the Trilobite group; the head and tail well developed, but not extravagantly so; the former with a three-lobed glabella, very convex and narrowed in front, and with prominent supine eyes, which have evidently a very thin cornea, in which, only very rarely, the lenses are visible;[1] a thick margin to the head, the suture being in front submarginal and subtending a broad rostral shield. This bears a notched labrum, with a central gibbosity, conspicuous in all the species. Body of thirteen rings, the axis convex, and the pleuræ facetted and rounded at the ends. The tail, of about five lateral segments duplicated throughout, and the convex axis reaching to the very end in all the species, and with eight or ten rings to it.

The range of the genus is from the Arenig Group to the Ludlow Rocks. In Bohemia alone is *Calymene* known to rise somewhat higher; the upper limestones of that country, containing *Calymene*, being generally considered to be the lower beds of passage in the Devonian strata.

[1] Hall, in his 'Palæontology of New York,' has figured the lenses. I have never seen any traces of them.

C. TUBERCULOSA, *Salter*. Pl. VIII, figs. 1—6.

CALYMENE BLUMENBACHII, *var.* TUBERCULOSA, *Dalman*, p. 36? (not tab. 1, fig. 2). 1826.
— — — *Hisinger*. Lethæa Suecica, p. 10? (not tab. 1, fig. 3), 1837.
— — *Murchison*. Sil. System, pl. vii, fig. 5 only, 1839.
— TUBERCULOSA, *Salter*. Mem. Geol. Survey, vol. ii. pt. 1, pl. xii, 1848.
— — *Id.* Decade 2, Geol. Survey, pl. viii, 1849.
— — *Id.* Morris Catal. 2nd edit., p. 102, 1854.
— — *Id.* Siluria, 1st ed., pl. xviii, fig. 11, 1854 (by accident not introduced in the 2nd edit., 1859).
— BLUMENBACHII, var. *Hall*. Pal. New York, vol. ii, pl. A 66, fig. 6? 1852.

C. lata biuncialis depressa alutacea, nec tuberculosa; margine frontali capitis valde producto recurvo; genis gibbosis, glabellâ brevi depressâ. Thorax axe angusto, pleuris planis usque ad fulcrum, quod anticè ad dimidium, posticè ad tertium positum est. Cauda lata depressa, lateribus abruptè deflexis, axe conico subplano 7—8 *annulato; costis lateralibus* 5 *planis, sulcis acutis haud interlineatis.*

If species be anything more than confirmed varieties, this is a good species; and being common in Shropshire, there is sufficient material to judge from. It is entirely different in aspect from the common Dudley Trilobite, being greatly more depressed, and having a projecting recurved snout, which distinguishes it at a glance.

Our specimens are not more than $2\frac{1}{2}$ inches in length, and in breadth $1\frac{1}{2}$ inch. Whole surface equally and minutely scabrous. General form broad for the genus, not much attenuated posteriorly, depressed. Head short, wide; the glabella not more prominent than the cheeks, and much narrower, contracted in front, and separated by a deep furrow from the front margin: it has three lobes on each side, the basal one large, the middle one nearly spherical, the third minute; the forehead-lobe is small, the neck-lobe large and prominent. The neck-furrow is continued nearly to the posterior angles, which are rounded. Cheeks gibbous, often more elevated than the glabella, bearing the small eyes on their most prominent part. These are placed opposite the middle lobe of the glabella, and at some distance from it. A strong deep furrow separates the cheeks from the glabella, except opposite the eye, where a buttress is thrown across from the cheeks touching the middle glabellar lobe. The wings are strongly bent downwards, and even inwards on the under surface of the head, and the anterior margin is much recurved, and produced into a snout. On looking at the under view of the head, the margin appears greatly bent, and in the angle so formed, the curved rostral shield, half as long as broad, is inserted; beneath this is attached the hypostome, which is squarish-oblong, with the terminal angles rounded; it is strongly convex forward, the convexity terminating in a compressed tubercle;

one or two concentric lines, as if of growth, mark the surface, which is also scabrous, like the general crust of the body. The axis of the thirteen body-rings is convex, but narrower than the pleuræ, and constantly tubercular on the sides. The pleuræ are horizontal half-way, and then strongly decurved; their ends rounded posteriorly, and bent forward. Fulcrum distant from the axis, about half-way from it near the head,—at one third, or rather less, behind. Pleuræ sharply furrowed, the forward or fulcral half somewhat narrower than the posterior. Tail nearly semicircular, with the front angles truncated; evenly and gently convex, the axis not prominent, the sides decurved strongly towards their edges. Axis not percurrent, narrow, conical, with seven rings and a terminal boss. Lateral ribs flattened, separated by sharp, narrow furrows, starting at a wide angle from the axis, and curved back on the sides, simple, or but rarely marked by a central line near their ends,—not bifurcate, as in *C. Blumenbachii.*

—— *Junior.*—The proportions of the axis to the sides, and the structure of the pleuræ are similar; but the glabella is more cylindrical, not widened below; the tail is proportionately smaller, has the axis wider and more convex, with fewer ribs; and there are but four distinct ribs on each side.

Variations.—In some the axis is a little more prominent; in others a greater or less depression of the glabella occurs, and apparently the production forwards of the snout is not always in the same degree. But these variations are within narrow limits, and our species never seems to approach *C. Blumenbachii* in convexity, especially with regard to the glabella and caudal axis. The front is constantly produced, the surface minutely scabrous, not covered with scattered tubercles; but this last character occurs in some varieties of *C. Blumenbachii*, which is more variable than we formerly believed.

This really beautiful species shows in the reduction of the glabella a tendency towards the characters so strongly displayed in the Lower Silurian forms. In the projecting buttress which stretches from the region of the eye towards the second lobe of the glabella, an approach is made to the very curious *C. camerata* of Conrad, in which the processes from behind the eye form projecting wing-like covers to the axal furrows. And in the slight furrows of the arched lateral lobes of the tail, several foreign species are imitated.

Localities.—WENLOCK SHALE, Burrington, Shropshire, abundant. LUDLOW ROCKS, Underbarrow, Westmoreland.

Foreign localities.—Clinton Group (May Hill Group). Hall's figure probably represents this species.

C. Blumenbachii, *Auctorum*. Pl. VIII, figs. 7—16. Pl. IX, figs. 1, 2.

Lyttelton. Phil. Trans., vol. xlvi, pls. i, ii, p. 598, 1750.
Mortimer. Ib. p. 600. *Mendez da Costa*, ib., vol. xlviii, p. 296, 1753; also *Guettard, Wilckens, Klein, Walch, Beckmann*, &c., 1757 to 1773.
C. platys, Green's Monograph; cast No. 4, 5, (not C. Blumenbachii, *id.*) 1832.
Trilobites tuberculatus, *Brunnich*. Nye Samml., &c., i, 389, 1; 1781.
— — *Blumenbach*. Abbild. Naturh. Gegenst., i, t. 50; 1810.
Entomolithes paradoxus, *Parkinson*. Org. rem., III, pl. xvii, figs. 11, 13, 14; 1811.
— tuberculatus, *Wahlenberg*. Nov. Act. Ups., viii, 31, 6; 1821.
Calymene Blumenbachii, *Brongniart*. Crust. foss., ii, 1, pl. i, fig. 1 a—c; 1822.
— — *Dalman*. Palæad. 35, 1, tab. i, figs. 2—3 a—c; 1826.
— — *Payton*. Trilob. of Dudley, fig. 14 (plate only) 1827.
— — *Murchison*. Sil. Syst., pl. vii, figs. 6, 7 (not fig. 5) 1837.
— *var*. Niagarensis, *Hall*. Pal. N. York, vol. ii, p. 307, pl. lxvii, figs. 11, 12; 1852.
Calymene Niagarensis, *Hall*. Geol. Report 4th district, p. 101, fig. 3; 1843.
— subdiademata, *M'Coy*. Pal. Foss. Woodw. Mus., pl. i f, fig. 9 only; 1851.
— spectabilis, *Angelin*. Palæont. Suecica, t 19. f. 5; 1852.

C. magna 3—5 *uncialis, elongata, valde convexa,—per totum tuberculosa. Caput glabellâ magnâ longâ, utrinque trilobâ, marginem frontalem compressum crassum haud productum attingente; genis declivibus, vix glabellâ latioribus. Thorax axe quam pleuris lato, convexo ferè gibbo; pleuris deflexis nec planis, fulcro approximato, anticè ad tertiam, posticè ad quintam partem latitudinis posito. Cauda angusta, subtrigona, axe gibbo, lateribus declivibus* 5-*costatis, costis omnibus bifidis, sulcis profundis.*

It is necessary to be diffuse in the specific character; for in this genus it is only by contrasted differences of proportion that we are able to distinguish the species. *C. Blumenbachii*, like all common Trilobites, has a considerable range of variation; but the greatly extended large glabella is after all the best character of the species. It occupies fully a third of the head in width, and is so long as to touch the thickened front margin, which is neither produced nor much reflexed. In this all the specimens agree, while they differ in points of mere proportion. The glabella is sometimes a little wider, especially at the base; sometimes less convex, but always more prominent than the cheeks, and even overtopping the eye-tubercles. The tail varies in width, but is always trigonal with deflexed strongly ribbed sides; and the pleuræ, always strongly decurved, are sometimes flatter, and sometimes very convex, *i. e.*, curved very strongly down, as in our figure 15.

The species being so common, it is only necessary to describe the points relied on for its distinction from others; for the Dudley fossil has been much confused with kindred forms, and these differ sufficiently from it when closely examined.

C. Blumenbachii is the largest of the genus; our fine central specimen, the one figured in the 'Silurian' System, being 4 inches by $2\frac{1}{2}$ broad. Swedish specimens sometimes reach 5 inches. Form ovate oblong, not much pointed in front, and obtuse behind. The head occupies more than a fourth the whole length, and is semicircular,—our fig. 8 giving the true shape, while fig. 7 is more pointed than usual. The glabella is more than equal to the width of the cheeks; and reaches quite to the thickened front margin. It is very convex, especially in front, and is bell-shaped, the base expanded, the large, round, lower lobes fully $\frac{2}{5}$th the whole length, the middle ones about half this length, the upper minute. All are convex and well circumscribed. The marginal furrow and the axal furrows deep; the cheeks convex, but not nearly so much so as the glabella; and the eyes, placed two thirds up, are not very prominent, nor is there any strong buttress connecting them with the central glabella-lobe. The neck- and marginal furrows strong, and all but complete, sometimes (fig. 18) quite so.

The margin is very thick, and seen on a front view (fig. 16) not so much bent upward as in the last species. The rostral shield is a good deal wider than long, and the sutures converge much toward the labrum, which (fig. 10) is squarish-oblong, with parallel sides beneath the broad ascending processes. The centre is gibbous, with a strong tubercle; and there is a strong concentric furrow separating a broad margin. The tip is emarginate, with obliquely truncate angular lobes.

The axis of the 13 body-rings is very convex, and as wide as the deflexed pleuræ, from which it is abruptly separated, but not by a furrow. The axis tapers very little backwards. The pleuræ, horizontal for a third of their width, bend strongly down at the fulcrum, placed at one third near the head, and in the last joint at $\frac{1}{5}$th the width of the pleuræ. Pleural groove strong, the forward half of the pleura being the smallest.

Tail roughly trigonal, with the front greatly arched, and the base-line very little curved. The convex axis occupies one third the width, converges quickly near the tip, and is thence continued by a short appendix to the very end of the tail. It has six or seven rings, and a smooth, convex, terminal portion, beyond which is a smooth appendix. The sides have five strong bifid ribs, very little arched, subparallel, the hinder ones becoming quite longitudinal in direction. The margin is strongly bent inwards at right angles to the surface, but not further incurved. All the tail is granulated. The granulation of the margin is very close and fine, (fig. 18 *b*) but over all the rest of the body tubercles are mixed with the granules (fig. *a*) often conspicuously. (See also Pl. VIII, figs. 7, 15, &c.)

Fig. 19.

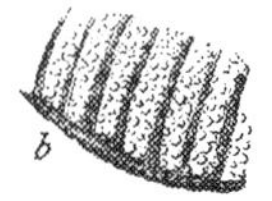

Body-ring and tail of *Calymene Blumenbachii*, showing mixed granules.

The following variations may be noted. Our Pl. VIII, fig. 7, has a somewhat more pointed head and longer glabella than usual, while fig. 16 has it rather shorter than ordinary, and more expanded at the base. Fig. 8 has the normal shape of the glabella-lobes; the basal ones being pyriform and somewhat wider than long. In fig. 14 the length is greater. Fig. 16 has the large lobes triangular. Fig. 15, in the 'Mus. Pract. Geology,' has the granulation stronger than in most specimens we have seen,[1] and is also deeper in the body; the vertical height of the very convex form being exactly half the width. Few specimens have the height greatly more than one third. In some, fig. 16 for instance, the marginal furrow is continuous all round the front. Fig. 14 has scarcely any interval behind the thick front margin and the glabella.

Var. pulchella, Dalman; Palæadæ, p. 35, &c., has the glabella narrow in the young state; our fig. 19 may very well represent such a form. The head is more triangular, and the glabella narrower than usual. Fig. 11 has the same character. Fig. 16, on the contrary, has a wide short glabella. But all these variations are within narrow limits.

There are differences of proportion in the tail. Some have the length as ten, to a breadth of eleven. A second has it in the proportion of seven to ten. In others it has a length of five, to a breadth of seven. And there are all intermediate proportions; I figure a variety from Mr. Allport's collection, which differs more from the ordinary form of the species than any other Upper Silurian specimen I know. I call it

Var. ALLPORTIANA;— *capite trigono, fronte productâ, glabellâ breviore.*

In this variety the knots on the axis are very strong; the fulcrum is further out than usual, the glabella more sunk, and the front so much produced as to suggest the idea of a strictly intermediate variety between this and the preceding species. Such specimens as this tend to shake our faith in species, and make us ready to believe that they are after all only confirmed varieties of some more common type. *Even if this be true,* they are not, for this reason, of less consequence either to the naturalist or to the geologist: nor is it necessary to extend the idea indefinitely to genera and families.

FIG. 20.

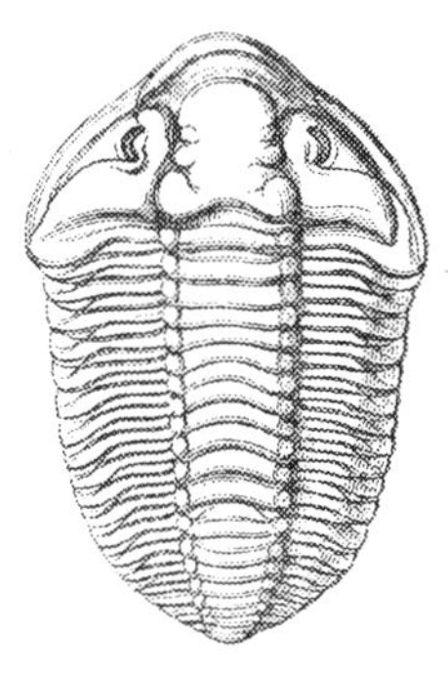

C. Blumenbachii, var. *Allportiana*, from Dudley. Mr. Allport's collection.

Fig. 12 shows decided knots down the sides of the axis, and this is very probably Dalman's variety *tuberculata*. Figs. 15, and in part 16, also show them; they are absent in the majority of Dudley specimens. But these variations are all slight compared with the amount of difference seen on comparing the Dudley fossil with the Lower Silurian forms distinguished under the name *C. senaria*.

The internal casts figured from the Woolhope beds, Pl. IX, fig. 1, and from the Llandovery beds, Pl. IX, fig. 2, show a tendency to a narrower and shorter form of the glabella; and some Dudley specimens also tend this way. And it will be well in comparing *C. Blumenbachii* with its var. *brevicapitata*, to bear in mind that we must compare casts with casts in order to understand the amount of difference. For instance, Pl. IX, fig. 1, is that of

[1] Another specimen in the Mus. P. Geology, imperfect, but as large as our central figure, has the granulation remarkably close, covering the whole surface. And, whenever the true surface is preserved by careful cleaning of the specimens, this character is more or less conspicuous.

a very thick old crust, and that partly accounts for the space between the glabella and the front. In fig. 2 the crust is much thinner, and the space is consequently not so great. But in both these the glabella reaches nearly to the front margin; and is bell-shaped, not triangular. In the species which follow, the glabella is more and more reduced in size, till it reaches its minimum in the *C. parvifrons*, Pl. IX, fig. 21 We shall take these species in descending order.

Localities.—Caradoc, Ireland (Prof. McCoy), Lower Llandovery, Carmarthenshire; Radnorshire; Pembrokeshire; Mullock, Girvan, in Ayrshire; Mayhill Rocks of Norbury and the Longmynd; Malverns, &c.; also Dingle, Ireland. Woolhope beds, Bogmine, Shelve, Shropshire. Wenlock Shale and Wenlock Limestone, everywhere in Great Britain and Ireland, where these rocks or their equivalents occur; the chief specimens from Dudley and Walsall. Lower and Upper Ludlow Rocks, Shropshire (rarely). Marloes Bay, Pembrokeshire. Aymestry Limestone—Leintwardine (*C. subdiademata*, McCoy).

Foreign Distribution.—Sweden and Norway, in Upper Silurian. *Niagara Group* of New York and Pennsylvania, &c. *Etage E.* Bohemia; *Barrande*. (This last locality is somewhat doubtful.) Kindred species are found all the world over in rocks of Silurian age.

C. Blumenbachii, var. CARACTACI.—*Glabellâ angustiore breviore, caudâ normali, axe lato, lateribusque deflexis.* Pl. IX, figs. 3—5.

[C. selenecephala, *Green's* Monogr. Cast No. 3, 1833?]
C. subdiademata, *M'Coy.* Pal. Foss. Woodw. Mus. pl. i f, (fig. 10 only); 1855.
C. Blumenbachii, *var.* brevicapitata, in part. *Salter* in Mem. Geol. Survey, vol. iii, pl. xvii, fig. 9 only, 1865.

Not above 1⅓ inch long, and fully 1¼ broad, ovate, the head semicircular, with the glabella equal in width to the cheeks, and in length rather greater than its breadth. The glabella is parabolic, blunt in front, and has the three side-lobes well developed,—as in the type-variety. The eye is rather forward, and there is no buttress opposite the middle lobe. Except in this reduced size of the glabella, and in a slightly narrower axis and more arched side-ribs to the tail, there is no essential distinction between it and the ordinary form. It has been, however, occasionally (being always in the state of casts only, confounded by myself and others with the true *C. brevicapitata*, which, indeed, being the *C. senaria* of Conrad, the name had better be extinguished. I would call the present variety Var. *selenecephala*, were I sure of Green's name; but his cast is very imperfect.

Locality.—Caradoc Rocks of Shropshire; abundant. (Mr. Edgell, Mr. Lightbody, Geol. Surv. Collections, &c.) Bala, Llanwyddyn, Llangollen, Snowdon, &c., in N. Wales. Desertcreat, Tyrone; Col. Portlock (Mus. Pract. Geology). Coniston, Westmoreland.

Foreign localities.—North America, if *C. selenecephala* of Green be the same.

Subspecies I.

CALYMENE SENARIA, *Conrad.* Pl. IX, fig. 5—11.

C. BLUMENBACHII, *Green,* in part, No. 1 cast, 1832.
C. SENARIA, *Conrad.* Ann. Geol. Rep., N. Y., p. 49, 1841.
— *Hall.* Palæont. N. York, vol. i, pl. lxiv, fig. 3, 1847.
— *Emmons.* Geol. Rep., p. 390, fig. 2, 1842.
C. BREVICAPITATA, *Portlock.* Geol. Report, pl. iii, fig. 3 only ?, 1843.
— *Salter.* Mem. Geol. Surv., vol. ii, pt. 1, pl. xi, figs. 1, 2, 1848.
— „ „ vol. iii, pl. 17, figs. 10—12 (not fig. 9). 1865.
— „ Morris Catal., 2nd ed., p. 102, 1854.
— *M'Coy.* Pal. Foss. Woodw. Mus. 165, tab. i F, figs. 4—6, 1855.
C. BAYLEI, *Id.* *Op. cit.*, tab. i F, fig. 8 (not of Barrande).
C. FORCIPATA, *Id.* Sil. foss. Ireland, pl. iv, fig. 14 (head only), 1846.

C. modica vix biuncialis, elongata, alutacea; glabellâ brevi trigonâ, trituberculosâ, lobis rotundis. Gena convexa, absque processu oculari, oculis submedianis. Frons producta recurva, longitudine tertiam partem glabellæ æquans. Thorax axe convexo, pleuris valde deflexis, fulcro approximato. Cauda trigona, axe lato conico; lateribus deflexis 5-*costatis, —costis furcatis.*

Under the name *brevicapitata* of Portlock have been commonly included two marked Caradoc varieties or subspecies, in one of which the glabella, though nearly resembling that of the ordinary form, is narrowed in front, and has a smaller proportion as compared with the cheeks. This I have already distinguished in the previous page by the varietal name, *C. Caractaci.* It is, in truth, not far removed from the Dudley variety. The other there will be much less hesitation in admitting as a distinct species. It has so short a glabella as to have suggested the specific name to Gen. Portlock, and so cylindrical a form as to have formerly induced me to rank with it *C. parvula,* Barrande, a truly distinct fossil. And by the convexity of the form, and the breadth of the axis, it is distinct enough from our next variety or subspecies, while the finely granular surface easily distinguishes it from the "Dudley fossil."

C. senaria is a rather small species, the fragments seldom indicating a size of more than two inches in length. It is elongate-oval and very convex, the width of the head being less than half the whole length, of which the head itself occupies two-sevenths. It is semicircular, with the front produced and elevated; and the triangular glabella is fully as wide at the base as the whole length from the neck-furrow to the front margin: it is also as wide as the cheeks in their full measure. The lobes are rounded, and all three quite distinct, the basal ones not so wide as the central portion of the glabella. The axal furrows deep, but not broad; and in the cast the attachment for the ascending processes is well marked (as in figs. 5—7). The cheeks gibbous anteriorly, but without any buttress opposite the middle lobe. The thorax convex, its axis as broad as the sides; the fulcrum (as in *C. Blumenbachii*) approximate. The tail is trigonal; but its sides slightly arched,—the axis convex, not much tapering, abrupt at the end, and with six or seven rings. The sides are deflected, but a little arched, and have five ribs, forked thoughout.

FIG. 21.

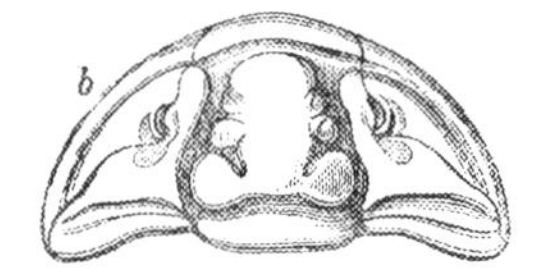

Calymene senaria. a is from British Lower Silurian, a cast. *b* from Ohio, with the crust preserved.

The whole surface is granulose (not roughly tubercular as in *C. Blumenbachii*), with larger and smaller tubercles. I think *C. callicephala*, Green, Cast No. 2, is too extreme a form to be reckoned with this subspecies, though closely allied. Hall says it is identical, but the *C. Blumenbachii* of his work (*C. senaria*, Conrad) is quite our species, and is as common in the Lower Silurian Limestones of the Western States as in our own Caradoc and Welsh Slate Rocks. See woodcut, fig. 21, *b*.

Localities.—CARADOC or BALA ROCKS of Shropshire and North Wales—everywhere; Westmoreland; North and South Ireland, &c. Figs. 5 and 6 are the original specimens of General Portlock's work; and fig. 8 is the perfect young specimen given in the 'Memoirs of the Geological Survey.'

Foreign Localities.—*Trenton Limestone* of the United States, abundant (under the name *C. senaria*). (*Calymene nivalis*, Salter, is the representative species in the Thibetan range of the Himalaya. See Col. Strachey's collection, now in the Mus. P. Geology.)

Subspecies II.

CALYMENE CAMBRENSIS, *Salter*. Pl. IX, figs. 12—14.

C. BREVICAPITATA, *Salter*. Mem. Geol. Surv., vol. ii, pt. 1, pl. xi, figs. 3—5, 1848.
C. BREVICAPITATA, *M'Coy*. Pal. Foss. Woodw. Mus., pl. i F, figs. 4, 5, 1855.
C. BLUMENBACHII, *var.* CAMBRENSIS, *Salter*. Mem. Geol. Surv., vol. iii, pl. xvii, figs. 13, 14, 1865.

C. modica, alutacea, fronte subrectâ; glabellâ vix trigonâ brevi trituberculatâ. Cauda axe angusto, lateribus valde arcuatis, multisulcatis,—sulcis profundè interlineatis.

This marked species or subspecies has the glabella of the same shape with the Caradoc fossil *C. senaria*, last described; but is flatter, and the upper glabella-lobe is almost obsolete. The front margin is produced, but not recurved. The body is elongate, and not so convex as in *C. brevicapitata*, and the tail especially is more expanded,—its sides less deflected and more arched, resembling in this respect our *C. tuberculosa*. The axis is narrow, compared with the width of the sides, and seven- or eight-ribbed. The sides have six ribs, strongly interlined throughout, and arched outwards; and the general shape of the tail is more transverse, ovate, and less triangular than in either the Dudley or the Caradoc species.

Locality.—LLANDEILO FLAGS of S. Wales, abundant at Llandeilo; at Lann Mill, near Narberth, &c. Fig. 13 is from a very perfect specimen observed by Sir H. De la Beche, and of which a cast only of the external surface is preserved in the Mus. P. Geology. The original is, I believe, in the rich cabinet of Mr. J. E. Lee, of Caerleon; and is the finest known.

CALYMENE TRISTANI, *Brongn.* Pl. IX, figs. 15—18.

Tristan, Journ. des Mines, tom. xxiii, page 21. 1807.
CALYM. TRISTANI, *Brongniart*. Cr. foss. 12, pl. i, fig. 2, A—K. 1822.
— — *Schlotheim*. Nachtr. ii, 14, 2, 23, 2, and 40, tab. xxii, fig. 5, 1823.
— — *Dalman*. Palæad., 62, 3. 1826.
— — *Emmerich*. Dissert, 39, 4. 1839.
— — *Milne-Edwards*. Crust. iii, 320, 5. 1840.
— — *Burmeister*. Org. Trilob., tab. 2, figs 7, 8, 1843, and Ray Ed. p. 40. 1846.
— — *Salter*. Quart. Geol. Journ., vol. xx, pl. xv, fig 5. 1864.

Calymene Tristani is a fossil belonging to the French, and not the British, Silurian fauna. But imperfect heads of this species are certainly found in the quartzose strata of Gorran Haven, near Mevagissey, Cornwall; and the tail, fig. 17, is found in company with several other French fossils in the now famous Budleigh Salterton pebble-bed.

Our specimens from Gorran Haven, figs. 15, 16, show only the glabella and front of the head. The former is short, and wider than long, with the base broad, so as to form a nearly equilateral triangle. All the lobes deeply marked, and oblique forwards; the basal lobes triangular, the middle ones nearly linear; the upper ones strong for the genus. The cheeks highly convex, separated from the glabella by broad but shallow axal furrows; which at the base expand over a triangular flattened depressed space (fig. 18) analogous to the corresponding portion in *Homalonotus*, see p. 104, line 6. The eyes are placed forward opposite the upper glabella lobes. The cheeks strongly margined exteriorly. The front is produced, very convex and recurved; and the space from the front of glabella forwards is nearly equal to the length of the glabella itself. We have not the body,—which in foreign specimens has a broad axal lobe and greatly decurved convex pleuræ. The axis of the subtrigonal tail, fig. 17, is highly convex and broad, of six or seven rings, and an appendix which reaches the end. The sides have five strong ribs interlined for nearly their whole length. I do not consider our British specimens sufficient to afford a regular diagnosis of the species.

Calymene Arago, a kindred species found with *C. Tristani* in the French and Spanish deposits, has a smooth trilobed tail, with scarcely a trace of ribs.

Localities.—LLANDEILO (or ARENIG) ROCKS, of Gorran Haven, Cornwall, figs. 15, 16, Mr. Edgell's cabinet. Budleigh Salterton, fig. 17. [Fig. 18 is added from a specimen from Néhou, Normandy, where the species is very common.]

CALYMENE DUPLICATA, *Murchison*. Pl. IX, figs. 19—24.

ASAPHUS DUPLICATUS,	*Murch.*	Sil. System, pl. xxv, fig. 8, 1839.
CALYMENE DUPLICATA,	*Id.*	Siluria, 2nd edition, pl. iii, fig. 6, 1859.
— —	*Salter.*	Mem. Geol. Survey, vol. iii, pl. xvii, figs. 15—20, 1865.

C. modica, rarissime biuncialis, alutacea, depressa, fronte paullum productâ recurvâ. Glabella parallela, longa; oculi antici. Axis corporis angustus, caudæ longus præangustus (*nec quartam partem latitudinis caudæ subplanæ efficiens*) 9—10-*annulatus. Costæ laterales* 7—8, *valde interlineatæ, arcuatæ, haud deflexæ.*

An inch and a half long,—seldom more, and fourteen lines broad,[1] greatly depressed, with a scarcely produced front not above one third the length of the glabella; a very narrow axis, and a many-ribbed flattened tail. The head is wide, semicircular, and depressed, with a strong but very narrow margin all round.

The glabella rather parallel-sided than parabolic (in the ♀ form, figs. 19, 20, somewhat broader and more arched on the sides), and not occupying above one fourth the width of the head. It has three well-marked lateral lobes, the lowest not greatly projecting beyond the others. The glabella is depressed, and does not rise above the level of the broad gently convex cheeks; on which the small eye is placed very forward, and distant from the glabella about half its width. The neck-furrow is sharp, narrow, and strong; the axal furrows narrow and deep. The surface of the head, and indeed of all the body, is finely granular.

The thorax has a very narrow convex axis, in some specimens less than a fourth the whole width. The pleuræ are very flat at first, and then, beyond the fulcrum, which is placed far out (at one half in the front rings, and one third posteriorly) they are strongly and vertically decurved, but not abruptly bent.

The shape of the tail is semioval, depressed above, but convex around the margin. The form is remarkable for the genus, being so flat as to be more like the tail of an *Ogygia* or *Asaphus* (see fig. 21); and the numerous ribs, 9—10 on the axis, and 18 on the sides, heighten the resemblance. The axis is barely more than one fifth the width of the tail in the ♂ (fig. 19), and about one fourth in the ♀ form (fig. 24);—long conical, and rather abruptly contracted about the middle. The end of the axis is abrupt, and

[1] Mr. J. Lee, of Caerleon, has one specimen with the head 1½ inch broad.

there is no appendix beyond it (see fig. 21). The side ribs arch widely out, and are so strongly interlined throughout, as to give the appearance of having twice the proper number of ribs.

I have ventured to call fig. 24 the ♀ form, or it may stand for variety β, *fœmina.* All the characters are the same as those of the typical variety *a*, except the somewhat broader axis. And such a difference we should be prepared to expect in Trilobites, for nearly all, as Barrande has shown, have a " forme longue " and a " forme large."

Accordingly, in our *var. fœmina* the general shape is rounder, and the glabella broader and less pointed; the axis of body and tail broader, and the proportion borne by them to the sides consequently greater. But these differences do not constitute a distinct species, and the general aspect is much alike in both forms.

The labrum is oblong, twice as long as broad, with parallel sides. It has a narrow, not extended base, and is concave under the front border, then very convex, and with a blunt tubercle above the centre. Two small tranverse lobes lie beneath this, forming a nearly continuous ridge, and the limb is tumid beyond as far as the base of the short furcate terminal lobes.

C. Baylei, Barrande, is a closely related species, the tail is especially like; but Barrande's figure and good Bohemian specimens in the British Museum show a species with more remote eyes, and the axis of the tail yet narrower than in the British form (Barrande, 'Sil. Syst. Bohême,' Pl. XLVIII, fig. 49). It is an Upper Silurian fossil.

Localities.—LLANDEILO FLAGS, near Shelve and Wilmington, Shropshire; Builth, Radnorshire; abundant. Lann Mill, near Narberth; Abereiddy Bay, Pembrokeshire, abundant. The species is not known beyond the British Isles, and appears to be very restricted in its range. It has been found at Waterford by Major Austin.

Such species as the above help to conduct us to the *Conocephalidæ*, and show the near connection of *Calymene* with that group.

CALYMENE PARVIFRONS, *Salter.* Pl. IX, figs. 25—28.

CALYMENE PARVIFRONS, *Salter.* Appendix A, p. 3 of Pal. Foss. Woodw. Museum, 2, pl. i F, fig. 7. 1855. *M'Coy*, ib., p. 167, pl. i F, fig. 7. 1855.
— — *Salter,* in Morris' Catalogue, 2nd ed., p. 102, 1854.
— — *Id.* Siluria, 2nd. ed., p. 53, fig. 4, 1859.

C. biuncialis et ultra, convexa, sublævis; capite subtrigono lato, fronte valde productâ tumidâ elevatâ, ⅔ glabellæ brevissimæ æquali,—hâc parabolicâ quam longâ latiore, lobis tribus elongatis subradiatis. Genæ valde convexæ, glabellam latitudine ⅓ superantes; oculis prominulis remotis, subcentralibus. Thorax axe angusto, pleuris valdè deflexis. Cauda lata, axe convexo, lateribus valde decurvatis 5*-costatis, omnino interlineatis.*

We formerly only knew the head of this curious species (see fig. 25). It is 16 lines wide, and only 7 lines long. And of this length the tumid and produced front is two thirds as long as the small parabolic glabella. The latter is very little convex, a good deal wider than long, truncate in front, and with the three lateral lobes all distinct, oblique forwards, and reaching well into the glabella, especially the basal ones, which are subtriangular and much contracted at their inner edge. The cheeks are large and tumid, and the small eyes are placed far forwards, on a level with the upper glabella-lobe, about as remote from it as the width of the glabella. The neck-furrow is strong beneath the glabella, but broad and shallow underneath the capacious cheeks, and the lower margin of the latter is much curved.

Fig. 22.

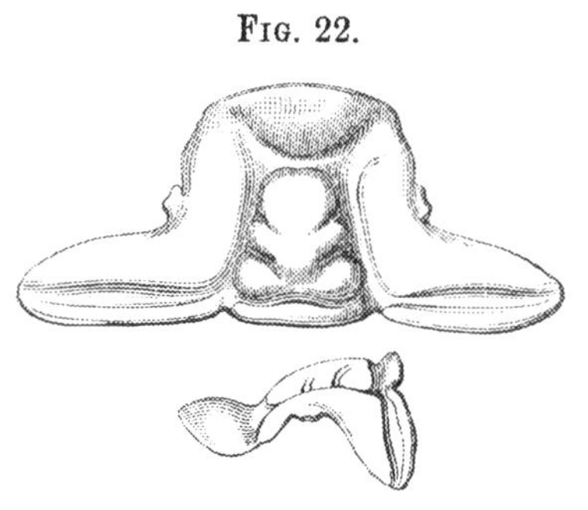

C. parvifrons, nat. size and magnified.

Prof. M'Coy's figure is very inaccurate in proportion. I have only lately been able to see the specimen, and give a more correct outline in our woodcut than fig. 25 in the plate, which was unfortunately copied from the Cambridge Synopsis.

Var. Murchisoni. Pl. IX, figs. 26—28.—*Glabellâ latiori, oculis vix remotis.*

Imperfect as our materials are, it is clear this variety must be distinguished from the typical *C. parvifrons*. The var. *Murchisoni* has a broader glabella, and the eyes less remote, and larger. Such differences may be due to sex rather than variety, and there is reason to suspect sexual variations to a small amount in this genus, as I have above shown (p. 101).[1] This variety, *Murchisoni*, is much more like *C. Tristani*, from which, indeed, except by the more truncate and less triangular form of the glabella, and far less strongly ribbed tail, it is somewhat difficult to separate it.

Locality.—Arenig or "Skiddaw" group. Tai hirion, on the road from Bala to Ffestiniog; collected in 1844, by Prof. Sedgwick and J. W. Salter (Woodw. Mus.).

Of the *var.* Murchisoni;—the Stiper Stones; viz. at Lord's Hill, and other neighbouring localities. Also at Cae Glyd, under the Manod Bach, near Ffestiniog—and at Ty-obry, near Garth, Portmadoc (Mus. P. Geology; and the cabinets of Messrs. Ash and Homfray).

[1] Prof. Thomson, who has observed hundreds of specimens of the *Calymene Blumenbachii*, thinks there is no difference of sex observable in that species. But that is no reason for doubting it in others.

CALYMENE? DAVIESII, *n. sp.* Fig. 23 (Woodcut).

C. caudâ latâ subtrigonâ, depressâ, anticè subrectâ, posticè rectangulâ: axe angusto longo, 9-annulato, percurrente; lateribus 7-costatis-costis radiatis, valde duplicatis: sulcis profundis, fasciâ caudali angustâ convexâ. Lat. $1\frac{1}{2}$ *unc. Long.* 1 *unc.*

The general character of this curious specimen forbids us to associate it with any other genus, as it is unlike the usual Upper Silurian forms.

FIG. 23.

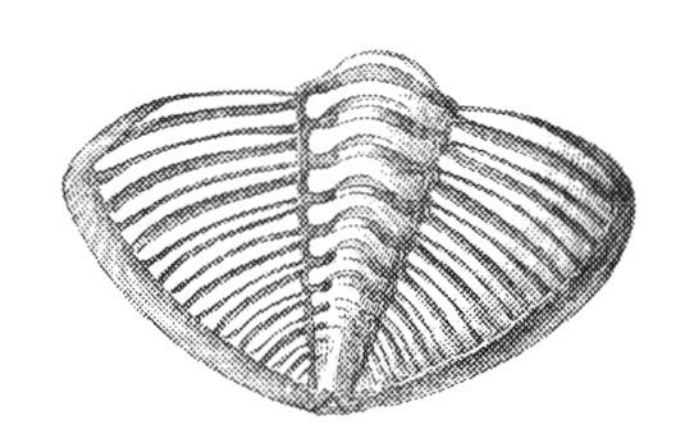

C. ? Daviesii, from the Wenlock Shale.

The tail is much depressed for the genus, wide, and rectangular behind, the sides very little curved, the apex obtuse. The length is rather more than two thirds the width, which is $1\frac{1}{2}$ inch; the axis occupies only one fourth of the whole width, and is regularly conical, convex, and reaching the very end of the tail. It has eight distinct rings and a short terminal portion. The axal furrows are strong, the side-lobes convex along the middle, but not towards the margin, which is only slightly decurved. There are seven distinct lateral ribs reaching almost to the margin, and strongly duplicated throughout. The direction of the ribs is outwards, not much backward. The specimen is only an interior cast, and the actual margin is not seen. But the caudal or subcaudal fascia is very distinct, convex, and narrow all round, and the furrows, both of the axis and the sides, are strong and deep. The fulcral point is so remote, that if, after all, this should prove to be an unusual form of *Phacops,* I should not be greatly surprised. Still I think it is a *Calymene.*

Locality.—WENLOCK SHALE of Glan Wye, 3 m. N.W. of Builth, in Radnorshire. (Cabinet of Griffith Davies, Esq., of Islington.)

HOMALONOTUS, *König.*

Dipleura, Trimerus, Green, &c.

A natural genus, which has been admitted by every writer on Trilobites, since our old friend Dr. König gave us the figure in the 'Icones Sectiles.'

There is no pretence for separating any of the forms except as convenient subgenera. From *Calymene,* its near ally, as, indeed, from most Trilobites, it is distinguished at once by its want of distinct trilobation; and this peculiarity is only less distinct, not absent, in the earlier species. In the characters of the thorax, eyes, labrum, and even in the contour of the tail, the resemblance is so close to *Calymene* that it is often difficult to distinguish fragments of one genus from the other. In general, the very obscure glabella, scarcely lobed; and the broad, highly convex, and slightly trilobate thorax, distinguish it.

Homalonotus is—Elongate, convex, with steep sides and a very broad axis, scarcely distinguished from the pleuræ. There are thirteen body-rings, deeply grooved, and the fulcrum is close to the axis in most of the species. The head with an obscure quadrate glabella, slightly lobed; a rostral shield; and a quadrate labrum, tuberculate and gibbous in the middle, and with a bilobed tip. Surface of the body scabrous, occasionally spinous. Internally the cheeks have at their base a broad flat space next the glabella.

Range.—Upper Silurian, and Lower (and Middle?) Devonian.

The genus has not yet been satisfactorily divided, but the species may be conveniently arranged in five, or possibly six, groups or subgenera, as follows:

1. *Brongniartia,* Salter, 1865. Depressed, with broad rounded head, remote eyes, well-defined lobeless urceolate glabella, and many-ribbed rounded tail. Lower Silurian.

§ 1. Body scarcely trilobed; the axis broad. (*H. bisulcatus* is the type of the subgenus, and of this section.) Lower Silurian.

H. bisulcatus, Salter.
H. Sedgwickii, id.
H. Edgelli, id.
H. platypleurus, Green.

§ 2. Body strongly trilobed; the axis narrow. (Type, *H. rudis*; this leads off directly towards *Calymene.*) Lower Silurian.

H. rudis, Salter.
H. Bohemicus, Barr.
H. rarus, Corda.
H. brevicaudatus Deslongch.
H. Vicaryi, Salter.

2. *Trimerus,* Green, 1832. Elongate, convex, with triangular head; eyes not remote; a defined, but obscurely lobed, broad glabella. Thorax slightly lobed; tail many-ribbed, pointed, often acuminate. (Type, *H. delphinocephalus.*) Upper Silurian.

H. delphinocephalus, Green.
H. Johannis, Salter.
H. cylindricus, id.

3. *Kœnigia,* Salter, 1865. Convex; head wide transverse, with concave and tricuspidate front. Glabella subquadrate, well defined. Eyes rather approximate, on gibbous cheeks. Tail pointed, many-ribbed. (Type, *H. Knightii.*) Upper Silurian.

H. Knightii, Murch.
? H. ludensis, Salter.

4. *Dipleura*, Green, 1832. Convex; head wide, semi-oval, or subtriangular, with somewhat pointed front. Glabella narrow, well defined. Eyes rather remote, on gibbous cheeks. Thorax slightly lobed. Tail obtuse, hardly ribbed. (Type, *H. Dekayii.*) Upper Silurian; Lower Devonian.

H. Dekayii, Green.
H. sparsus, Eaton.
H. obtusus, Sandberger.
? *H. crassicauda*, id.
? *H. Ahrendi*, Roemer.
&c. &c.

5. *Burmeisteria*, Salter, 1865. Elongate, convex; head triangular; eyes approximate on gibbous cheeks. Glabella distinct, lobeless, spinous. Thorax slightly lobed and spinous, as is also the many-ribbed pointed tail. (Type, *H. Herschelii.*) Devonian.

H. Herschelii, Murch.
H. armatus, Burm.
H. Greenii, Goldfuss.
H. elongatus, Salter.
H. Pradoanus, De Vern.

I have been obliged to include some foreign species, but these are inserted in italics. The genus *Homalonotus* seems to have been chiefly a northern one in Silurian times, but during the Lower Devonian epoch it was common both north and south of the equator. After this period the whole group was extinct.

Section—Brongniartia, § 1.

Homalonotus bisulcatus, *Salter*. Pl. X, fig. 2—10.

Homalonotus bisulcatus, *Salter*. Appendix to Sedgwick and M'Coy, Synopsis Brit. Pal. Foss., pl. 1 G, figs. 24—31, 1851.
— — *M'Coy*. Ibid., fasc. 1, p. 168, 1851.
— — *Salter*. In Morris' Catal., 2nd ed., p. 109, 1854.
— — Ib. Siluria, 2nd ed., p. 74, foss. 12, fig. 2, 1859.
— — Ib. Memoirs Geol. Surv., vol. iii. Appendix, pl. xvi, figs. 1—8, 1865.

H. 7-uncialis ovatus lævis, capite caudâque semiovalibus, rotundatis. Glabella depressa. Genæ latæ, extûs deflexæ; oculi submediani, valdè remoti. Frons unà cum genis continua, plano-concava. Thorax vix lobatus. Cauda semiovata, obtusa, margine angusto recurvo; axe conico, 11—12-annulato, per $\frac{4}{5}$ longitudinis caudæ extenso, et appendice conico; lateribus 8-sulcatis, sulcis duobus anticis valde profundis, haud interlineatis.

Long-ovate, but broader than in many of the genus, with a blunt semi-ovate head and rounded tail, ribbed strongly throughout. Trilobation conspicuous, but not deep.

Glabella without lobes. These general characters will enable us to recognise the species, which is a very common one in the Caradoc sandstones and slates of Britain, and especially of Shropshire and North Wales. The finest head is from Mr. Lightbody's cabinet (fig. 9); the most complete body and tail are from the Geological Society's Museum (fig. 4). The species is a large one, and, judging from fragments, must have been fully ten inches long. It is ovate and blunt at both extremities.

Head semioval, of which the glabella occupies three fifths; it is pyramidal, but with sinuous sides, and truncate above; and at its base is exactly as broad as the cheek. It is rather abruptly raised above these, in a step-like manner. The cheek itself is gently convex, but not abruptly so towards the eye, which is placed about centrally[1] and outside the highest point of the cheek, but does not nearly rise to the level of the glabella. This gives a peculiar character to the present species, for in the genus *Homalonotus* the cheek is often very convex. Young specimens (fig. 7) have the glabella greatly more distinct and more separated from the cheeks than in the adult, and the proportions are not quite the same, the glabella being longer.

In front of the glabella the head is slightly concave only; and the cheeks are as gently convex, without any central gibbosity or any strongly distinct margin. The neck-furrow is distinct all along, especially in the cast (fig. 9). It is strongest beneath the glabella, and then descends to a lower level beneath the cheek, along the base of which it is continuous almost to the rounded angle. The facial suture is vertical in front of the eye, and beneath it curves boldly outwards. The eye is small and apparently ovate (not globular, as in the last species). The free cheek is tumid just outside the eye, and much deflected.

Fig. 24.

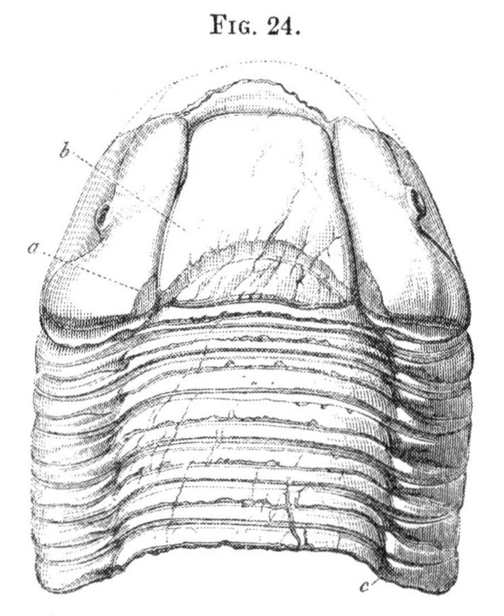

H. bisulcatus, Marshbrook, Shropshire.

The body, of thirteen rings, has its axis suddenly and greatly wider than the base of the glabella, as indicated by the diverging lines on the neck-segment (woodcut). It is not actually a great deal wider than the pleuræ; but, from the general and regular convexity (the back is not depressed as in *H. delphinocephalus*), it appears much more on a front view. The axal line is less strong in the adult than in younger specimens (fig. 7), and is marked out chiefly by a sudden angular thickening of the pleuræ over the fulcral point (see figs. 4, 5). (Inside the crust this point is marked on the cast by an oblique furrow, indicating, of course, an

[1] The deflected cheeks give the eye in our figure 9 too exterior a position; and the eye is really placed a little too far out in the figure.

internal ridge.)[1] The pleuræ are themselves regularly convex and not abruptly bent down. Their facets are large, their ends expanded, and not much bent forward. The pleural groove is narrow.

The tail is semi-oval, its end obtusely rounded, its margin all round slightly concave; the anterior edge not greatly curved, and the facet very strong. The axis is gradually tapering, and reaches four fifths of the length; and it is then continued by a conical appendage nearly to the margin. It is not very prominent or strongly marked out, but is nearly as wide as the side-lobes, with eleven or twelve segments, of which the two anterior are much stronger than the rest; sides gently sloping, with about seven broad flat ribs, separated by narrow sulci, the two anterior of which are disproportionally deep, the rest very shallow. These side-furrows are straight, and have a marked interval between their origin and the furrows on the axis. The contrast between the two deep upper furrows of the tail and the remaining faint ones is very striking, and suggested the name.

Variations.—The young have the head-furrows somewhat stronger. The tail-axis varies a little in width, and has nine to ten or eleven rings; and the sides have stronger or fainter furrows.

Trimerus platypleurus of Green, ('Silliman's Amer. Journ.,' vol. xxxii, p. 169), nearly resembles ours in the caudal shield, but differs in its proportionate length and breadth; in the short-conical form of its axis, and the stronger lateral furrows. *H. Jacksoni,* Green, is another allied American form.

Localities.—Arenig group, Tremadoc, N. Wales. Llandeilo flags, Anglesea? Caradoc sandstone, everywhere in N. Wales and Shropshire.

Homalonotus Sedgwicki, *n. sp.* Woodcut, fig. 25.

Fig. 25.

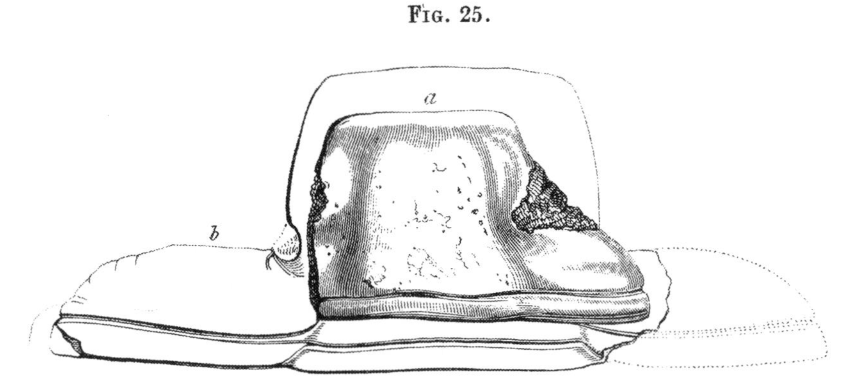

Homalonotus Sedgwicki, n. sp. From Caradoc Rocks in Westmoreland.

[1] The condition of the fulcrum is rather complicated in Homalonotus, from its coinciding with the axal line, and it is difficult at first to interpret the casts of exterior and interior.

H. magnus, 7-9 uncialis; capite latissimo, longitudine haud dimidium latitudinis attingente; fronte latè truncatâ rectâ. Glabella subquadrata, sulcum obscurum marginalem frontis petens, ad basin latitudini genæ convexæ æqualis. Oculi post medium capitis positi, subremoti. Sutura facialis margini postico parallela.

Heads only are known of this splendid species. It must have grown eight or nine inches long, and has, strangely enough, been quoted by M Coy, in the Cambridge Collection, not only as *H. bisulcatus*, from which the wide transverse head separates it, but even as *Asaphus Powisii*.

The head is transverse, more than twice as wide as long, with so truncate a front as to have that edge parallel to the hinder one. The front margin has a slightly concave furrow within it, up to which the glabella reaches. This is subquadrate, its sides a little contracted, its base not much the widest, but still broad; the front emarginate, and defined by the front furrow, as before stated.

The cheeks are as wide as the base of the glabella, convex, almost gibbous at the eyes, which are somewhat remote. The facial suture curves out boldly from the eye, parallel to the hinder margin, and cuts the outer edge in advance of the blunt angle. The neck-furrow is strong in the cast, and reaches all across; it is sharp under the glabella, and broader and blunter beneath the cheeks, separating there a broader and more convex neck-segment than that beneath the glabella.

The eyes, in large fine specimens in the Cambridge Museum (fig. 25) show a small eye-lobe. I only know two specimens; and yet I expect the species to be a common one, and its body and tail-portions should be looked for in the following—

Locality.—CARADOC SLATES of Ravenstone Dale, Westmoreland. Both specimens are in the Woodwardian Museum.

HOMALONOTUS EDGELLI, *n. sp.* Pl. X, fig. 11 (and fig. 10?)

H. minor, H. bisulcato *simillimus, nisi caudâ (solùm cognotâ) angustiore, axe angusto conico, 8-annulato, â lateribus declivibus bene sejuncto. Costæ laterales 7 planæ, haud interlineatæ, ad marginem angustum ferè tractæ, duo superiores profundè exaratæ.*

It is possible that the small head (fig. 10) from Horderley may belong to this species, but I only describe the tail.

A neat species, which has so strong a resemblance to the preceding that it may be difficult to distinguish it when more specimens are found; and it may possibly, though, I think not probably, be the ♂ form of *H. bisulcatus*. Nevertheless the proportions are so different that we cannot well mistake it. The tail is very convex: the axis is decidedly narrower,—hardly more than half the breadth of the side-lobes, and the ribs fewer: and though these are only differences of proportion, and we must wait for better materials to be certain about the species, they are probably sufficient in a genus like the present, where the forms are so closely allied.

Locality.—CARADOC BEDS. Acton Scott, Shropshire. Mr. H. W. Edgell's cabinet.

Section—BRONGNIARTIA, § 2.

HOMALONOTUS RUDIS, *Salter*. Pl. X, figs. 12—14.

HOMALONOTUS RUDIS, *Salter*. In Appendix to Sedgwick's and M'Coy's Synopsis. Foss. Woodw. Mus. pl. 1 E, fig. 20, 1851.
— — *M'Coy*. Ibid., fasc. 1, p. 168, 1851.
— — *Salter*. Morris' Catal., 2nd ed., p. 109, 1854.
— — „ Siluria, 2nd ed., p. 539, 1859.
— — „ Mem. Geol. Surv., vol. iii, plate xvi, figs. 9—11, 1865.

H. maximus, pedalis et ultra, valdè trilobus, costatus et corrugatus. Glabella convexa benè distincta. Oculi antici. Thorax axe lato convexo, pleuris usque ad fulcrum planis, dein deflexis. Cauda semicircularis et latior, axe convexo conico appendiculato, 7—8-*sulcato, vix* $\frac{3}{4}$ *caudæ efficiente, convexo noduloso; lateribus costis* 6—7 *duplicatis, haud marginem lævem attingentibus.*

A still larger and coarser species, in all respects, than the last. Our largest specimen must have been a foot long when perfect. The several specimens now collected together show that the lower furrows of the tail are all nearly equal in strength to the upper two, contrary to what is known in *H. bisulcatus*. Moreover, *H. rudis* is trilobate, while the other follows the usual form of the genus.

We have but one specimen of the head, and that, belonging to the specimen fig. 14, is so crushed that it is not worth while to figure it. But it shows that the glabella is convex, and well distinguished from the cheeks, which bear a small eye rather forward; the neck-furrow is distinct.

The body has a narrow axis for the genus, and the axal furrows lie some distance within the fulcrum. The axis is convex, as in ordinary genera, and the axal furrows deep. The posterior half of each segment is convex, the anterior, separated by the groove, is flattened. The pleuræ are flat as far as the fulcrum, which is placed about one fourth out (as seen in fig. 14 *a*), and has the usual angular character; the pleural groove is deepest beneath this point. In all these respects *H. rudis* resembles a *Calymene*, and differs from the ordinary forms of the genus.

The tail is semi-oval, the front much arched; the axis prominent, conical, and well raised above the sides, which slope gently down, and have seven nearly straight furrows, of which the two front ones are the strongest. The axis has seven or eight rings, faintly nodular on their sides and central part. It extends three fourths the length of the tail, exclusive of its pointed appendix. The side-ribs, six or seven in number, are, at least in the front ones, faintly interlined.

If fig. 13 be of the same species, of which there is some little doubt, the whole upper surface is finely tubercular. Fig. 12 shows the great size attained by this fossil.

Locality.—CARADOC ROCKS of North Wales and Shropshire; Cader Dinmael, near Corwen, Denbighshire (fig. 14); of Nantyr, near Llanarmon, Denbighshire (fig. 12): Capel Garmon, same county (Woodw. Mus.): Cressage, Shropshire; fig. 13 (Mus. Pract. Geol.).

HOMALONOTUS BRONGNIARTI, *Deslongsch.* Pl. X, figs. 15—17; Pl. XIII, fig. 9.

ASAPHUS BRONGNIARTI, *Deslongschamps.* Trans. Soc. Linn. Calvados, vol. ii, p. 301, pl. xix, xx, (fig. 12?) 1825.
HOMALONOTUS BRONGNIARTI, *Marie Rouault.* Bull. de la Soc. Géol. France, 2me Sér. vol. vi, p. 379, 1849. Id. in vol. viii, p. 370 (not of De Verneuil, id., vol. xii, pl. xxiii, fig. 1).
— — *Salter.* Quart. Geol. Journ., vol. xx, p. 290, pl. xv, fig. 1, 1864.

P. 9-*uncialis et ultra, oblongo-ovatus, capite caudâque minoribus. Glabella parabolica, vix lobata, sulco distincto circumdata, á fronte curvâ benè sejuncta. Genæ elevatæ, oculi retrorsi. Thorax trilobus, axe lato. Cauda rhomboidea, ut longa quam lata, axe distincto* 8—9-*annulato; lateribus subplanis lævibus, sulcis angustis* 7—8; *margine serrato.*

Although the figures in the Calvados Transactions are rather out of date, yet the care which M. Deslongschamps took to figure all the varieties leaves no doubt as to what species he intended. It was rather a large fossil; some of the foreign specimens appear to have grown more than nine inches long, and it is rather conspicuous for the trilobation of the body. Deslongschamps has not shown the curious serration of the tail-margin, but except in that point his figure is exact. The swelled front border, the slightly lobed parabolic glabella, the backward eye; and the tail, with its shortened axis and almost smooth sides,—are very well figured; and his description is characteristic and complete.

I will only describe our own specimens, and add a note upon the body-rings from the French author, who had larger specimens than we possess.

Head an inch long, of which the parabolic glabella occupies seven lines, and is eight lines broad at the base, depressed, very slightly lobed, and well distinguished from the sides by a shallow furrow. The front margin ("chaperon") is much elevated, and the interval between it and the cheeks depressed. The latter are tumid and rather narrow; the small eyes placed considerably behind the middle: the free cheek triangular and smooth, except on the vertical outer margin, which is roughly granular all over.

The body in French specimens is very convex, but not semicylindrical, being depressed above. The sides bend down abruptly. The central lobe is well marked out, and is wider than the pleuræ in front, but narrower behind; the distinct trilobation takes place some distance within the fulcral points. The facet is large and distinct, and the pleural groove nearly reaches the broad tip. The separate figures of the pleuræ given by Deslongschamps much resemble our figure 16, *a b.*

The tail is regularly convex, ten lines long by eleven broad; its front margin greatly arched, and the sides retreating at almost a right angle from the centre; the axis broad, conical, nearly reaching the end of the tail, its tip blunt and prominent; it is marked by eight or nine rings (ten in larger French specimens), and the furrows which separate these are distinct though shallow,—several of the front ones showing a plane pseudo-articular surface between the rings (this character is well shown in the foreign figures).

The axal furrows of the tail are sharp, but shallow. The sides slope regularly outwards, and have a very smooth appearance. They are scored by seven or eight narrow sharp lines, which reach the sharply incurved margin and decussate it so strongly as to produce sharp serratures, a character not observable, so far as I know, in any other species. The incurved edge is very abrupt, but this is usual in the genus.

H. Brongniarti is described by Rouault as having a slightly prominent glabella, narrowed in front, and showing two lateral lobes; and the tail with a shortened axis, marked by ten ribs, and an equal number on the side-lobes. It is somewhat doubtful if M. Rouault's be the true species, but ours is clearly the form described by Deslongschamps. The fossil quoted by De Verneuil from the Sierra Morena has a longer and more triangular tail, with interlined ribs, and the head so figured has wider cheeks. The two are nevertheless allied forms.

Locality.—Lower Silurian pebbles, in New Red Sandstone! at Budleigh-Salterton Cliff, South Devon.

Foreign localities.—May, Normandy; also Gahard, near Rennes (Rouault), and possibly Vitrè (Rouault).

Homalonotus Vicaryi, *n. sp.* Pl. XIII, fig. 10.

This small species, of which we have only the caudal portion, is quite distinct from any other, and I wish to distinguish it by the name of the gentleman who has paid so much attention to the fossils of the locality in which it is found. The shape is blunt-triangular. There are seven nodular rings on the axis, and six ribs partly interlined on the convex sides, the outer margin strongly incurved, the apex notched beneath. The regular conical axis, distinct all round, resembles that of the *H. Brongniarti*, with which it is found. But the smooth blunt incurved edge has none of the serrations seen in that species, nor do the furrows quite reach the margin. It is, moreover, quite distinct from the unnamed species in our Pl. X, fig. 18.

Locality.—Llandeilo (or Arenig ?) Rocks; found in the pebbles of Budleigh Salterton, South Devon: (cabinet of W. Vicary, Esq., who first drew attention to the fossils of this remarkable pebble-bed, and has laboriously collected and investigated them, thus adding to the British series a new fauna, which is identical with that of Normandy).

Foreign Distribution.—I have figured (fig. 10 *b*) a Normandy specimen of the species. It is from the May Sandstone of Caen (Mus. Geol. Society).

HOMALONOTUS, *sp.* Pl. X, fig. 18.

H. sp. caudâ latâ brevi, profunde sulcatâ, axe angustiore, lateribus convexis curvis (nec subplanis declivibus); sulcis lateralibus 6—7 profundis vix marginem granulatum attingentibus, et ad apices paullulum interlineatis.

There is some indication in Deslongschamps' figures, pl. xx, figs. 1, 2, of a distinct species, with a squarer front to the head than in his *H. Brongniarti*, figured above. But these indications are obscure: they certainly do not represent our second species, for this has fewer and stronger, not more and fainter, ribs to the tail.

The tail was broad when perfect, probably $2\frac{1}{4}$ inches by less than $1\frac{1}{3}$ in length. Of this width the axis can scarcely be one third, and it is well marked out and scored with several rather deep furrows. The sides are convex, not steeply sloping off, and deeply grooved by at least seven curved furrows, which do not quite reach the granular margin, and are interlined near their tips. The facet is very broad and distinct, and the upper furrow strongest, but not greatly so.

This I do not name, for it is possibly a described foreign species. But it is clearly not the *H. Barrandei*, of Rouault, a species with only four lateral ribs to the tail, and a very strongly marked axis, which reaches the border.

Locality.—Budleigh Salterton (Mr. Vicary's cabinet).

HOMALONOTUS, *sp.* Woodcut, fig. 26.

? H. BISULCATUS, *M'Coy.* Quart. Journ. Geol. Soc., vol. viii, p. 13.

FIG. 26.

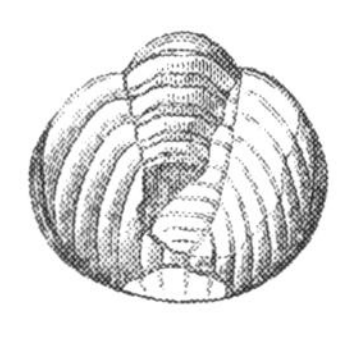

Homalonotus, sp., Lower Silurian, Gorran Haven, Cornwall.

I figure here a small species, which, by its distinct ribs on the axis and sides, appears to differ from the others, and which may possibly be the species catalogued by Prof. M'Coy as *H. bisulcatus.* It differs from that fossil in several points; having a greatly more arched anterior margin, and strong ribs on the narrower axis, with numerous lateral ribs. We wait for better specimens.

Locality.—LOWER SILURIAN quartzites of Gorran Haven, S. Cornwall; (cabinet of Mr. H. W. Edgell). These beds appear to me to be identical with those which supplied the pebbles for the Budleigh Salterton bed above referred to. They contain at least some of the same fossils. (See 'Geological Magazine' for July, 1864, vol. i, p. 9.)

HOMALONOTUS VULCANI, *Murchison.*

Siluria, 2nd edition, pl. ii, figs. 3, 4, 1859. (*Asaphus* of the original work.)

(The description and figures will appear in the appendix.)
Locality.—LLANDEILO FLAGS, Corndon mountain, Shelve, Shropshire.

Section—TRIMERUS, *Green.*

HOMALONOTUS DELPHINOCEPHALUS, *Green.* Pl. XI, figs. 1—11.

ASAPHUS CORNIGERUS, *Brongniart.* Crustacés Foss., pl. iv, fig. 10 (from a drawing of a Dudley specimen) not of pl. ii, fig. 1, 1822.
TRIMERUS DELPHINOCEPHALUS, *Green.* Monog. Tril. N. America, pl. i, fig. 1; model No. 32, 1832.
BRONGNIARTIA PLATYCEPHALA, *Eaton.* Geol. Text-book, pl. ii, 1832 (fide *Hall*).
HOMALONOTUS DELPHINOCEPHALUS, *Murchison.* Sil. Syst., pl. lxxvi bis, figs. 1, 2, 1837.
— — *Murchison.* Siluria, 2nd ed., p. 123, Foss. 16, 1859.
— — *Milne-Edwards.* Crust., iii, 314, 1840.
— — *Hall.* Geol. Report, N. York, 1843. Palæont. N. York, vol. ii, pl. lxviii, figs. 1—14; 1852.
— — *Emmrich.* Dissert. 41-7, 1839. Neues Jahrbuch, 1845.
— — *Goldfuss.* Syst. Uebersicht Tril.; Neues Jahrb., p. 559, 1843.
— — *Bronn.* Lethæa Geognostica, 1, 112, tab. ix, fig. 5.
— GIGANTEUS, *Castelnau.* Essai Syst. Sil., p. 20, pl. iii, fig. 1, 1843. (Also *H. Atlas* and *H. Herculaneus*, ib., p. 20. All from Lockport, N. York.)

H. 6-*uncialis, depresso-convexus, alutaceus fere scaber, capite caudâque æqualibus acutis, hâc acuminatâ, illo triangulato. Glabella undulata punctata, vix lobata, anticè truncata, bis quam fronte angulato plano longior. Genæ declives angustæ, margine vix distincto, angulis obtusis. Oculi convexi, præ medio genarum positi. Thorax haud lobatus. Cauda brevis trigona, convexissima, acuminata; axe haud distincto, per totam caudam extenso per-annulato, annulis* 10—11; *lateribus declivibus* 8—9-*sulcatis, sulcis obscuris.*

It is a curious synonym we have to head the above list withal. Mr. Charles Stokes was very liberal in communicating English specimens and drawings to Brongniart, whose artist has in this case faithfully copied the course of the facial suture, and enabled us to recognise the species. The reference by Brongniart is evidently an inadvertence; at least, the species has no resemblance to the Russian *Asaphus cornigerus*, figured correctly in his plate ii. Brongniart made other mistakes with his English materials.

Elongate, broadest in front and regularly tapering backwards, pointed at both ends; semicylindrical, with strongly deflected sides and flattened back; the head depressed, truly triangular; the tail regularly convex, and rather rhomboidal than triangular, about as long but not nearly so broad as the head. Young specimens seem to have the same general proportions. We may now consider the details, and they are taken from the magnificent central specimen long known as the ornament of Mr. E. Blackwell's collection, and which we are enabled to figure afresh (the specimen served for Murchison's very accurate figure), together with others in the collections of Mr. E. J. Hollier, jun., of Dudley, of Mr. Ketley, and of Mr. H. W. Edgell, of Sandhurst College.

The largest specimen known, in Mr. Hollier's cabinet, is not given on the plate; it measures fully six inches by about three inches broad at the base of the head, which forms nearly a right-angled triangle, and is shaped in front like a pointed gothic arch. Behind it is nearly straight, and the angles are obtuse, but not rounded.

The glabella is less than half the width of the head, and is four-sided, exactly as long as broad, but narrowed rather suddenly towards the front, and therefore apparently longer than broad. It is distinct all round, and shows faintly a central ridge, and traces of three lateral furrows. It is also slightly emarginate in front, and opposite the notch there is a depression in the rather concave front border, which is less than half as long as the glabella. Cheeks moderately convex, and most so toward the outer angle; the border distinct, except in advance of the angle, where it is fused with the cheek Neck-furrow plainly marked, as a narrow sharp line beneath the glabella, and a broad furrow beneath the cheeks. The eye is round, prominent, and placed opposite the middle of the glabella, in advance of the centre of the cheek. All the head is covered by large and small puncta, the former equally spaced over the surface; and these on the outer margin pass into and are mixed with squamous granules and short lines. The facial suture is very prominent, running close along the front border for a short distance; it describes a narrower arch than the front of the head to reach the eye, and thence curves largely out in advance of the angle.

The thorax is half as long again as the head, and is very regularly arched from side to side, scarcely a slight indentation marking the place of the fulcrum, which is in this genus the boundary of the axis.[1] Beyond this fulcral point the rings arch down a little, and then suddenly descend in a vertical line, the broad ends of the segments curving forward and much expanded at their tips, especially in the front rings. The facet is very large, commencing at the fulcrum. The pleural groove, as in all the genus, bisects the pleuræ very unequally, the forward half being narrow, the hinder broad; the groove is deep, but narrow, ends abruptly before quite reaching the broad tip, and inwards continues quite across the axal lobe, separating a narrow, flat, articular band in front of each segment,

[1] In nearly all other Trilobite genera the fulcrum is beyond the axis; in the section ***Dipleura*** of *Homalonotus* it coincides with it. Some few other genera have it nearly in the same place: ***Æglina***, *Barrandia*, and ***Remopleurides*** are instances of this character.

which lies at a lower level than the body of the ring, and is often quite covered in the act of bending (see fig. 1). The hinder rings gradually decrease in width towards the tail, but very slowly so.

The tail at its origin is rather less than three fourths as wide as the base of the carapace. It is highly convex, trigonal, but with the front so much arched forward as to give a sub-rhomboidal shape to the whole. The width and length are equal; the axis, more than one third the whole width, is convex above and highly prominent behind. It extends nearly to the pointed tip, and is annulated for the greater part of its length by ten or eleven rings, bent forward a little in the middle. The front furrow is sharper than the rest, and is continuous with its pleural groove; the others diverge more or less from the corresponding axal furrows, and divide the lateral lobes into fewer than ten or eleven rings, seldom more than eight ribs being visible on the sides. These do not quite reach the margin, which is curved down and vertical, and is not produced or at all flattened. The apiculus in most cases is but short; its difference in length may indicate the sex.

The cornea of the eye has never yet been described for this genus, which is generally supposed to have hiant or hollow eyes, *i. e.* with so thin a cornea that it is not preserved. This, however, is not the case. It is very convex and finely reticular, covered with minute lenses (see fig. 6). But it appears to be but loosely connected with the free cheek, a very unusual character, and of course this would make it easily separable, for there is no real connection with the upper eye-lobe in any Trilobite possessing a facial suture. Mr. Hollier's little specimen figured above in the plate, fig. 3, shows the cornea apparently connected with the upper lobe. The very perfect young head (fig. 4) (Mr. Edgell's) has the same appearance. The cornea seems to be soldered to the upper lobe, and divided from the supporting free cheek. But this is an appearance only, as fig. 6 plainly shows.

Locality—WOOLHOPE LIMESTONE, [Woolhope Valley, Herefordshire, ?] Malvern; Dudley, abundant. WENLOCK or LUDLOW ROCKS. Marloes Bay, Milford; Llandeilo. I am not quite sure if this or the following species be the one found in numerous fragments along the South Welsh Silurian border.

Foreign.—Niagara Limestone of North America; abundant.

HOMALONOTUS CYLINDRICUS, n. sp. Pl. XI, fig. 12, and woodcuts, figs. 27, 28.

H. 6-uncialis, convexus, minutè granulosus, (capite triangulato?) caudâ longâ productâ. Thorax convexissimus, insuper planatus, pleuris longè deflexis, axe angusto posticè contracto. Cauda longa trigona, axe longiconico 10-costato, et apice læve circumscripto; mucrone terminali elongato cylindrico distincto vix elevato; costis lateralibus 7 convexis, haud marginem verticalem angulatum attingentibus.

I have only lately obtained a good specimen, which is given in the woodcut; and ascertained that this Woolhope species is quite distinct from the *H. delphinocephalus*, of which it was formerly thought a variety.[1] It is a far more convex species, and might easily be mistaken for the *H. Johannis*, to which it is indeed nearly allied (we have not the head as yet), and differs only, so far as I can see, in the want of trilobation, more tapering body-axis, and rather longer pleuræ; but especially by the cylindrical convex tail, with a much flatter axis, and less distinct side-ribs than in that species. The

Fig. 27.

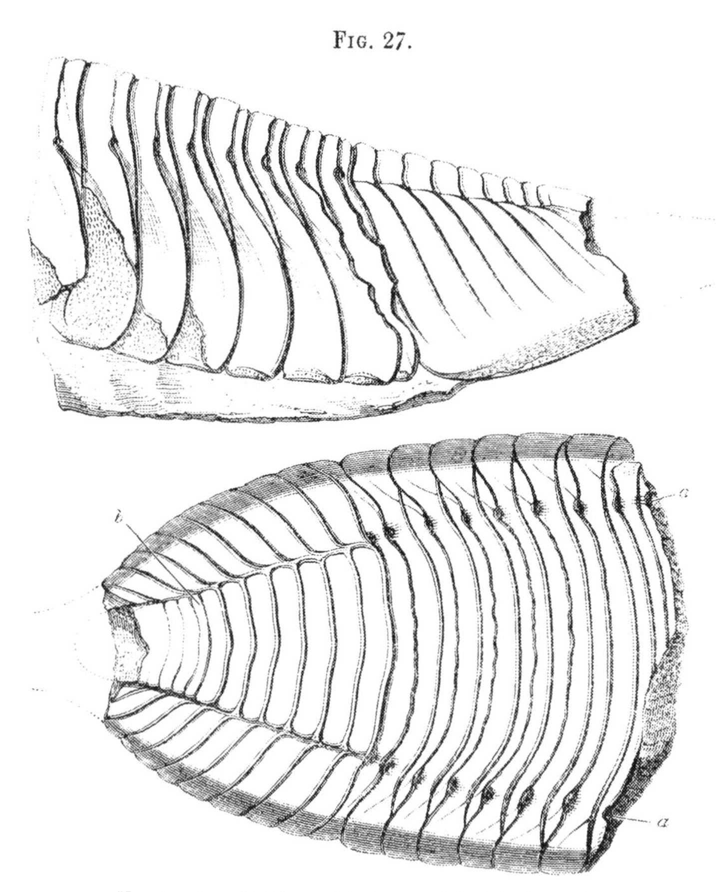

Homalonotus cylindricus; Woolhope. (Mr. Edgell's cabinet.)

differences in the tail are indeed so marked, that I cannot doubt its being a distinct form, and it has probably been always quoted as *H. delphinocephalus* from Woolhope.

Rather a large species. It must have been full $5\frac{1}{2}$ inches long when perfect, and $2\frac{1}{2}$ inches broad. The sides are very parallel as far as the root of the tail; the outline thence curves rapidly inwards towards the end of the tail-axis, and is again much produced into a straight and somewhat cylindrical point. The depth is very great for a Silurian

[1] *H. cylindricus* is probably the common Woolhope species. It is now distinguished for the first time.

species, being exactly half the breadth. The axis is broad in front, but narrows much behind towards the origin of the tail; the trilobation being quite obsolete in the body-rings, while in *H. Johannis* it is distinct.

The tail is a long triangle, nearly half as long again as the breadth; and of this the long-conical axis is at first more than half the breadth of the whole tail, and thence tapers less slowly than in *H. Johannis*, but not so rapidly as in *H. delphinocephalus*. The axis is, moreover, very flat, even quite to the end, and is well distinguished, but not deeply so, from the side-lobes. It is annulated by eleven furrows, of which the uppermost is much the strongest; all are angulated in the middle, not straight across as in the Wenlock fossil. The sides are very convex, almost gibbous outwards, and turning down quite vertically; on their lower third they are again sharply angulated, leaving a broad rough-sculptured area above the final incurvation on the lower surface, a character common to all the Upper Silurian species. The side-furrows are curved, shallow even in the cast, and reach only three fourths to the margin: and, except the upper three, they do not coincide with the axal-furrows, while in *H. Johannis* at least five of them do so.

It is unfortunate that we do not possess the head; but I figure one found at Woolhope, which certainly differs in proportion from that of *H. delphinocephalus*, and was found by the late Hugh Strickland, Esq., at Woolhope. The glabella is very flat, and this coincides with the character of the axis in our species. It is, moreover, very pyramidal, almost triangular, but truncate in front. The eyes are somewhat more approximate than in *H. delphinocephalus*, if I may trust in this respect a careful sketch made some eighteen years back (fig. 28). The cast is punctate all over the glabella, and has larger puncta over the lower portion of the fixed cheeks.

Fig. 28.

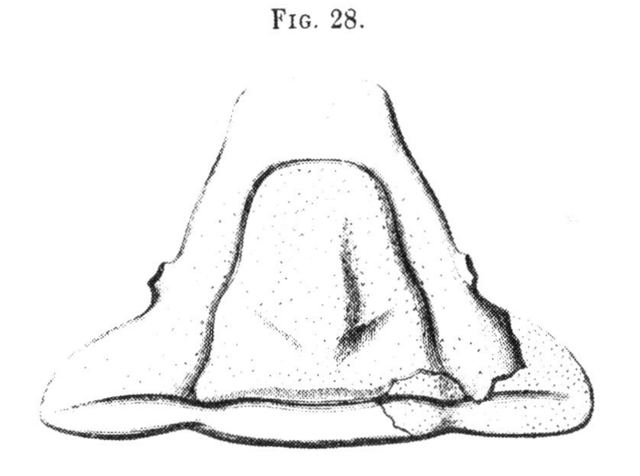

Head of *H. cylindricus?* Woolhope Limestone. Collection of the late Mr. H. Strickland.

Locality.—Woolhope Limestone of Woolhope. Plate XI, fig. 12, is in Mr. Edgell's collection. Our fig. 27 is from a fine specimen in the Woodwardian Museum. Fig. 28 was in the late Mr. Hugh Strickland's cabinet. I do not know where that collection now is, and shall be glad of the information.

Homalonotus Johannis, n. sp. Pl. XII, fig. 11; and Pl. XIII, figs. 1—7.

H. modicus elongatus, 6-pollicaris, scaber, cauda acutâ, quam capite triangulato paullo longior. Glabella valde distincta, pyramidata, subcarinata, anticè truncata, lobata,—lobis basalibus tumidis. Frons[1] *concava, subacuta, scuto rostrali quam lato longiori. Genæ convexæ: oculis ante medium positis. Thorax benè lobatus, granulis posticis spinoso-pectinatis. Cauda triangularis convexa acuminata; axe prominente* 12-*annulato, lateribus declivibus* 8-*sulcatis, sulcis profundioribus.*

With great pleasure I recognised this new form in the very choice cabinet of my friend John Lee, Esq., of Caerleon, and beg to dedicate it to him. It seems not to be a rare species; but has hitherto been confounded with the *H. delphinocephalus,* from which it differs in every particular. More strongly trilobed, more roughly tubercular, with a more triangular and far more deeply lobed head, a narrower tail-axis, and deeply sulcate side-lobes to the tail, it is rather surprising the species had not attracted attention before.

The length of large specimens is just 6 inches, the base of the head, which is the widest part, being there $3\frac{1}{2}$ inches. This gives a size all but equal to the great species just described. The head itself is trigonal, $3\frac{1}{2}$ inches wide at base, and $2\frac{1}{4}$ long, with a blunt-pointed front and similar posterior angles. A broad neck-segment is deeply divided from the tumid four-sided glabella, and these taken together are rather longer than the concave front. The glabella is twice as wide behind as in front, trapezoidal in shape, the angles rounded, the sides tumid, and the lateral lobes very distinct for the genus; the centre is raised and somewhat carinate, as in its ally *H. delphinocephalus,* but the glabella is greatly wider behind. A distinct, wide, and rather strong sulcus borders it all round, separating the tumid cheeks with their sub-central eyes. The course of the facial suture is the same as in the last species. One specimen (fig. 2) shows the rostral shield; it is longer than broad, and is of a nearly rectangular shape. The whole head is covered with a coarse and fine granulation; but, as we have only internal casts, we cannot give further details.

Thorax depressed; the rings convex, well separated from each other, and projecting posteriorly more than usual; the hinder half large, the anterior portion separated by the pleural groove narrow, as in *H. delphinocephalus.* The trilobation is distinctly marked; the axis thus marked out being narrower, and more tapering backwards, than in the last-mentioned species. Internally the pleural groove is very strong (see fig. 6). The pleuræ are square at their ends (fig. 5).

The tail is strongly ribbed. The axis, not much broader than the sides, is tapering, long-conical, and distinctly ribbed all down by nine strong ribs, while the sides have seven strong furrows. The axis is prominent and pyramidal at their tip, and does not fade off into the broad-pointed mucro, which is more than one third the whole length of the tail, acute, not recurved; it is much longer than in the allied species so often quoted.

[1] I use the term *frons* for the produced front of the head, in front of the glabella. Some authors use it for the glabella:—it is best to keep the terms distinct.

The whole thorax and tail are sparsely covered with rather strong granules, which become (see fig. 5) spinous combs along the hinder edges of the thorax-rings, but are not so produced in the tail-portion.

Altogether the species is very distinct from the Dudley fossil with which it has hitherto been confounded. It appears to be abundant at the following locality.

Locality.—WENLOCK SHALE (soft mudstone) of Craig y Garcyd, in the bed of the Usk, Monmouthshire, whence a plentiful supply has been obtained by Mr. J. E. Lee, of Caerleon, whose fine series of specimens we have figured. Mr. Nichols, jun., late of Usk, now of Cowbridge, has also a fine series. Also Golden Grove, Llandeilo (Mus. Pract. Geol.).

Section—KOENIGIA, *Salter.*

HOMALONOTUS KNIGHTII, *König.* Pl. XII, figs. 2—10, and Pl. XIII, fig. 8.

HOMALONOTUS KNIGHTII, *König.* Icones Sectiles, pl. vii, fig. 85, 1825.
— — *Bronn.* Lethæa, 1, 119, tab. ix, fig. 14, 1835; also in Leonh. und Bronn's Neues Jahrb., p. 445, &c., 1840.
— — *Murchison.* Sil. Syst., pl. vii, fig. 1, 2, also *H. ludensis,* ib., figs. 3, 4, 1837.
— — *Milne-Edwards.* Crust., iii, 315, 1840.
— — *Goldfuss.* Neues Jahrb., p. 559. *H. ludensis,* ib., 560, 1843.
— — *Emmrich.* Dissert. 41, 8, 1839. Neues Jahrb., 1845.
— — *Burmeister.* Org. Trilob., Ray ed., p. 86, 1846.
— RHINOTROPIS, *Angelin.* Palæont. Suec., t. xx, fig. 1, 1855.

H. fere 6-*pollicaris, depresso-convexus, scabriculus, capite acuto, caudâ multo breviore, truncatissimo. Glabella haud convexa seu lobata, a genis latis distincta; margine frontali brevissimo tricuspidato. Oculi antici. Thorax omnino homogeneus, nec lobatus. Cauda triangulata acuminata, apice obtuso vix recurvo; axe distincto contracto, per* $\frac{4}{5}$ *longitudinis annulato, sulcis* 9 *intùs profundis, et cum tot sulcis lateralibus profundè exaratis omnino continuis. Ad apicem caudæ sulci axales et laterales subitò absunt. Margo abruptè incurvus angulatus, ad latera contractus, scabrosus.*

Of this fine species little need be said, as it has figured in every book of fossils since König instituted the genus and gave the appropriate name. It is one of the commonest and yet most choice of the Ludlow rock fossils. Fragments are found in every locality. The barren mountains of Radnorshire present us with fine and nearly perfect specimens. Others have been disinterred by the patient labour of the Shropshire geologists, and we are indebted to Mr. G. Cocking, of Ludlow, especially, for saving every specimen that could be preserved when the Ludlow railway was being made. Figs. 2—4 are from his specimens, the crust being beautifully preserved in fig. 3. But the largest and finest we have yet seen is our fig. 7, from the cabinet of Dr. Grindrod, of Malvern. With his usual kindness he has sent me every specimen that could illustrate this or other species.

A large convex species, 5½ inches long, and nearly 3 inches wide at the shoulders. Foreign specimens occasionally attain the length of 9 inches? The form is oblong, pointed behind, and abruptly truncate in front, the head being nearly a parallelogram, twice and a half as wide as long, and divided by the distinct axal-furrows into three nearly equal parts. The glabella is urceolate, a good deal widest below, where it is wider than the cheeks, and rather abruptly contracted halfway-up; the front is straight-truncate, and slightly emarginate. The cheeks, separated by broad axal furrows, are decidedly gibbous, the greatest convexity towards the front in advance of the eye, which is placed two thirds up the cheek and not quite halfway-out. It is small, oval, prominent, and subtended by a flattened base. The facial suture runs nearly direct to the front margin from the eye, and beneath it turns abruptly outwards, gaining the outer margin considerably in advance of the facial suture. [On the cast the cheek is both granulate and punctate. I do not know the exterior; but it is figured as granular by Angelin.] Neck-furrow strong and broad, but not reaching the angle. No outer marginal furrow to the gibbous cheek.

Fig. 29.

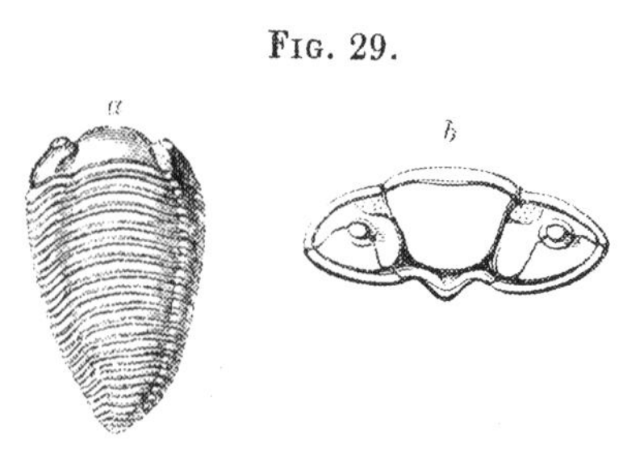

a. H. Knightii, young (Mr. Edgell's cabinet).
b. Front view of the head enlarged.

The front margin is of most singular structure, and may be described as tricuspid. The narrow edge is so deeply indented, and at the same time folded, that the front portion overhanging the rostral shield (fig. 2 *b*) forms one projecting angle, flanked by two smaller projections opposite the axal-furrows, exactly like the salient and re-entering angles of a fortification.

The rostral shield can only be indistinctly seen in fig. 2, at *b*. It is pointed in front. But the labrum (fig. 10), from a good specimen of Mr. Edgell's, is a flattened quadrate plate, deeply bilobed at the apex, and with parallel sides, no lateral wings, a convex centre separated from the margin by a shallow furrow; and with a pair of lateral tubercles well developed (as in *Asaphus*, to be described hereafter).

The thorax-rings are convex on the back, but descend so abruptly at the sides as to give a square appearance to the body. The convexity of the back is greater than in *H. delphinocephalus*. The axal furrows are but very slightly marked; the fulcrum a very short distance outside them. The facet large, but not flat, a depression followed by a longitudinal ridge occurring near the apex, which is rather abruptly truncate (fig. 8 *a*).

The pleural groove is narrow externally, very deep and strong on the inner surface (see figs. 5 and 8). It separates the ring into two unequal parts, of which the front is less convex than the hinder part, and is smooth, while the latter is squamous near the edge. The pleuræ are strongly granular, the granules taking the form of closely placed squamæ over the whole surface.

On the tail the granules are rounder and less closely placed, but they cover the whole surface. The shape of the tail is truly triangular, as wide as long, contracted near the apex, and then pointed, the tip somewhat recurved.[1] The axis is rather narrow, and not much distinguished from the sides, the furrows being quite continuous across. There are eight of them, the upper not stronger than the rest; and all are deep furrows, only rather less broad than the intervening ridges. They arch across the axis, which is only slightly convex, and pass direct across the sides, which are gibbous externally, but rather flattened above.

The eight furrows abruptly cease at rather beyond halfway down the tail: on the sides they cease entirely; on the axis there are one or two faint ribs beyond them. The incurved under margin of the tail is very remarkable. It is sharply incurved, strongly concave, and roughly granular, and there is besides a strong lateral contraction about halfway down the tail (see fig. 9).

Localities.—Abundant in the UPPER LUDLOW ROCKS of Shropshire, and all the border counties; also of Radnorshire, and other places in South Wales; Llangollen, North Wales; Westmoreland, near Kendal, abundant; and under the name of *H. rhinotropis* it occurs in Gothland. Dr. Honeyman has found it at Arisaig, Nova Scotia.

H. ludensis, of Murchison, is now well understood to be a synonym;[2] but nevertheless there are two species in the Ludlow rocks, even in England. And to preserve this old name, I beg to apply it to the very unpretending looking head, fig. 1 (lent to me by Mr. H. W. Edgell, jun.). That species has not the tricuspid front of *H. Knightii*, but a smooth regular contour to the margin; and is, indeed, more nearly like the *H. bisulcatus;* but I cannot be sure to what sub-genus it should be referred—possibly to *Dipleura.* It may be shortly described as follows:

HOMALONOTUS LUDENSIS, *n. sp.* Pl. XII, fig. 1.

(Not of Sil. Syst., pl. vii, figs. 3, 4, for which see *H. Knightii.*)

H. capite transverso semicirculari convexo, glabellâ brevi distinctâ, fronte concavâ subplanâ angustâ, genis modicis, oculis haud remotis, sulco cervicali perdistincto.

[1] Angelin's figure does not show this. Our figures 9 (and fig. 5, copied from the 'Silurian System') show it best. It is not very conspicuous.

[2] The original specimen of *H. ludensis*, 'Sil. Syst.,' pl. vii, fig. 3, was in the cabinet of the late Mr. J. Evans. It is now in the Museum of the Worcestershire Nat. Hist. Society.

Distinct at once from *H. Knightii* by the form of the front margin, which is not sinuous and tricuspid. I do not know the body or the tail, and have much doubt if the species belong to the section *Kœnigia*. It may be referable to *Dipleura*, and I think it is so.

Locality.—UPPER LUDLOW ROCK, Ludlow (Mr. W. H. Edgell's cabinet).

Section—BURMEISTERIA, *Salter.*

HOMALONOTUS ELONGATUS, *n. sp.* Pl. X, figs. 1, 2.

HOMALONOTUS HERSCHELII, *Phillips.* Pal. Foss. fig. 253 (not of *Murchison*), 1841.
— sp. *Salter.* In Morris' Catal., 2nd ed., p. 109, 1854.

H. caudâ 2¾*-unciali, vix* 2 *uncias latâ, longè trigonâ, valde convexâ fere gibbâ. Axis dimidium latitudinis efficiens, convexus, et ad apicem obtusum longum gibbus,* 12*-annulatus, annulis perdistinctis,—primo, secundo, quarto, quinto bituberculatis. Latera gibba, costis circiter* 8 *valdè obliquis, secundo quintoque tuberculatis, tuberculis magnis.*

Mr. Townshend Hall, junr., of Wadham College, Oxford, has made a fortunate discovery, for we have long wanted to know what spinose species of *Homalonotus* furnished the fragments described from South Devon. The Lower Devonian in every country yet examined shows some species or other of this peculiar group—the *H. armatus* on the Rhine, the *H. Pradoanus* in Spain, the *H. Herschelii* in South Africa. In South Devon we now have a new form, remarkable even in this elongated genus for the lengthened shape. We have only the tail as yet; but a diligent search at Meadsfoot Sands will surely discover all the portions of this fine species. And I beg of our friends to make the inquiry. Prof. Phillips has figured some fragments.

Caudal portion 2¾ inches long, and only 2 inches wide at the top, and the convexity about 1¼ inch. Of this form (a frustum of a cone) the axis occupies fully one half the width, and is well marked out, convex in front, and very prominent below. It is annulated nearly the whole way down, eleven or twelve rings being distinctly marked, and there is but a small terminal smooth portion. Of these rings, the first, fourth, and fifth bear a pair of approximate spinous tubercles, and the second ring a pair wider apart. The side-ribs are about eight on each side, very oblique, broad, with narrow intervening furrows; the first and the fourth rib bear each a strong tubercle (probably spinous), directed outwards, and placed less than halfway out. There is a narrow, recurved, lateral margin. The apex of the tail is not recurved or mucronate, but obtuse, the axis being prominent.

This may be compared with the *H. Herschelii*, Murchison, a common South African species; but ours has a much more elongate and more nodular tail than that fossil. The

H. armatus, Burm., from the Lower Devonian of Germany, hardly needs comparison, as it has a minute tail for the size, with but a single lateral spinous tubercle. The species figured, but without a name, by Steininger in the 'Mém. Soc. Géol. France,' vol. i, pl. xxi, fig. 8, has a large spinous tubercle on the first and fifth lateral ribs; but the tail is truly triangular. I think that probably this species does occur in the German Devonian; and that there is a fragment of it in the Geol. Soc. Museum from the Rhine Provinces, (marked as from the Eifel; Drawer 40); but I cannot identify it with any of the published forms.

Locality.—LOWER DEVONIAN; Meadsfoot Sands, S. Devon. (Mr. Hall's cabinet).

FAMILY—ASAPHIDÆ, *Emmrich*, 1839 (?).

If the *Calymenidæ* are but a small group, the *Asaphidæ* are at present an unwieldy one. This is to a great extent a natural family, and I am unwilling to divide it; though we may well distinguish in it several sub-groups, hereafter, perhaps, to be called families—*Ogygides, Asaphides, Illænides, Æglinides,* &c.

Taken as a whole, the *Asaphidæ* comprehend the most bulky and expanded forms of the Trilobite family, together with some which are of much less stature. The head and tail are in general of great comparative size, the latter often free from visible segments, though really composed of many. The thorax is reduced in size, and generally of eight segments, but not always—for instance, *Illænus* and *Stygina.* Respecting *Æglina,* which has only six rings, and which genus I have kept apart in the preliminary sketch, p. 2, there may be reason to suspect it to belong to a distinct family when we know more of its relations. We must exclude it from the character of the family, which, as above said, has in general eight, more rarely nine or ten, body-rings.

Taking the general characteristics, we may say that the *Asaphidæ* are—Large Trilobites, generally of an oval form, without any ornaments except a close striation (which is very common), and totally destitute of tubercles or spines on the surface. The head and tail are large and well developed, and the former has usually an obscure glabella, with but indistinct lobes. Yet in this family we have Trilobites possessing distinctly four lateral lobes to the glabella, *Ogygia, Barrandia,* and *Niobe* having that number.

The facial sutures end on the posterior margin. The eyes are smooth and large; occasionally, but rarely, very prominent (*Asaphus*); usually conical or even depressed (*Illænus, Ogygia,*) &c. The head-angles are more often obtuse than spinous, and never greatly prolonged.

The thorax of eight rings (rarely more, as in *Illænus* and *Stygina*), always facetted, and, with few exceptions, grooved throughout: the apices blunt, or only shortly pointed.

The large tail is of many segments, and there is every gradation in it from a flat expanded form, with the segments all distinct, through others which have them more or less obscure, to such convex forms as *Nileus* and *Illænus*; where they are obsolete, or only visible on the interior of the crust. The margin of the tail is not unfrequently sharply striated, and the incurved portion (caudal fascia) is broad, distinct, and strong, and often convex beneath.

The hypostome varies in character. In most of the genera it is continuous, or in some species of *Asaphus* it has a vertical suture only; the rostral shield being in this case atrophied. In *Illænus* the rostral shield is very large and transverse. Indeed, this genus is abnormal in many characters, and leads the way to the next family, No. 12, in which the form is more compact, the body-segments more numerous, and the rostral shield a necessary part of the structure. The labrum varies much in shape in the different genera, and only in this group is it occasionally forked: but it is constantly so in *Asaphus*.

Having said so much on the general characters, we may proceed to describe one of the most typical and constant of the genera. *Ogygia* is a very limited genus; it has all the elements of the *Asaphidæ*, and has a great resemblance to some of the sub-genera of *Asaphus*, which genus has, on the contrary, the greatest variety of forms within a single generic group.

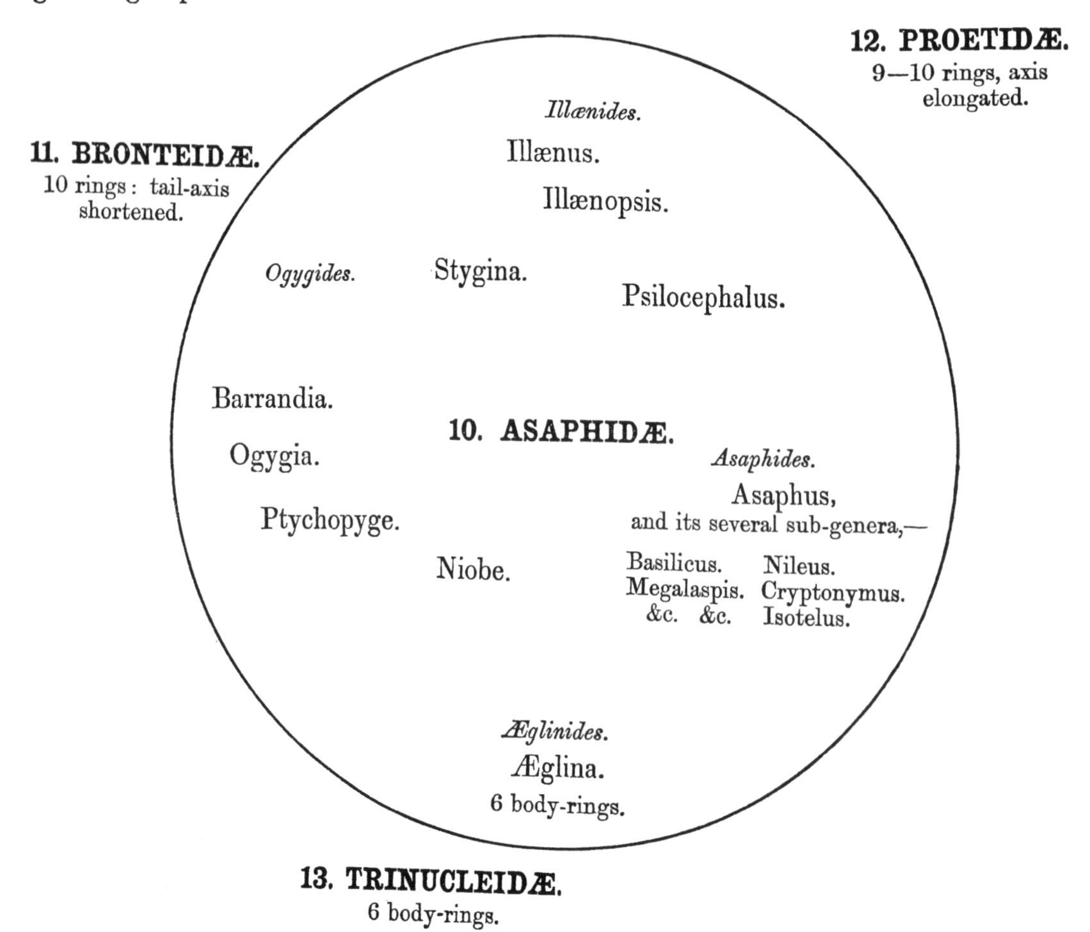

Genus—OGYGIA, *Brongniart,* 1822.

Flat or very slightly convex, with wide semicircular head, large many-ribbed tail, and slightly pointed body-segments. The glabella well marked, wider in front, and with four lateral furrows. The eyes large, lunate, depressed, and smooth. Facial suture generally marginal in front. The hypostome continuous, the labrum pointed.

The thorax has a narrow very distinct axis, and the pleuræ are pointed, strongly grooved, and have a very obscure and remote fulcrum, so that the animal could only bend, not roll up completely (and it is probable that some of the species never did so). The tail has many segments: these are very distinct, and generally strongly duplicate.

RANGE.—*Lower Silurian only.* North and central Europe: not yet known in America (or the southern hemisphere ?). The species are rather numerous, and often attain a great size, but not so large as that of *Asaphus.*

OGYGIA BUCHII, *Brongn.* Pl. XIV, figs. 1—7; Pl. XV, figs. 1—6.

Dr. Llhwyd, in the Philosoph. Trans., vol. xx, 279, tab. addit., fig. 15, 1698.
Buglossa curta strigosa, ib., Ichnograph. Brit. Epist. 1, t. xxii, fig. 2, 1690.
Trilobite from Llanelly, *Parkinson,* Org. Rem., vol. iii, t. xvii, fig. 13, 1811.
ASAPHUS DE BUCHII, *Brongniart.* Crust. Foss. t. ii, fig. 2, 1822.
TRILOB. DE BUCHII, *Schlotheim.* Nachtr. ii, 34 (excl. the locality "Norway"), 1823.
ASAPHUS BUCHII, *Dalman.* Paleadæ., 68, 1826.
— — *J. D. C. Sowerby.* Loudon's Mag. Nat. Hist., vol. ii, fig. 27, 1829.
— — *Buckland.* Bridgwater Treatise, t. xlvi, fig. 7, 1836.
— — *Murchison.* Sil. Syst., t. xxv, fig. 2, 1837.
— — *Quenstedt.* Wiegm. Archiv, vol. iii, 1, 346, 1837.
— — *Emmerich.* Dissert. 28, 1839.
— — *Milne-Edwards.* Crust. 3, 309, 1840.
— — *Burmeister.* Org. Trilob., t. i, fig. 2, 1843, Ray edition, p. 59, (facial suture wrongly figured), 1846.
— — *Corda.* Böhm. Tril., t. iv, fig. 39 (bad), 1847.
— — *Barrande,* Sil. Syst. de Bohême, pl. ii A, f. 25, 26, 1852.
OGYGIA BUCHII, *Salter.* Decade II, Geol. Survey, pl. vi, 1849.

Junior. Pl. XIV, figs. 5, 6.

TRINUCLEUS? ASAPHOIDES, *Murchison.* Sil. Syst., t. xxiii, fig. 6, 1837.
— — *Milne-Edwards.* Crust. v, iii, 333, 1840.

O. ovalis 5—7-*uncialis, capite semicirculari, glabellâ angustâ utrinque quinquesulcatâ, oculis ad medium capitis; spinis posticis brevibus; labro obtusè acuminato; thoracis axe bis pleurâ angustiore,—caudæ longo abrupto; sulcis lateralibus* 12—14 *perduplicatis, propè marginem angulatìm deflexis.*

Length occasionally near seven inches, generally three or four; proportion of length to breadth in some 7 × 5, in others 10 × 8, a considerable variation, to which we must refer by and by. "General form a broad oval, depressed; the head nearly semicircular, and as long as the thorax; the tail semi-elliptical and longer than either; glabella at its base occupying one fourth or less of the width of the head, broader and more convex above, and a little pointed in front; it extends nearly to the front margin, and is marked a little within each side by a longitudinal depression, which is curved, with the convex side inwards. Along this hollow lie five pits, which represent the glabella-furrows: the basal one is the neck-furrow; that above is deepest, and placed nearer towards the centre of the glabella; the other three recede more,—the fourth is opposite the top of the eye, and the uppermost just above it. Eyelid large, semicircular, and not touching the glabella. The facial suture above the eye is curved a little outwards to reach the margin, along which it runs for some distance; beneath the eye it curves largely outwards, and cuts the posterior edge just within the border. The latter is broad and striate, more or less concave, and marked off by a distinct marginal-furrow. Eye placed half-way up the head, smooth; the lentiferous surface broad; its minute structure is shown in Pl. XIV, fig. 7. Head-angles short-spined, the spines not produced below the third body-segment. Labrum broad above, narrowing below, where it is suddenly and obtusely pointed (Pl. XV, figs. 2, 3); the centre is gently raised, and a furrow runs down each side; there are two transverse furrows near the apex, with compressed tubercles between them.

"The cheeks appear to be nearly smooth, but the glabella is covered by fine, short, broken, almost microscopic lines. The labrum is strongly and concentrically striate.

"Thorax with the axis gently convex, variable in proportion, but seldom more than half the width of the pleuræ. In the broad form it is not so much as this The pleuræ are flattened, furrowed along the upper margin as far as the fulcrum, at which there is an obtuse bend backward and very little downward; the pleural groove diverges from the margin here, and does not quite reach the recurved and pointed extremity; it becomes widest immediately beneath the fulcrum; and from the lower margin of the pleuræ at this point a narrow ridge runs obliquely upwards and meets it. The curved sabre-shaped tips are strongly striated transversely.

"Tail varying from little more than a semicircle in the broad form, to a half ellipse in the narrow form; very slightly convex; the axis gently tapering, ribbed by about thirteen or fourteen furrows, which have a slight angular downward bend in the middle; its obtuse tip projects abruptly. Lateral furrows twelve, or in some specimens fourteen, at nearly right angles to the axis, angularly bent downwards at the place of the fulcrum,[1] interlined all the way by fainter furrows, and leaving but a narrow caudal fascia. The interlining furrows nearly meet the primary ones at the end, and here the ribs are sud-

[1] The tail, consisting of anchylosed segments, preserves traces of nearly all the characters of the thorax; the strong furrows are analogous to the furrows of the pleuræ, the fainter lines between to the joints. This is, of course, universal, but *Ogygia Buchii* shows it more clearly than most species.

denly tumid, so as to undulate the flat marginal band. Incurved under portion of the tail (caudal fascia) narrow, closely striate, not indented by the axis; the upper side has a striate band of the same width,[1] which is indented by every rib of the tail. The fascia is continued (of the same width) along the ends of the pleuræ, and the striæ run in the same direction, while on the upper surface they are transverse, as above said, and reach further inwards. Numerous fine striæ, arched upwards, cover the axis.

" *Variations.*—This fine Trilobite varies a little in convexity, but greatly more in proportionate length and in the width of the axis; this variation appears to be chiefly due to sex. If we are right in referring the long form to the ♂, we must allow the Pl. XIV, fig. 3, and Pl. XV, fig. 1, to be males, and the remainder female specimens. The difference is rather extreme in this species. But there is also another variety. Pl. XV, fig. 5 (a ♀ form), has a pair of tubercles on each segment of the axis; they are distinct and strong in this specimen, and are faintly seen in some others, both ♀ and ♂ forms. It might be called var. *tuberculata.*

"*O. Buchii* is known from the French species *Asaphus Guettardi* by its less elongate and more truly oval shape, and the greater width of the pleuræ in proportion to the axis. In *A. Guettardi* they are but half as wide again. The axis of the tail is longer; has thirteen instead of nine ribs; and is abrupt, not attenuated, at the tip. In *A. Guettardi* it extends but three quarters the length. Our fossil has also duplicate, bent, and more numerous side-furrows; the French fossil has but eight or nine straight simple ones; the labrum also is subconical, not dilated laterally. But the general aspect of the two fossils is much alike.

"*O. dilatata*, Brünnich (not of Portlock), which seems to have been frequently esteemed a variety of *O. Buchii*, has the glabella short, with its lobes crowded down towards the base and the eyes remote. The facial suture behind the eyes appears much less arched in Sars' figure. He also describes the tail as with ten ribs, separated by broad furrows. The labrum of *O. dilatata*, according to Sars' figure, is but slightly different from that of the British species; but an important difference resides in the facial suture, which in *O. dilatata* is within the front margin on the upper side, but in *O. Buchii* is along the edge itself," as in *Asaphus tyrannus*, soon to be figured.

" The earliest mention we can find of Trilobites is concerning this species, and is that which all writers on these fossils have quoted. Dr. Edward Llhwyd, in a letter to Dr. Martin Lister, of the Royal Society (1698), writes ' concerning several regularly figured stones lately found by him.' ' The fifteenth,' he says, ' we found near the Lhan Deilo, in Caermarthenshire, in great plenty; it must doubtless be referred to the sceleton of some flat fish;' but he remarks a few lines after, ' Not that these or any other marine terrestrial bodies were really parts or exuviæ of animals; but they bear the same relation to them as fossil shells to marine ones,' &c. This latter opinion he maintains in his '*Lithophylacii*

[1] In many of the *Asaphidæ* the caudal fascia is much wider than the upper striated band.

Britannici Ichnographia' (1699), where he again says the specimen represents only the sceleton of a sole fish, and wants the tail; and he marvels that the 'Piscis Icon' should be raised above the surface of the stone, 'ac si verus piscis esset.' It is curious that Brongniart should have placed this species in his heterogeneous group *Asaphus*, at the very time he was founding *Ogygia*, as he appears to have recognised the latter genus more by its marked habit than by any positive characters; and the principal species, *O. Guettardi*, turns out to be an *Asaphus*. He probably meant to unite *A. dilatatus* with *A. Buchii*. Dalman distinguished them, but with doubt, in 1826, and gave a figure of the Norwegian fossil from a plaster cast; and had not Sars, in 1835, given a complete description and a good figure of the head, the identity might still have been maintained. But, though often quoted, *O. dilatata* does not occur in Britain. Dalman described a specimen of it with seven body-segments; and Prof. Quenstedt, in 1837, two specimens of *O. Buchii* with seven rings; whether an accidental variety, or, as Burmeister thinks, occasioned by the slipping of one ring under the others, it is difficult to say. Quenstedt, however, relied on these and Dalman's seven-ringed specimen of *O. dilatata*, and asserted the same number for *Asaphus* (*O.*) *Guettardi*. But the error in both cases led him to see the generic affinity between the two former species, and he distinctly says that their union with the typical *Asaphi* is unjustifiable. Burmeister, in his first edition, set the number of rings right, but confounded *O. dilatata*, which Sars had well distinguished in Oken's 'Isis' (1835), with our species; and adhered to this view in the Ray edition (1846). Emmerich had, in the mean time, spoken of them as different species, and figured *O. dilatata* in Leonhard and Bronn's 'Neues Jahrbuch' for 1845;" and Corda, in his notoriously incorrect work, in 1847, maintained the error of the seven rings, after everybody else had clearly understood that all the *Asaphi* and *Ogygiæ* had eight segments.

Localities.—LLANDEILO FLAGS only. Shelve and Hope Mill, Shropshire; Rorington, Middleton, and Meadowtown, Shropshire; Builth, Radnorshire; Llangadoc and Llandeilo, Caermarthenshire; near Haverfordwest; Musclewick Bay, Pembrokeshire; and Abereiddy Bay, Cardiganshire, where it is abundant.

It is not known in any foreign country; nor, indeed (though often so quoted), has it been found out of South Wales. It must be esteemed an extremely local and very abundant species, and not very variable, except as regards those differences referable to sex.

Sub-species I.

OGYGIA ANGUSTISSIMA, n. sp. Pl. XIV, figs. 8, 9.

O. ovata, anticè obtusa, parva, 17 *lineas longa,* 13 *lata, capite transverso, axe angustissimo. Omnino* O. Buchii *simillima, nisi glabellâ tertiam partem latitudinis genæ vix efficiente; margine antico capitis obtuso truncato: axe thoracis caudæque pleuris ter angustiore.*

I cannot believe this to be a mere variety of *O. Buchii*, though it does approach very near in character to young male specimens of that species. *O. angustissima* has so extremely narrow an axis, and so broad, transverse, and truncate a head, that, whatever may have been the 'origin of the species,' there is sufficient difference in this to afford us a good diagnosis. The specimens are in the choice cabinet of Mr. Griffith Davies, of 17, Cloudesley Street, Islington.

Locality.—LLANDEILO FLAGS. Gilwern, near Llandrindrod, Builth.

To show how these Llandeilo trilobites are occasionally maltreated by cleavage, I give a woodcut from a specimen of *O. Buchii* in Mr. Henry Hicks' cabinet. It was found at the east end of the section of vertical slate-rocks in Abereiddy Bay, Cardigan-

FIG. 30.

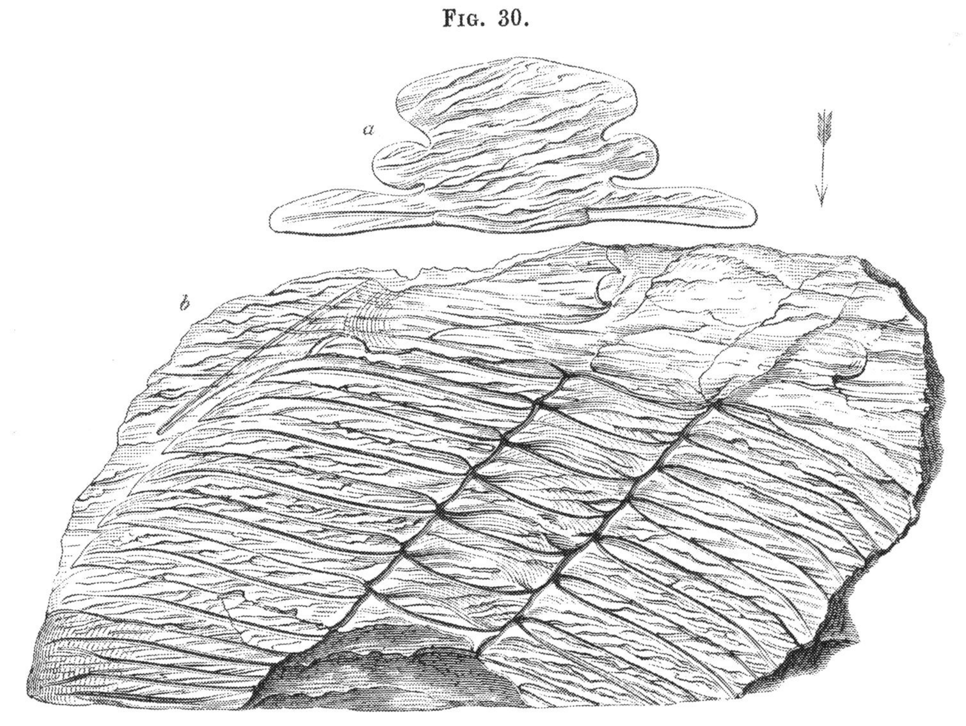

Ogygia Buchii, distorted by cleavage; from the vertical black slates of Abereiddy Bay, Cardiganshire.

shire. The arrow indicates the direction of the pressure, which has shortened all the parts; and the wrinkles on the surface are in the mean direction of that force, which was undoubtedly, as Prof. Phillips, Prof. Haughton, Mr. Sharpe, and others have shown, the cause of this common phenomenon in the slaty rocks. Lateral pressure, acting on a confined plastic mass, is quite sufficient to produce such effects; and has, indeed, been proved by experiment to be equal to the result. Mr. H. C. Sorby has philosophically illustrated this change of dimensions in the rock, which is noted here the more particularly, as our future plates will show several species of trilobites affected by it.

OGYGIA? (vel PHACOPS) SUBDUPLICATA, *Salter*. Pl. XV, figs. 7, 8.

O. caudâ latâ semiovatâ, subtrigonâ, $1\frac{1}{2}$ *unciam longâ, fere planâ, costis perduplicatis. Axis conicus, per* $\frac{5}{6}$ *caudæ extensus, apice prominulo 13-annulato, annulis anticis abruptè angulatis. Costæ laterales 9, marginem angustissimum ferè attingentes, ad apices decurvatæ, omnino interlineatæ, sulcis posticis conniventibus, ultimis profundis parallelis nec abbreviatis.*

Although placed in *Ogygia*, from its general resemblance to that genus, I confess to a strong suspicion that this is a *Phacops* of the section *Chasmops*, and as such it should have been previously described. The strong posterior furrows, with their connivent intermediate sulci (or rather it is the intermediate ones which are the longest and deepest), strongly remind us of some *Ogygiæ;* while the character of the axis, with its arched and duplicate anterior rings, is much more like that of *Chasmops* (see Pl. IV, fig. 19). Moreover, it is more likely that this should be a lingering species of the latter subgenus, than that *Ogygia* should reappear in the Lower Llandovery Rocks, while it is absent from the Caradoc.

Locality.—LLANDOVERY ROCKS of Haverfordwest, Pembrokeshire. The specimens are in the small but excellent collection of Mrs. Breawell, of Cambridge Terrace, Brighton; relict of H. Day, Esq., formerly of Hadlow, Uckfield.

OGYGIA (PTYCHOPYGE[1]) CORNDENSIS, *Murchison*. Pl. XVI.

ASAPHUS CORNDENSIS, *Murchison*. Silurian System, pl. xxv, fig. 4, 1839.
— — *Salter*. Morris's Catalogue, 2nd ed., p. 100, (List at end of *Asaphus*,) omitted in p. 112; 1854.
— — *Id*. Siluria, 2nd ed., pl. iii, fig. 4, 1859.

O. obtusè ovata, 3—4-uncialis, capite lato semicirculari, spinis parallelis. Glabella

[1] *Ptychopyge*, Angelin, consists of those species of *Ogygia* which have the facial suture *within* the margin in front. *O. corndensis* certainly belongs to it. I do not know sufficiently the structure of other species to subdivide the genus.

urceolata, anticè clavata, lobis obscuris. Oculi modici, circumscripti. Thorax articulis rectis, fulcro remoto, sulco pleurali profundo, apicibus obtusis. Cauda axe primùm conico, dein parallelo, 7-8-costato, lateribus 8-costatis, costis rotundis rectis simplicibus, ad marginem striatum concavum undulatis.

The cabinet of Mr. Griffith Davies, of Islington, has for several years contained a fine series, from Builth, of this Trilobite, which has been so long obscure, from the very imperfect specimen figured in the 'Silurian System.' Mr. James Sowerby's figure showed, indeed, that it was distinct from the *Ogygia Buchii;* but its characters could not be fully known from the fragment figured. The exact age of the Corndon mountain, relatively to the trappean beds of Builth, may now probably be determined from this single fossil.

It is a large species; occasionally $4\frac{1}{2}$ inches by $3\frac{1}{2}$, depressed, but not flat; of an obtuse oval shape, with rather blunt extremities in the ♀ form (figs. 9, 11), or with the head somewhat pointed in the ♂ varieties (figs. 1, 7). I will describe the broader form first.

Head a broad segment of a circle, the length being to the breadth nearly as 2 to 5. Of this breadth the glabella occupies more than a third and less than a fourth; it is urceolate in shape, being broad below, contracted at the sides rather below the eye, and then again clavate above. Traces of two short furrows only on each side (the three upper ones are obsolete in the adult), the lower one oblique;—it is the neck-furrow, incomplete in the middle, but marking off a much broader segment beneath the glabella than below the cheeks, where the furrow is strong and continuous.

In front the blunt glabella invades the striated margin, which is here narrow, but very broad round the cheeks, and produced behind into an abruptly narrowed spine, which reaches the sixth thorax-segment in the adult; it is shorter in the young. The eye is rather small for the genus, placed fully half-way up the head, near the glabella, and over-hung by it; a depressed space surrounds it. The facial suture cuts the posterior margin rather more than half-way out, and in front it circles round the glabella (intra-marginal). The labrum (fig. 10) is pointed, with two pairs of furrows, and a broad arched base. It is also visible in fig. 9.

The glabella-furrows are strongest in young specimens, and get partly obliterated in age. In some young ones four pairs of furrows show clearly; the two upper ones directed obliquely backwards towards the eye, the middle one obliquely forwards, the basal one (neck-furrow) direct across. Besides these there is the narrow marginal furrow frequent in the genus, and which probably marks out the articular area on the under surface of the crust. All the glabella-furrows are short and shallow, except the hindermost pair, which are deeper, and show in every specimen of all ages.

Thorax with a wide axis, less than one third the whole width, the rings straight across, and ornamented with coarse arched striæ. The pleuræ straight, scarcely bent at the remote fulcrum, very deeply grooved and strongly facetted; the apices blunt and rather tumid. The fulcrum is about half-way out in all but the front rings.

Tail semicircular, but with rather straight sides and a broad, coarsely striate margin, which is waved by the strong lateral furrows. The axis narrows abruptly from a conical base, and is then parallel-sided; reaching nearly to the end of the tail. It is 9-ringed, and has, besides, a blunt terminal portion, and no appendix beyond. The sides have eight furrows, strong and deep, and rounded at their ends, reaching to the striated border, but not the margin; the upper or marginal one much the strongest and deepest. The ribs are rounded, very convex, quite straight, radiating, and without any duplications. In the tail of young specimens, however, a faint duplication is observable in the forward ribs, and only on the forward edges of these ribs, making them angular instead of rounded. The whole surface of the tail is minutely granular; of the glabella, apparently punctate.[1]

Variations.—There seem to be the usual "*forme longue*" and "*forme large*" in this as in other species of *Ogygia.* Our principal specimens (figs. 9, 11) are of the broad form, and the smaller ones (figs. 1, 7) of the narrower one. Figs. 1 and 7, which are probably ♂ forms, have the head much more pointed, forming a gothic arch, and the margin in front of the glabella much broader than in the female form; the glabella is more pointed and clavate in front. The thorax, too, is relatively narrower, and the tail a semicircle, but with somewhat straight sides. Young specimens do not differ much in proportion or shape from older ones, except in having the tail more triangular (fig. 7), and with a narrower axis, the number of ribs in the tail remaining the same. But very young ones (figs. 5, 6) show the metamorphosis. Fig. 5, which is only two lines long, has clearly but six rings, and fig. 6 has only seven rings, besides having the cheeks contracted at the base of the spine, which is also more divergent than in the older form. This last character may be accidental in this specimen, for somewhat older ones (figs. 2, 3, 4) show the spines more parallel to the body, and the cheek not contracted at this part.

The appearance of a reduced number of body-rings is often due to accident. In fig. 4 the body has slipped under the carapace, so that on the left side there are seven rings, and on the right hand only four. But this is not the case in the two youngest specimens just quoted, which have six and seven rings respectively.

Locality.—LLANDEILO FLAGS, Gilwern, near Llandrindod, Builth. All the specimens figured are in the collection of Mr. Griffith Davies, of Islington, except fig. 8, which is Murchison's original specimen from the Corndon Mountain W. side (Mus. Geol. Soc.)

[1] A delicate lineation occupies the whole of the axal-lobe of the thorax, and the outer half of the pleuræ is much more strongly marked with it than the inner. The concentric lineation of the tail-border is very sharply defined, in a rather narrow band, the lines abutting obliquely, as usual, against the inner edge.

OGYGIA SCUTATRIX, *Salter*. Pl. XVII, figs. 9—13 (9, 10 are wrongly marked as *O. peltata* on the plate).

OGYGIA SCUTATRIX, *Salter*. In Siluria, 2nd ed., p. 53, woodcut 9, fig. 1, 1859.
— — *Id*. Append. Ramsay, Geol. N. Wales; Mem. Geol. Survey, vol. iii, 312, pl. ix, fig. 1, and pl. viii, fig. 8, and plate ix, fig. 1, 1866.

O. septuncialis, fere rotunda! vel latissimè elliptica, depressa. Caput?—. Thorax caudâ brevior, axe lato; pleuris longis axin 1½ *superantibus, profundè sulcatis, á fulcro (ultra dimidium posito) paullo deflexis. Cauda quam semicirculari latior, axe lato* 8—9-*annulato; limbo sulcis primariis* 7—8, *ad apices fractis; secondariis rectis profundis haud axin attingentibus.*

I well remember finding this large specimen in 1853, together with the *Asaphus affinis* hereafter to be described. They were on the face of a good slate fence in the valley behind Garth, Portmadoc; and were the harbingers and first fruits of the fauna, then unexplored, of the Tremadoc Slates.[1] The large size and round contour of *O. scutatrix* distinguish it easily from any other of the genus. The specimen is in the Museum of Pract. Geology: it may have been a little altered in form by the pressure accompanying cleavage; but in the figure some allowance is made for this. If not sufficiently so, still the form must have been a very wide and rounded one, and can scarcely ever have had the shape of the species next described, although it is evidently closely allied to it.

Length 7 inches; the breadth fully 6 inches. Of the head we have only a single specimen, in Mr. Lee's collection (fig. 10); and it is from a different locality and formation. It appears to be very short and broad, with a wide urceolate glabella, a broad striated margin, a small forward eye, and a wide labrum, with a narrow arched base.

The body is four fifths the length of the tail, and of eight widely transverse joints, with a broad axis, which is not convex, but sharply defined, and is more than two thirds as wide as the flat pleuræ. These are deeply grooved the groove much curved at its origin and bent (as are the pleuræ) at the fulcrum-point. This occurs at about two thirds out; it is rather more remote in the front rings. The anterior or fulcral half of the pleuræ is much broader and less convex than the hinder part, which is angular, almost gibbous, beneath the fulcrum. From this point a strong oblique striation covers all the tips of the pleuræ, corresponding to the broad-striated area which runs round the tail-border. The tips of the pleuræ are truncate, the hinder end produced a little into an acute angle.

[1] I am afraid I replaced the gate-post with somewhat insufficient material. A new *Ogygia*, and that in a new and unexplored formation, was not to be resisted. The owner of the field must accept my apology; and will oblige me much by searching for more complete specimens.

The tail, nearly semicircular in outline, is more than twice as wide as long. The axis broad, quickly tapering, reaching less than four fifths the length of the tail; depressed, and with (apparently, for the extreme tip is lost) a blunt termination. It consists of eight strong rings, and a semicircular terminal portion. The broad side-lobes are grooved by eight strong primary furrows, interlined from near their origin by furrows as strong as themselves, but which commence very abruptly at a short distance from the axis, and radiate straight out, not bent (as are the primary furrows) in a line corresponding to the fulcrum of the pleuræ. A broad band of very oblique and wavy striæ ornaments the border of the tail above, and beneath it the concentric striated fascia (fig. 12) has the striations wide apart.

The labrum (fig. 13) is pointed abruptly, and seems to have two pairs of concentric furrows, but it is very imperfect in our specimen (fig. 13) from the Tremadoc Rocks.

In fig. 10, which is from the Arenig group, the labrum is more perfect. It is as broad as long, and subquadrate in outline, the front much arched; the base of attachment scarcely wider than the broad-winged sides; the notch above the short auricle somewhat acute. The concentric furrows are strong and double, the central part enclosed by them roundish and convex, and the apex (broken off) is connected with this central part by a raised space, and this elevation is sharply divided from the flat wings by longitudinal furrows.

The large size and round form of this fine species much recals that *O. Desmaresti*, Brongn. (the *O. Brongniarti* of Rouault), and well figured by the latter authority in the 'Bull. Soc. Géol. France,' vol. vi, 2nd ed., pl. i. From that species the less width of the axis, the straight, not curved, pleuræ, grooved nearly to their ends, distinguishes our rare fossil. From *O. Edwardsii* (pl. ii, fig. 1, of the same work) it is distinguished by the fewer joints of the tail, and the less number of lateral furrows, all of which are duplicated. The pleuræ, too, are curved in *O. Edwardsii*, Rou., which in many respects closely resembles *O. Desmaresti*. I do not know any species with which ours need be compared, except the following one, which at one time I thought identical ('Siluria,' l. c., p. 53).

Localities.—Lower and Upper Tremadoc Rocks. Lower Tremadoc.— Carnarvon Road, 1½ miles W. of Tremadoc (Mr. Homfray). Upper Tremadoc.—North face of Garth Hill, at the mouth of the Traeth Bach, Merionethshire, opposite Portmadoc—in company with *Angelina* and many other fossils (Mus. Pract. Geology). Also Portmadoc, quarries in the town. The labrum (fig. 12) is from thence, and is in Mr. Ash's cabinet. Arenig group.—North-east corner of Whitesand Bay, St. David's, Pembrokeshire, figs. 9, 10 (in Mr. Lee's cabinet).

OGYGIA PELTATA, *Salter*. Pl. XVII, fig. 8.

OGYGIA SCUTATRIX, *Salter*. Siluria, 2nd ed., p. 53, Foss. 9, f. 1, 1859.
— PELTATA, *Id*. Append. Ramsay, Geol. N. Wales; Mem. Geol. Survey, vol. iii, p. 313, pl. xii, fig. 8, 1866.

O. modica, vel caput, thoracem, caudam æqualis. Glabella lata, lævis, sulcis nullis. Thorax axe depresso, pleuris angustior; his usque ad apices falcatos nec decurvos planissimis; fulcro obsoleto. Cauda quam semicirculari longior, axe lato 8-*annulato; limbo sulcis primariis subrectis, secondariis abbreviatis.*

We have only portions of this fine species, but these enable us to reconstruct a fossil certainly four inches in length, of a broad-oval shape. Our fig. 8 is a copy of the figure in the 'Survey Memoir' above quoted. It is probably not of the full breadth, but nearly so; and I have no specimen which more completely shows the true shape.

The head, thorax, and tail, are nearly of the same length; the head is somewhat longer than a semicircle, with very short broad head-spines. It is widely margined; and the glabella is well distinguished from the cheeks, and is of the same width with them, parallel-sided, reaching five sixths the length of the head, and without any distinct lobes except the neck-segment, which is strongly marked out. The eyes are large?, placed half way up the head. The portion of the neck-furrow beneath the cheeks is much nearer to the posterior margin than that beneath the glabella, and abruptly so (see fig. 8). The striated margin to the cheeks is very broad towards the front.

Thorax with a wide axis, not quite so wide as the pleuræ, and somewhat narrower behind. The rings of the axis are very flat, and scalloped out at their junction with the pleuræ, so that the axal-furrows in this (as in the preceding species) present a set of re-entering angles, with concave arches between them. The pleuræ are flat, straight as far as the fulcrum, which is placed at two thirds out. The pleural groove is sharp and sigmoid in its curve, and reaches nearly to the falcate tip without any angular bend. It is parallel to the front edge, apparently more so than in *O. scutatrix*, and the hinder half of the pleuræ is rather the largest.

The tail is a semicircle, the axis occupying rather more than a fourth of the entire width, and tapering backwards for nearly three fourths of the length. It is annulated by seven furrows, which, except the two or three last, stretch right across. The tip is blunt, and rather obscure. The sides are radiated by eight strong grooves, including the upper or submarginal one, which are all slightly bent downward near their ends, and are interlined by similar but much shallower furrows, which start abruptly (as in the last species) from a point near, but not close to, the axis. A plain smooth (?) border to the upper surface, and the incurved fascia narrow and closely striated.

The specimens described are all from the north-west angle of Whitesand Bay, St. David's Head, and differ from the N. Welsh species last described as follows:—The form more elongate; the pleuræ flatter, much shorter, and more pointed; the tail with a narrower axis, and with less distinct interlineations to the sides. From *Barrandia Portlockii* (see Pl. XIX), the much greater length of the axis, and narrower glabella, will easily distinguish it.

Locality.—"Arenig group" of dark earthy slates at Whitesand Bay, St. David's Pembrokeshire, north of the reef called *Trwyn hwrddyn*, which is probably of the age of the Tremadoc Rocks.

Ogygia Selwynii, *Salter*. Pl. XVII, figs. 1—7.

Asaphus Selwynii, *Salter*.		Reports Brit. Assoc. Trans. of sect., p. 57, 1852.
— —	*Id.*	In Morris's Catal., 2nd. ed., p. 100, 1854.
Ogygia Selwynii,	*Id.*	Append. Ramsay, Geol. N. Wales; Mem. Geol. Survey, vol. iii, 313, pl. ix, figs. 2—6, and pl. 11 b, fig. 5, 1866.

O. elliptica plana, triuncialis, capite brevispinoso; glabellâ oblongâ, genis angustiori. Sulci glabellares brevissimi. Oculi magni. Thorax pleuris a fulcro remoto decurvis. Cauda axe angusto, longiconico, 6, 7-sulcato, apice prominulo; lateribus 7-sulcatis, sulcis brevibus, obscuris, interlineatis. Fascia latissima, striis confertis.

A species of more than ordinary interest, since it was the first true indication of the fauna of the great group now distinguished in Britain under the name of the "Arenig or Skiddaw group." A solitary fossil, found by Prof. Sedgwick and myself in 1844, proved different from all other Lower Silurian species, and characterised a particular set of strata, which were successfully compared with the rocks west of the Stiper Stones, by means of a specimen of this same species. Such exact indices are our friends the Trilobites of age in the Palæozoic Rocks.[1] We have specimens from three localities, viz., Hengwrt Uchaf, four miles north-east of Dolgelly, and we figure the original small specimen (Fig. 31); from S. Carnarvonshire; and lastly from near Chirbury, Shropshire. The latter are by far the finest specimens, and our larger figures are taken from them. Mr. Lightbody has lent some fragments of the head, and our restored figure (Pl. XVII, fig. 1) is partly made from these, and partly from others in the Museum of Practical Geology.

Fig. 31.

Ogygia Selwynii. Original specimen from Dolgelly.

The head is broad, semi-oval, an inch and a half broad or more in Shropshire specimens, and broadly margined. The oblong glabella is distinct all round, depressed, gently clavate in front, and contracted at the sides, but otherwise parallel-sided; with a large upper lobe and three very short lateral lobes, which do not reach even one third across the glabella (fig. 2), the lower pair largest and somewhat oblique. Eyes large lunate, placed less than half-way up the head. The facial suture cuts the margin more than

[1] The "Skiddaw group" has since proved rich in *Graptolites*, and these enable us to compare it with the Quebec group of America and the black slates of Melbourne. But *Ogygia Selwynii* was the first step in the comparison for the whole series.

half way out, but not very near the short spinous angles. The glabella reaches and invades considerably the front margin, so as to leave it very narrow in front; on the sides it is rather broad. The neck-furrow is indistinct. Labrum as broad as long, obtusely pointed, with strong concentric furrows.

The thorax, of eight narrow well-defined segments, is not quite so long as either head or tail, which are equal. The axis is well defined, gently convex, not quite two thirds the width of the pleuræ. The latter are flat as far as the fulcrum, which is placed half way out, then bent down; the pleural groove very shallow, except beyond the fulcrum, where it is deeper.

The tail is rather more than a semicircle, ten lines long by sixteen broad. The axis narrow, and tapering, not reaching quite three quarters the length of the tail (in the smaller Welsh specimens it reaches fully three fourths). It is annulated by six or seven furrows above, the upper one often strongest, the rest faint. The sides of the tail are marked by seven shallow short furrows, which do not reach to the depressed border, and the upper one of these also is the strongest. The broad margin is flat or slightly concave. The incurved fascia is very broad, folding over the tip of the axis, and closely striated. Young specimens (fig. 3) show fewer ribs to the axis and sides, and the upper lateral furrow is stronger in proportion.

Localities.—Arenig group.—Hengwrt Uchaf, four miles north of Dolgelly. (Woodwardian Museum, woodcut fig. 31): Llanfaelrhys, Aberdaron, Pl. XVII, fig. 3 (Museum Pract. Geology); plentiful at White Grit Mine, near Shelve, Shropshire, figs. 1, 2, 4 to 7 (Mus. Pract. Geology; and Mr. Lightbody's cabinet); Mytton Dingle, Stiper Stones, abundant (Mr. Homfray's cabinet).

Barrandia, *M'Coy*, 1849.

A group of well-marked species, characteristic of the Llandeilo Flags. The three forms known have the habit of *Ogygia*, but differ in such obvious characters, that there can be no doubt Prof. M'Coy was correct in separating the typical species *B. Cordai* as a distinct genus. The other two species, here arranged under *Barrandia*, are not very closely connected with it; but I do not wish hastily to give a new generic name to them, though very little doubtful of its propriety. The characters of the glabella, and the position of the fulcrum in the pleuræ, are too unlike to render it probable that *Homalopteon* and *Barrandia* are merely subgenera. But I shall leave them so for the present.

This small group is a remarkable one, and tends to render the passage of the *Ogygides* into the Bronteidæ so complete, that we may well hesitate about drawing a line between the two families.[1]

[1] But the Llandeilo Flag genera have a thin crust and a wide expanded shape, and are, moreover,

Generic Character.—Ovate, greatly depressed; head with short spines. Glabella widely clavate, the axal-furrows nearly parallel below, and sometimes obsolete above. Eyes large, depressed, anterior in one subgenus, subcentral in the other. Facial suture marginal in front, nearly vertical posteriorly. Labrum pointed? Pleuræ falcate. Tail large, fan-shaped, with a shortened axis, and few lateral furrows; the caudal fascia occupies a large part of the limb beneath.

Subgenera.—1. *Homalopteon,*[1] Salter, 1865. Glabella with complete axal furrows, widely clavate above, more or less distinctly lobed transversely by four transverse furrows. Eyes anterior. Pleuræ with remote fulcrum, grooved, but scarcely facetted. Tail with a distinct short axis of several rings; the sides few-ribbed. *Range.*—Llandeilo Flags.

2. *Barrandia,* M'Coy, 1849. Glabella with incomplete axal-furrows and no distinct lobes. Eyes subcentral. Pleuræ falcate, with a fulcrum close to the axis, grooved, not facetted. Tail with a short ribless axis and smooth sides, the strong articular furrow only present. *Range.*—Llandeilo Flags.

Subgenus—HOMALOPTEON.

BARRANDIA (HOMALOPTEON) PORTLOCKII, *Salter.* Pl. XIX, figs. 6—10.

ASAPHUS DILATATUS, *Portlock.* Geol. Report of Tyrone, &c.; p. 293, pl. xxiv, fig. 2 (not of *Dalman*) 1832.

OGYGIA DILATATA, *Salter* and *Phillips.* Mem. Geol. Survey, vol. ii, pt. 1, p. 239, 1848.

— PORTLOCKII, *Salter.* Decades Geol. Survey, No. 2, pl. vii, figs. 1, 2, 6, 7 (not figs. 3—5, for which see next species), 1849.

B. (*Homal.*) *ovali-oblonga,* $3\frac{1}{2}$ *uncialis, capitis margine angusto. Glabella genis latior, anticè gibbosior, lobata, lobis utrinque* 4, *distinctis. Thorax axe lato nodoso; pleuris per-sulcatis. Cauda axe lato* 5-*annulato, apice acuto; lateribus semicostatis, costis* 4-*radiatis divisis subrectis.*

The length of this, the largest of the three species known, is full four inches, judging by fig. 8, which is from the original specimen in Portlock's plate. It is nearly flat, ovate, widest in front; the head forming rather more than a semicircle, and considerably wider than the tail, but about as long as the thorax. The glabella is as wide as the cheeks below, and separated by narrow but distinct and complete axal-furrows from them. It widens and overhangs the eyes above, and is there somewhat abruptly bent down, the margin being over-

closely allied to *Ogygia* and *Stygina* in habit. The *Bronteidæ* have, on the contrary, a thick calcareous shell, and often a highly convex, almost gibbous form, while the axis is so greatly reduced as to be almost obliterated. I shall return to the Bronteidæ as soon as the Asaphidæ are completed. It is impossible to take the affinities in a linear form, because that is not the order of nature.

[1] ὁμαλος, *planum,* πτεον, Att. for πτυον, *flabellum;* from the flat, radiated character of the tail.

hung by it. For nearly two thirds its length it is marked by four lateral furrows, irregular in direction, which divide it into four lateral lobes, exclusive of the forehead- and neck-lobes. The furrows reach far inwards, leaving only a narrow median space unoccupied. The front ones are straight across, the second pair curve from without downwards, the third is a short deep impression, the fourth or lowest pair are nearly straight, but so oblique towards the narrow neck-furrow as to separate a pair of triangular lobes. None of the furrows reach the outer margin.

In the Decade, from which this description is partly drawn, I have inaccurately described the lower glabella-lobes as part of the neck-lobe. It is not so. There are four pairs of lobes to the glabella.[1]

Eyes rather large, placed very high up towards the termination of the facial suture on the margin. The eyelid semilunate, but not constricted above or below. The facial suture is nearly vertical above the eye, and below it turns very little outwards, cutting the posterior margin at about half the cheek's width. Free cheeks moderately large, with a narrow border and a small spine. Labrum with the centre gently swelled, and with concentric furrows and the usual pair of tubercles. The apex is broken off; it was probably obtusely pointed.

Thorax with a broad axis, not quite so wide as the pleuræ, and tapering backwards. The rings are nodular, with bilobed tubercles on the sides, the hinder rings with a central tubercle; but all rather faintly marked. The pleuræ are directed backward, and are somewhat sigmoid in outline; the fulcrum rather faintly marked at about half-way along them. The divisions between the pleuræ are not nearly so conspicuous as the oblique furrow that reaches nearly to the end of each, separating a broad and rather tumid posterior portion. The tips are recurved and pointed, the foremost ones, perhaps, not quite so strongly as in our large figure; the hinder ones more so, as in fig. 6 *a*.

Tail a semicircle, with the upper angles rounded off; the axis hardly more than two thirds the length, flattened, conical, the ends a little pointed. There are about five distinct ribs on the axis, each obscurely trituberculate. The axis is most convex behind; it reaches two thirds down the tail, with a rather large terminal portion connate with the triangular appendix, making the axis a pointed one (see fig. 8). Lateral furrows four or five, very oblique and short, the upper ones reaching two thirds—the lower not half across the broad striated limb. The three upper ones (see fig. 8) are strongly duplicate throughout.

Whole surface of tail covered with a concentric lineation, which runs across the limb transversely, not parallel to the margin. The striæ are continuous, but mixed with smaller interrupted ones. The incurved caudal fascia is very broad, but its margin is not distinctly seen. An equally broad fascia runs all up the pleuræ, the sharp lineation crossing them at right angles. We do not know the striæ on the head; but the labrum is strongly and coarsely striated (fig. 9).

[1] *Ogygia*, *Niobe*, the present genus, and *Cromus*, a Bohemian form, are examples of this large number of glabella-segments. But it is extremely rare in the Trilobite group.

Originally identified by its discoverer Portlock with the little known *Asaphus dilatatus* of Dalman. But Dalman's figure is far from good (that by Sars, in Oken's Isis, shows clearly that it is not even of the same genus), and Portlock's specimen was incomplete. Captain, (now Sir Henry) James afterwards obtained a noble series, which were figured in Decade 2 of the Geol. Survey. They are all from one locality: viz.—

Locality.—LLANDEILO FLAGS. Schists of Newtown Head, Waterford (Mus. Pract. Geology). Museum of Irish Industry. I do not know that this fine species exists in other collections, except that of Major Austin, Clifton.

BARRANDIA (HOMALOPTEON) RADIANS, *M'Coy*. Pl. XIX, figs. 1—4.

OGYGIA PORTLOCKII, jun., *Salter*. Decades Geol. Surv., No. 2, pl. vii, figs. 3, 5, 1849.
— RADIANS, *M'Coy*. Annals Nat. Hist., 2nd ser., vol. iv, p. 408, 1849.
— — *Id.* Synopsis Pal. foss. Woodw. Mus., p. 149, pl. i F, fig. 2, 1851.
BARRANDIA, sp. *Id.* l. c., p. 149.

B. (HOMAL.) *minor*, 1½ *unciam longa, latè ovalis. Glabella genis angustior, anticè haud gibba, illobata vel lobis obscuris. Thorax axe lævi, pleuris longè sulcatis. Cauda axe conico* 4—5-*annulato, apice rotundato: axe bis latiore limbo ferè, utrâque sulcis tribus curvis interlineatis. Sulci longi, primùm recti, dein profundiores, abruptè recurvi.*

My friend Prof. M'Coy, while animadverting, in the excellent work above quoted, (p. 149), on my mistake in placing this as the young of *B.* (*Ogygia*) *Portlockii*, apparently did not perceive that he also was referring to *Barrandia* the species he had in the same page described as an *Ogygia*. The truth is, my mistake and his own arose very naturally from the paucity of specimens. And as Burmeister had six years before predicted the occurrence of the metamorphosis among Trilobites, I was not so far out as my friend and critic supposed. With few exceptions, and those not very clearly made out, or even founded on mistake,—the increase of the number of segments with advancing age was not known till 1849, when M. de Barrande published his beautiful series of observations on the genus *Sao*; and at the same time, without knowing of his work, my own figures of this species were given. In 1851 M. Barrande examined the British specimens, and satisfied himself of the true occurrence of the metamorphosis; and that in a genus to which he had not access in Bohemia. *Barrandia* and *Ogygia* are northern types.

As I am not sure that I shall be able to show this change of form with age in any other British species, I would refer particularly to it here.

M. Barrande, to whom the real credit must belong of working out intentionally a point only accidentally observed by others, has established the fact that Trilobites—*probably all of them*—undergo a distinct and regular series of changes in the young state, comparable very nearly to that observed in the other Crustacea. The development of the eyes and facial suture takes place simultaneously with the increase of the body-segments. And the very young Trilobite (Barrande has traced some of the species from the egg) looks like a totally different genus from the adult; it is a simple disk, which gradually only becomes lobed, segmented, and provided with organs, as its life goes on and its size increases. The Zöe of the crab is now familiar to us. But its discovery hardly excited more interest among naturalists, than did Barrande's observations on the young Trilobites among palæontologists, when first announced in May, 1849, to the French Academy.' My own far less complete description was published in June of the same year.

It seems that the additional segments always come in between the thorax and the tail, while the latter also receives at each fresh moult an accession of segments till it attains its full number. What becomes of the moulted crusts of Trilobites? From all analogy they should be cast off, slit open in some part of the dorsal region. And such specimens should be looked for in our shaly rocks; for it is impossible that they should be much more destructible than the newly-formed crust.

Description.—I have obtained from Mr. J. E. Lee's cabinet a much larger tail of this species than we before possessed. And we may therefore reckon the species to have been fully an inch and a half long, widely oval, with obtuse ends; the head forming a semicircle, longer than the body, and larger than the tail.

The axal-furrows are parallel below, somewhat abruptly divergent above, and reach the very narrow margin in front. The glabella is without lobes, or very obscurely lobed. The cheeks, wider than the glabella, are gently convex, narrowly margined, and bear the lunate eye so forward as nearly to touch the front margin, a character by which *B. radians* is easily distinguished from *B. Cordai*, next described. Head-angles very shortly spinous.

The axis of the body is much narrower than the straight pleuræ, which are grooved three fourths of their length, and have the scarcely visible fulcrum placed at more than half-way out from the axis. The tips are sabre-shaped.

The tail is nearly a semicircle, and all but flat. And our largest specimen shows a length of six lines by a breadth of ten lines. Of this length the conical axis occupies less than three-fourths; and but for the triangular appendage (appendix) beyond its bluntish tip, does not reach more than half-way down. It has four ribs straight across, divided by shallow furrows.

The sides have three strong furrows, which at first run directly out, and then suddenly bend down, expand, and nearly touch the margin. They are interlined by three others, which are shallow, and cease at the straight portion. The rest of the tail is bare of ribs, and apparently smooth; but the incurved caudal fascia beneath is very broad and straight-edged anteriorly. It just reaches the tip of the axis (see fig. 4).

The metamorphosis above referred to (figs. 1, 2) is very simple. Fig. 1, a very young specimen, has only four rings, and this is the true number in this specimen, and is not due to accident. The ribs of the tail are more conspicuous in proportion than in the somewhat older specimen, fig. 2, which has truly seven rings; while in fig. 3 the eighth thorax-segment is becoming visible (see fig. 3 *a*).

Locality.—LLANDEILO FLAGS. Two or three places near Builth, Radnorshire, viz.: Wellfield; (figs. 1, 3), and Pencerrig, (Mus. P. Geology fig. 2); also Gwernyfyd (Mr. J. E. Lee's cabinet, fig. 4).

Subgenus—BARRANDIA, *M'Coy*, 1849.

BARRANDIA (BARRANDIA) CORDAI, *M'Coy*, Pl. XIX, fig. 5.

BARRANDIA CORDAI, *M'Coy*. Annals Nat. Hist., 2nd ser., vol. iv, p. 409, 1849.
— — *Id.* Synopsis Pal. Foss. Woodw. Mus., p. 149, pl. i F, fig. 1, 1851.
OGYGIA PORTLOCKII, *Salter*. In Morris' Catalogue, 2nd ed., p. 112. 1854.

B. *uncialis, ovata, vel caput, thoracem, caudam æqualis. Oculi post medium capitis positi. Pleuræ retrocurvæ, brevisulcatæ. Cauda axe integro, lateribus unisulcatis.*

One specimen only of this most distinct and characteristic species is known, and it is a perfect one in the Woodwardian Museum. But I believe it is a common fossil, and it should be sought for at Builth. I had no excuse for uniting this very distinct form with the *B. Portlocki.*

Length 11 lines (our figure is enlarged to once and a half the natural size), of which the head, thorax, and tail are nearly equal parts. The semi-oval head is rather more than twice as wide as long, lunate, with a concave hinder border and obsolete neck-furrow. The axal-furrows are parallel below, slightly converging at less than half-way up the head, and then obsolete. The front margin concave, the border of the cheeks extremely narrow; the large slightly curved eyes placed much nearer the axis than the border, and very near the hinder edge.

The facial suture curves slightly out above the eye; and beneath it cuts the posterior margin below the eye, and very remote from the short spinous angles.

Thorax with its slightly convex axis very little less in width than the pleuræ, and tapering backwards; the rings a good deal arched forward, and strongly grooved across. The pleuræ are oblique backwards, and falcate at their tips; with a short, strong, oblique groove not half their length, and the fulcral point very prominent (as in some of the *Olenidæ*, for instance, *Remopleurides*).

The tail is half as wide again as long, and the front much arched to follow the form of the retreating pleural segments. The axis short-conical, extending less than two thirds down the tail, pointed, not prominent at the tip, and without furrows. It is well distinguished from the gently sloping sides, and the border is very gently concave. One upper marginal furrow is all that is visible, and that is very strong and deep. The upper surface is not well preserved; but the caudal fascia on the under side is so broad as to occupy the whole surface of the sides beneath, and the ornamental lines are rather close. The same width of fascia is continued beneath the pleuræ; and the lines on it run directly across them, as seen in our figure. The same sharp lines run round the under margin of the head.

Locality.—LLANDEILO FLAGS. Penkerrig, near Builth, Radnorshire; (Woodwardian Museum).

The passage from the Ogygides to the *Bronteidæ* is rendered still more easy by a MSS. genus, *Bronteopsis,* distinguished so far back as 1857, by my friend Prof. Wyville Thomson, who was at that time carefully studying the Caradoc fossils of the Girvan district. The genus is a very remarkable one, having all the characters of the *Bronteidæ,* and will be described under that family. But the crust appears to have been thin, not calcareous; and the habit of the tail is so much that of *Barrandia,* that it might well be mistaken for an extreme member of the group we are describing.

Again—*Stygina* leads from the *Ogygides* to the *Illænides,* besides having some affinities with *Bronteus.* But we must first describe the typical Asaphi; and therefore return on our road to them by means of the genus Niobe (see diagram on p. 124).

NIOBE, *Angelin,* 1852.

A genus instituted by Prof. Angelin to include a few flattened species of Asaphus, which have the body flat, the axis broad, and the labrum pointed like that of Ogygia. *Asaphus frontalis,* Dalman, is the typical species. The genus is strictly intermediate between Ogygia and Asaphus. Range, Upper Cambrian to Lower Silurian.

Broad-oval, depressed; with a distinct broad axis, and a scarcely clavate glabella which is slightly 4-lobed. Head-angles *obtuse.* Eyes approximate. The pleuræ facetted and grooved, not produced into points. Tail broadly margined, of a moderate number of segments. Labrum with a narrow base and parallel sides; the tip obtusely pointed or slightly emarginate, not forked. Hypostome (in *N. emarginula* at least) without a vertical suture.

N. læviceps, placed by Angelin with this genus, should, I think, rather be referred to *Nileus* among the subgenera of Asaphus. It has a perfectly smooth unribbed tail.

NIOBE HOMFRAYI, *Salter.* Pl. XX, figs. 3—12.

NIOBE HOMFRAYI, *Salter.* App. Ramsay Geol. N. Wales; Mem. Geol. Survey, vol. iii, p. 314, pl. vi, figs. 5—8, 1866.

N. ovalis lata, 3—4 *uncialis, depressa, axe subplano distincto, caudâ semicirculari semiradiatâ. Glabella genis æqualis, urceolata. Oculi præ medio capitis positi. Labrum acutum. Cauda sulcis lateralibus* 4—5 *abbreviatis, interlineatis.*

A most characteristic fossil, which rewarded a good day's hammering (in Mr. Homfray's company) at the head of the marsh, Penmorfa; and was afterwards obtained in great plenty by Messrs. Homfray and Ash of Portmadoc.

General shape a very broad oval, depressed, fully two thirds as wide as long; the length between four and five inches. The axis is flattened, yet distinct throughout, wide and tapering from the broad urceolate glabella to the blunt tail-axis. The head is smooth and very little lobed, and the sides of the tail imperfectly radiated.

The head is rather more than one third of the whole length, semicircular, with blunt outer angles; the glabella occupying one third the width of the head, and reaching to the sharp marginal furrow. It is urceolate, blunt and widest in front, then a little contracted, and thence widening again to the base, marked by a very distinct neck-furrow. The glabella-furrows are faint, four short ones on each side, somewhat radiating from the eye inwards, the lower one longest. The eye is very near the glabella, as in all the genus, and placed more than half-way up the head, semilunar, and rather large. The facial suture reaches the edge immediately over the eye, curves boldly out beneath it, so as to leave but a third part of the posterior margin of the cheek outside it; it is marginal in front.

The body-axis is broad, equal to the pleuræ in front, but narrower than those behind, and ornamented with arched striæ. The pleuræ are strongly facetted for rolling up, convex, blunt and rounded at their ends; the fulcrum close in toward the axis in the front rings, and gradually further out till it reaches the inner fourth of the pleura in the hinder ones. The pleural furrow deep, but rather short and diamond-shaped.

The tail, a true semicircle, is somewhat flattened; and has a short broad axis reaching three fourths the whole length, and about half as broad as the side portions, or a little more. It is marked by seven or eight distinct ribs nearly to the tip, which is prominent and bluntly pointed. The sides are scored by five short furrows, which (faintly interlined) only reach to the inner edge of the broad flattened margin, and there stop abruptly. The margin is concentrically striated, of equal breadth all round. Young specimens (fig. 6) differ but little in proportion, but there are some specimens (one particularly in Mr. Ash's cabinet) which are much narrower than usual, and may, as that gentleman suggests, be the male forms (see also figs. 6, 7).

The labrum is long, broader at the base, but still on the whole parallel-sided. The sides are not contracted, nor the front expanded, as in the *N. emarginula* figured by Angelin; and the apex is an obtuse angle, at about 100°, and the tip rather acute. There is a concentric furrow near the margin, and a strong pair of indentations near the tip.

Localities.—UPPER CAMBRIAN. LOWER TREMADOC SLATES of Penmorfa Church and Llanerch, Tremadoc; also Castle Deudraeth, near Maentwrog, N. Wales (Mus. Pract. Geology). The figured specimens are from the cabinet of our excellent friend Mr. David Homfray.

ASAPHUS, *Brongniart*, 1822.

Of so large a genus much might be said, and with advantage too, if our space permitted us to discuss largely its affinities. The diagram, however, on page 124, will give at a glance the relation which the chief subgenera of *Asaphus* bear to one another, and to the neighbouring groups, *Ogygides* and *Illænides;* and we can more easily refer to these affinities when the subgenera themselves have been defined. The genus is as remarkable among the smooth Trilobites of our right-hand division (see Preface, p. 2), as *Phacops* is of the left-hand group.[1] There is the same perfection of organisation and compactness of character, the same wide limits of variation within the genus, and as great importance numerically. While however *Phacops* and its subdivisions are more characteristic of the Upper Silurian rocks, *Asaphus* is strictly Lower Silurian,—scarcely ranging below, and never above that horizon. The *Phacopidæ* in the higher groups are remarkably convex; the *Asaphi* expanded. The *Phacopidæ* have, as a rule, the segments both of head and tail well marked out; the *Asaphi* have them obliterated in the same portions. The *Phacopidæ* are ornamented with a tubercular or granular coat; the genus we are describing has a smooth or only lineated one, a distinction seen and ably noted by Dr. Burmeister, who, however, attached too much value to the character, since both kinds of ornament coexist in several genera. The lenses of the great eyes in *Phacops* are very large—the equally prominent eyes of *Asaphus* have them very small,—and so of several other contrasting characters; and while the chief character of all—the relative possession of eight and eleven body-rings—keeps the two groups widely apart, they are no less sundered by the course of the facial suture beneath the eye, which in all the *Asaphidæ* is to the margin behind, and in all *Phacopidæ* to the outer margin.

For size, only the large *Paradoxides* among the *Olenoid* group (see Preface) can match the *Asaphi*, which include, if not quite the longest, certainly the bulkiest forms of Trilobites, and indeed, so many of these, as to make size an important character of the genus. Few of the species are less than three inches, many of them nine or ten in length, and a few range beyond a foot in extreme measure.

The subgenera include several extreme forms, but the essential characters of this large genus seem to be as follows:—

Form oval, without spines or tubercles on the surface, with the head and tail nearly

[1] It might, perhaps, be convenient to designate the Trilobites of the left-hand branch—which have a well-developed thorax, and a relatively small pygidium of few segments, *Micropygini*, and those on the right-hand *Macropygini*. But I do not think the terms are needed.

equal in size, and the thorax of eight rings. The surface smooth, or only covered with a close lineation intermixed with puncta, which does not interfere with the general smoothness of the exterior. The head is semicircular or half-oval, or pointed, with a glabella widest in front, and generally very faintly marked out, and with quite obsolete lobes.[1] Smooth prominent eyes; a hypostome without rostral shield (only a vertical suture being occasionally present), and a forked labrum.

The thorax-rings convex; the pleuræ rounded, grooved, and facetted for rolling. The tail large, of many segments; but these are not generally visible, except on the axis, and in some cases not even there.

The several subgenera are dependent chiefly on the degree of convexity, distinctness of the tail-furrows, course of the facial suture in front, and the degree in which the obscure glabella occupies a larger or smaller space of the head. I will proceed to give these characters as briefly as I can, the route being from the *Ogygides* last described towards the *Illænides*, which we shall take up afterwards. Some of these subgenera have been indicated by Goldfuss, Dekay, Dalman, Emmerich, Milne-Edwards, Burmeister, and others. None of them are satisfactory, the habit having been too much overlooked, while trifling differences in the glabella, &c., have been preferred to the more important characters of the facial suture. Corda's arrangement is perhaps the least worthy consideration; and Emmerich's and Goldfuss's seem to me the best. Professor Burmeister had not sufficient material for all the divisions, but his acumen detected the importance of the facial suture, and his arrangement is clear. Professor M'Coy's divisions are far too few. We are greatly indebted to Barrande for his laborious collection of the entire history of these groups, and are sorry not to be able to adopt his own subdivisions. Following Emmerich and Goldfuss, but subdividing their group of "Ogygiæ" especially, we have the following eight subgenera:

1. *Ptychopyge*, Angelin. Expanded, ovate, gently convex, or flattened, with narrow axis; short urceolate glabella, reaching more than half-way up the head, and lobeless; approximate elevated eyes; subangular tips to pleuræ; and tail with moderately long axis, and many faint ribs on axis and tail. Facial suture forming a long ogive in front. Hypostome entire. Labrum shortly notched. North Europe.

Types, *A. angustifrons*, Dalm.
A. latus, Angelin.

2. *Basilicus*, Salter, 1849. Flattened and expanded forms, with rather broad axis; clavate glabella reaching far up the head, with only obscure lobes, the basal pair most conspicuous; approximate depressed eyes, angular or even pointed tips to the pleuræ;

[1] Except within the crust, where they may be seen faintly marked out, or even strongly in some cases.

tail with many ribs on axis and limb; facial suture marginal in front; hypostome[1] entire. Labrum deeply lobed. *Range.*—North and South Europe.

Types, *A. tyrannus*, Murch.
A. Powisii, Murch.
? *A. ingens*, Barrande.

3. *Megalaspis*, Angelin, 1852. Flattened and expanded, with narrow axis; short urceolate glabella, reaching but a little way up the head, and lobeless; approximate depressed eyes, subangular tips to pleuræ; and tail with narrow long axis, and many ribs on axis and limb. Facial suture intramarginal, in a long ogive on the upper surface. Hypostome—? Labrum notched? North Europe.

Angelin figures a labrum of *M. planilimbata* as entire, but there is some imperfection in the look of the figures, as if the forked lobes were broken off.

Types, *A. gigas*, Angelin, not of Dekay.
A. heros, Dalm.
A. extenuatus, Dalm.
A. nobilis, Barr.

[I have inserted here *A. nobilis*, Barrande, chiefly to call attention to it. It is possibly a distinct subgenus. Its short glabella is lobed posteriorly; the hypostome appears to have a vertical suture; the pleuræ are recurved and pointed; the tail-furrows not duplicate; and the crust very thin. It seems intermediate between *Basilicus* and *Megalaspis*.]

4. *Isotelus*, Dekay, 1824. Convex, moderately expanded, with broad axis; obsolete glabella; remote, moderately convex eyes; rounded pleuræ, and no ribs on the tail. Facial suture intramarginal, ogived. Hypostome vertically divided. Labrum deeply furcate. North America, North-west Europe.

Types, *I. gigas*, Dekay.
I. platycephalus, Stokes.

5. *Cryptonymus*, Eichwald, 1825. *Hemicrypturus*, Corda, *Asaphus*, Angelin. Short, convex, with narrow axis, a clavate glabella reaching the front of the head, and with three pairs of lobes; elevated eyes (sometimes stalked in the ♂); rounded pleuræ; tail short, of few joints, very obscurely ribbed, except on the axis. Facial suture intramarginal in a short ogive. Labrum deeply lobed. North Europe.

Types, *A. expansus*, Linn.
A. raniceps, Dalm.

6. *Symphysurus*, Goldfuss, 1843.[2] Short convex, with narrow axis; the gibbous, lobeless glabella overhanging the front margin; large remote depressed eyes; rounded pleuræ;

[1] In former pages I have used Dalman's old term *hypostome* for the incurved under front margin of the head, which supports the labrum; and I adhere to this, although the term *epistome* is often adopted. Barrande calls it '*la doublure frontale*.'

[2] Better defined and restricted by Angelin, 1852.

and short tail, with axis of few joints; and sides without ribs. Facial suture marginal in front. Hypostome entire. Labrum—? North Europe.

Type, *S. palpebrosus*, Dalm.

7. *Brachyaspis*, Salter, 1866. Short, broad, expanded, with moderately broad axis; the head with obsolete glabella; eyes depressed but convex, remote; pleuræ rounded; tail short, with no ribs. Facial suture marginal in front. Hypostome entire. Labrum—? North Europe.

Types, *A. rectifrons*, Portlock.
A. lævigatus, Angelin.

[8. *Nileus*, Dalman, 1826. Very convex, scarcely trilobed, with very wide axis, and hemispherical lobeless glabella; eyes very large, remote, depressed, reniform; pleuræ rounded; tail short broad, with no trace of axis or ribs. Facial suture marginal. Hypostome entire. Labrum scarcely notched. North Europe.

Type, *N. Armadillo*, Dalm.]

I believe other subgenera might easily be formed, but these are perhaps sufficient for convenient reference. For instance, *Asaphus nobilis*, Barr., and the several Mid-European species connected with it, are very doubtfully forms of *Megalaspis;* and might with advantage be kept separate. *Asaph. ingens* of the same author does not quite fit with *Basilicus;* and so of several others.

I have noticed briefly the geographical range of each of these subgenera under their definitions. The species of all the groups except *Isotelus* are European; and even this reaches our north-western shores, but is not found on the Continent. *Nileus, Symphysurus,* and *Cryptonymus* are almost exclusively Scandinavian types; *Brachyaspis* is North-European only. *Basilicus* and *Megalaspis* range over North and South Europe, and *Ptychopyge* appears to have a wider range, extending to North America, and even to India.

I do not know to what group the smooth-tailed *Asaphi* of the Andes, described by D'Orbigny, belong. The genus is evidently world-wide.

Returning over the ground just trodden, it may well be questioned if *Nileus* should be included in the list of subgenera. It not only has the broad and lobeless glabella, the wide axis to the thorax, and the smooth unfurrowed tail of *Illænus,* but it has a nearly entire labrum. It evidently leads the way to that genus.

But then it is so near *Symphysurus,* which cannot be separated, except by its inflated glabella, from *Cryptonymus,* that it is difficult to see how to draw the line. It is not so far removed from *Cryptonymus* as this is from *Isotelus,* which last may perhaps be regarded as the central type of *Asaphus.*

Brachyaspis follows next; and from this point the subgenera diverge towards *Ogygia. Megalaspis* consists of expanded forms, which yet have, on the whole, more resemblance to the typical *Asaphi* than *Basilicus*, a subgenus having the facial suture marginal in front like *Ogygia*, but yet having the deeply lobed labrum of the typical forms.

Lastly *Ptychopyge*, of which we have no certain representative in Britain,[1] has the aspect of the *Ogygiæ;* and even its intramarginal suture finds its parallel in the *O.* (*Ogygiocaris*) *dilatata*, Dalm. Yet having the labrum strongly notched, it may for the present be safely included in *Asaphus*.

Subgenus 2.—Basilicus, 1849.

1. Asaphus (Basilicus) Tyrannus, *Murchison*, Plates XXI and XXII, figs. 5—12.

Asaphus Tyrannus, *Murchison*. Silurian System, pl. xxiv, fig. 4, pl. xxv, fig. 1, 1837.
— — *Milne-Edwards*. Crustacés, vol. iii, 310, 1840.
— — *Emmerich*. Leonhard und Bronn's Jahrb., 1845.
— — *Salter*. Decade ii, Geol. Surv., pl. v, 1849 (1848).
— — *Id.* Morris's Catalogue, ed. 2, p. 100, 1854.
— — *Id.* Siluria, ed. 2, pl. 1, fig. 5, pl. ii, fig. 1, 1859. (Not of Burmeister, Org. Trilob., t. 5, fig. 4, which is *A. heros*).
Ogygia Tyrannus, *Emmerich*. Dissert. (1839).
Asaphus Tyrannus, *Id.* Leonh. und Bronn's Jahrb., p. 42, 1845.
Isotelus Tyrannus, *M'Coy*. Synops. Woodw. Pal. Foss. Fasc. i, p. 171, 1851.
Asaphus Tyrannus, *Salter*. App. Ramsay, Geol. N. Wales; Mem. Geol. Surv., vol. iii, p. 312, pl. xiii, fig. 1—6 (1866 ined.)

A. magnus, interdum pedalis, ovatus ferè, per-sculptus; capite obtuso quam caudâ elongatâ breviore. Oculi depressi approximati, sese spatio glabellæ longitudinis remoti. Glabella brevis pyriformis, lobis lateralibus supra oculum distinctis. Anguli capitis breves. Cauda parabolica, apice truncato, axe multi-annulato, lateribus multicostatis, costis 12—13 *abbreviatis.*

Sir Roderick Murchison hit upon a happy name for this fossil, the largest then known among British Trilobites, and singularly plentiful in the region which he made his own by hard work along the South Welsh frontier. It is not only our finest British species, but is peculiarly British, and even restricted to the Welsh Principality and the border-counties. Often as it has been quoted out of Britain, it has never, except in one single locality (in North Wales), been really found outside the restricted area of Shropshire and South Wales, where, as above said, it may be counted in swarms. Earl Cawdor's magnificent specimen found in Dynevor Park, Llandeilo, is here again figured from the British Museum collection, with others, of smaller size, from the Museum of Practical Geology. It is the best of types, together with the *Ogygia Buchii*, for the Llandeilo rocks proper,

[1] It was an error to insert this subgeneric name under *Ogygia corndensis*, p. 130. That is a true *Ogygia*. It is just possible *A. radiatus*, p. 154, may belong to it.

and its place is taken in strata of the same age in Sweden by the equally large *A. heros*, Dalman, a species which Burmeister and others have confounded with it.

Besides, it belongs to a group of *Asaphi*, which, by their thinner crust, narrower axis, more expanded form, and strongly furrowed pygidium, conduct easily to *Ogygia*. This subgenus was first noticed in the 'Memoirs of the Geological Survey,' Decade 2, in 1849, and has been generally adopted.

Description.—Length of large specimens eleven and a half inches; breadth more than six inches. General form true ovate, more obtuse in front, moderately convex; the head semicircular, the tail much longer, more convex, and parabolic. Head sculptured, very gently convex, with only a very slight concavity marking off the margin. The cheeks are broad, and their hinder angles produced into short spines, which only reach the third thoracic segment. The glabella is convex in front, broad-clavate, well defined forwards, only slightly so on the sides, and as long as the breadth between the eyes. The forehead-lobe large, round, and rather protuberant, especially in old specimens. Beneath this lobe, and at about the level of the eye, is the uppermost of three short, obscure, oblique lobes, the basal one of which is largest. There is at the base of the glabella a tubercle, and beneath it a shallow transverse impression marking the place of the neck-furrow. Eyes rather large, depressed, much arched, placed behind the middle of the head, and close to the glabella in this species.

The base of the eye-lobe is constricted, the lentiferous surface probably narrow, and the numerous lenses beneath the smooth cornea are not very closely set (fig. 7). Facial suture curving widely out on the posterior margin to half the width of the cheek, and in front of the eye turned outwards in a sigmoid curve to cut the front margin beyond the parallel of the eye, and continue exactly along the front edge of the shield.

The hypostome is yet unknown; the labrum large, with a semicircular base, which is moderately broad only; a squarish slightly tumid centre, surrounded by a strong furrow, in which are set obliquely a pair of minute transverse ovate tubercles. The apex is deeply divided into two ovate-lanceolate forks, between which the shelly plate turns strongly inwards; the length of the lobes is greater than one third of that of the whole labrum.

Thorax of eight moderately arched rings; the axis is not strongly marked out, and is as broad as the pleuræ; these latter are grooved for more than half their length, the strong groove bounded behind and before by strong ridges, which form a prominent node at their point of junction beyond the groove; from thence the pleuræ rather flat. They are curved down, but scarcely backward, at the obscure fulcrum, which in the hinder rings is placed one third away from the axis, but nearer to it than in the forward rings. The anterior edge of each pleura is sharpened and coarsely striate, and its termination square.

Tail parabolic, a little truncate at the end; its width at the front one fourth greater than the length (though apparently not more than equal to it); not quite regularly convex, the axis being flattened in front, narrow, scarcely one fourth the width of the tail, and prominent behind, reaching seven eighths of the whole length; it ends somewhat abruptly in a swelled tip. There are eighteen or twenty rings marked out on it, not quite reaching the tip. The sides are convex, almost tumid at first, and ribbed by twelve distinct raised ridges for about two thirds their width. These end rather abruptly, and the concave limb is quite smooth and even, except for the imbricated lines of ornament which cover the whole surface in one form or another. The lateral ribs are as broad as the furrows between them, somewhat square in section, with tumid abrupt ends. They are as numerous in young specimens (see fig. 12), and these are nearly as convex as the older ones.

The caudal fascia is broad, considerably indented by the axis, and concave, so as to leave a space between it and the outer crust. The lines of growth are coarse, and rather distant, obliquely set against the margin of the fascia (see fig. 10), not parallel to it.

Of all the *Asaphidæ* this is the most highly ornamented species. Our largest figured specimen was, indeed, named, as a variety, *ornatus* by Sir R. I. Murchison ('Sil. Syst.,' pl. xxv, fig. 1), but it is the normal condition of the old specimens, sometimes a foot long.

The sculptured lines, which are veritable plicæ or folds of the crust, are wavy and interrupted, following a general curve all round the margin, but sinuous and inosculating with each other. They follow the broad head-margin, and are concentric, fainter, and closer round the forehead-lobe of the glabella; but they are far stronger and shorter, and form deep pits on the thorax-segments and tail, strongest on the axis of each, and occur only on the prominent parts of the side-lobes, the hollows being nearly free from them. They arch over the axis from side to side, almost obliterating the annuli of the tail-axis by their strength; while on the side-lobes they radiate from behind forwards, and decussate the ridges strongly. They become transverse lines on the outer margins of the thoracic rings, except in very old specimens (fig. 8), where they form pits and puncta to the very ends of the pleuræ. On the smooth concave limb of the tail they sweep forwards in sinuous lines, strongly imbricated forwards, and cut the margin obliquely.

On half-grown specimens they are much fainter, and, indeed, can only be rarely seen in young specimens.

The caudal fascia (fig. 10) has them distant, and branching, as it were, from an imaginary line along its middle portion. But on the corresponding part beneath the thorax-rings they are quite longitudinal instead of transverse. (See Decade 2, Geol. Survey, pl. v, fig. 8.)

There are few species to compare this fine fossil with, unless it be our next two species. It is most nearly like the *Asaphus Powisii*, which occurs in Caradoc, not in Llandeilo rocks. That species, however, shows no sculpture, and is, moreover, more convex; and it has very indistinct ribs on the tail, the axis of which is wide above, and then suddenly contracted for the rest of its length. The species figured by Prof. Burmeister as *A. tyrannus* has, indeed (in his figure), a strong resemblance; but that figure is a mixture of two species, for the Professor has transferred the sculpture of the British fossil to a worn Swedish specimen of *A. heros,* Dalman, which had lost the characteristic mucro of the tail. *Isotelus marginalis,* Hall, formerly supposed by me to be the same, is quite distinct.

In comparing with other species, the following must be particularly noted as a very close ally; indeed, it has hitherto always been confounded with it. Both occur at the same locality near Llandeilo.

Localities.—LLANDEILO FLAGS. Abundant in South Wales; Llandeilo and many localities in Carmarthenshire. In Pembrokeshire; Musclewick Bay, Llampeter-Felfrey, Narberth, Mydrim, and Lann Mill. In Shropshire, Meadowtown, Shelve, &c. In North Wales; Craig-y-glyn, near Llanhaiadr, in the Berwyns. But not in Scotland, Ireland, nor in any Continental locality.

2. ASAPHUS (BASILICUS) PELTASTES, spec. nov. Pl. XXII, figs. 1—4.

A. modicus, vix 4-uncias longus, convexus, ovalis, obscurè lineatus, capite caudæ æquali. Oculi elevati, sese spatio capitis totius longitudinis remoti. Glabella elongato-pyriformis, lobis basalibus subtrigonis isolatis. Anguli longè producti. Cauda axe 14—15-*annulato, lateribus* 9—10-*costatis.*

Certainly when I figured these specimens I did not expect to be able to determine two completely distinct forms, for the general aspect is exceedingly alike in both. Yet if form and proportion, difference in position of the eyes, in the form of the glabella and its lobes, shape and number of ribs in the tail, &c., be sufficient characters, we must reckon for the future two distinct Llandeilo fossils hitherto confounded under one name.

A. peltastes is probably as common as the better known *A. tyrannus.* It is easily distinguished by the more oval, (not ovate) shape, and the equality in length of the head and tail. But also notably by the small cheeks, remote and large eyes, long and broad head-spines, and the fewer ribs to the tail. The following description may serve:

Length, judging from large fragments, not more than six inches, the breadth two and a half. Strictly oval, convex; the half-elliptic head equal in length to the tail; the body shorter than either, not exceeding the glabella in length. The glabella elongate, narrower than the cheeks; a pair of very distinct oval basal lobes (really composed of the two lower pairs), isolated from the rest of the glabella, and set wide apart. A long oval fore-head-lobe, which reaches quite to the concave front margin, and is more than half the entire length of the head, from which it is not abruptly distinguished in front as in the last species.

The cheeks are very convex, steeply bent down towards the truly flat margin, and produced behind into broad blunt spines, which are long enough to reach to the sixth thorax-ring. This character at once distinguishes *A. peltastes.* The eyes are large and prominent, much wider apart than in the last species, the width from eye to eye being greater than the whole length of the head. The eye is broad, and almost as high as broad, and much curved, sunk in a depression of the cheek, and placed hardly its own breadth from the marginal furrow (see fig. 2). Facial suture beneath the eye reaching very little outwards, and cutting the margin beneath the outer edge of the eye. In front it cuts the margin more abruptly than in *A. tyrannus.*

The thorax is generally convex, but the axis is not very strongly separated from the sides, but more so than in the allied species; and is somewhat narrower than the pleuræ, which curve down steeply from the fulcrum, placed at about one third out, and they bend more sharply backward than in *A. tyrannus:* their ends are more oblique and less square, and the pleural groove is longer, much narrower, and somewhat deeper than in that species.

The tail is half-oval, rounded, not at all truncate at the end, and with a flat rim all round, not concave. as in the last. The axis narrow, more convex, shorter, and more regularly tapering, and with fewer (fourteen or fifteen) rings. The sides are convex, forming a channel against the flat margin, and with nine or ten ribs only (rarely eleven), and these are longer, more curved, and more prominent than in *A. tyrannus.* The caudal fascia has more closely set lines.

The sculpture, too, is different. The same in character, it is much less prominent and conspicuous. This may be due to the less size of the specimens, but there is a difference in the only part where I can find it conspicuous, viz., the border of the tail, where it is much closer, more thread-like, and directed more longitudinally than in the typical species. *A. peltastes* cannot be a simple variety of *A. tyrannus;* still less can the differences be regarded as mere sexual variations. It is a good and distinct species.

Locality.—LLANDEILO FLAGS. Llandeilo, South Wales; abundant. Cabinets of Mus. Prac. Geol. (figured specimens 1 to 3); of Mr. J. Lee, Caerleon (do. fig. 4); Mr. R. Lightbody; Rev. G. Smith, of Tenby; and Mr. Edgell, who finds it abundant in the flagstones, while *A. tyrannus*[1] affects the limestone. Also in volcanic grit, Builth, Radnorshire, and from Abereiddy Bay, Pembrokeshire (Mr. H. Wyatt Edgell).

ASAPHUS? (B.) HYBRIDUS, n. sp. Pl. XXIII, figs. 8, 9.

A. (*B*) *modicus,* 4—5 *uncias longus? complanatus, axe angustissimo. Cauda, solùm cognota, rotundata semiovalis* (*long. ad lat. ut* 8 : 11) *axe contracto quàm limbo ter angustiore, vix* $\frac{2}{3}$ *longitudinis caudæ superante, apice abrupto obtuso. Sulci axales anticè profundi,*

[1] The *A. tyrannus* figured by Hoffman in his summary of the Russ. Tribolites ('Verhandl. Russ. Kais. Min. Gesellsch.,' 1858, pl. vi, fig. 3) is a *Ptychopyge.* The *Ogygia Buchii,* fig. 4, is a *Megalaspis,* like *A. heros,* Dalm.!

limbum ad hoc elevatum ab axe depresso benè separantes. Limbus paullo convexus, costatus, margine lato plano seu concavo. Costæ axales angustæ, 7—8; laterales 7 abbreviatæ, curvæ, ad basin duplicatæ.

I do not feel quite sure that this caudal portion (we possess two specimens only) represents the tail of a *Basilicus.* It may be a fossil allied to the *Ogygia Selwynii* (Pl. XVII). But on close examination the axis is so much depressed in front, and is so short, that, as our fossil cannot possibly be that ancient species, it may as well be figured with *Basilicus.*

Tail half-elliptic, blunt at the end, the length being to the breadth as 8 to 11. Axis very narrow, not above one third the width of the limb, depressed at its upper portion, where, the axal-furrows being strong here, it is sunk between the convex upper portions of the limb. Behind, the furrows vanish, and the axis is prominent above the more depressed limb: its termination is blunt, and it reaches only two thirds down the tail, or but little more.

The axis is a little wider in front, then parallel for the rest of its length. The sides, on the contrary, are convex in front and for half their width, and then flat or slightly concave. The seven lateral curved ribs extend only over the convex portion, the ribs being broad and flattened and the furrows narrow. At the base the upper ribs are duplicate, and this is the chief reason for thinking the species may be an *Ogygia.*

Locality.—Llandeilo Flags? Henllan Amgoed, Carmarthenshire.

Asaphus (Basilicus) Powisii, *Murchison.* Pl. XXIII, figs. 2—7.

Asaphus Powisii, *Murchison.* Sil. Syst., t. xxiii, fig. 9 (not *a*, *b*), body and tail only, 1837.
— — *Burmeister.* Org. Trilob., Ray ed., p. 96, and note: *Isotelus Powisii*, p. 122 (not of Portlock), 1846.
Isotelus Powisii, *M'Coy.* Pal. Foss. Woodw. Mus., p. 170, 1851.
Asaphus Powisii, *Salter.* Mem. Geol. Surv., Decade 2, t. iii, p. 5, 1849.
— — *Id.* Morris's Catal., 2nd ed., p. 100, 1854.
— — *Id.* Siluria, 2nd ed., t. ii, fig. 2, 1859.
— — *Id.* App. Ramsay, Geol. N. Wales; Mem. Geol. Surv., vol. iii, p. 312, 1866, pl. xv.

As. (*B.*) *magnus, sæpe 6-uncialis, lævis, ovalis, convexus, oculis remotioribus; caudâ axe conico, lateribus vix costatis. Glabella ad frontem rotundata, posticè elobata. Caput immarginatum, spinis brevibus. Cauda thorace longior, axe ad basin lato, dein contracto, nisi ad apicem vix elevato, costis 8—9 obscuris. Limbus lentè declivus sulcis abbreviatis 9, primo solùm profundo, reliquis obscuris, margine concavo.*

The original name bestowed by Sir R. I. Murchison is still by common consent retained for this fine fossil, although the author confounded the head of *Phacops macroura* with the species, as Burmeister first observed in his Ray edition in 1846, and as has been already noticed in our Part I, p. 37. The species is very distinct, and has nothing to do with the form described by Portlock under the name *Isotelus Powisii* in 1843; nor is it easy to see why Prof. M'Coy retained it under that subgenus, since the facial suture is exactly that of *Basilicus*. It is one of the commonest of Caradoc fossils. From *A. tyrannus* the rounded shape, faint furrows on the pygidium, and quite different head, without any furrows to the glabella, easily distinguish it.

Rarely six inches long, but still a large species, of truly elliptic, very regularly convex form; the semilunar head much shorter than the thorax, and this shorter than the tail. The glabella is only convex and rounded in front; behind, it neither shows convexity, separation from the cheeks, nor lobes of any kind. The eyes are rather large and lunate, but not elevated, and, measured from their outside, are wider apart than the length of the head; they are not at all sunk in the free cheeks, and their curved lentiferous surface contains about 7000 lenses, according to observations of mine in 1849 (Mus. P. Geology, and Collection of the late Daniel Sharpe). The cheeks are very little convex, without any marginal furrows, and produced into very short head-spines.

Thorax distinctly three-lobed, the axis broad, wider than the pleuræ, but not strongly separated from them; a punctum in the cast marks the point of junction. The pleural groove is bordered, as in *A. tyrannus*, by strong ridges, meeting beyond it in a sort of node; and it extends as a broad, well-defined furrow for more than half the length of the pleuræ. These have the fulcrum near the axis, not above one third out, and at nearly an equal distance in all the rings. From the fulcral point the pleuræ bend down, but not backward, and are somewhat recurved at the tip.

The tail is the truly characteristic portion, and has a whimsical resemblance to that of the *Homalonotus bisulcatus* (Pl. X, fig. 3), with which it is so often found in company. It is semioval, rounded at the end, and regularly convex, the axis not greatly raised above the general convexity; broad at its base, where it occupies rather more than one third the whole width of the tail, but soon narrowing behind, and becoming more parallel-sided. It reaches quite to the inner edge of the concave margin, fully five sixths down the tail, and is there rounded and a little prominent, especially in young specimens (figs. 3 and 4). These indeed have the axis generally more strongly marked out than in the adult. The axis is marked by about eight or nine obscure ribs, and the sides by an equal number of rather short and obscure furrows; but the upper or border furrow, always the deepest in the *Asaphidæ*, is here distinctly and strongly so, and hence, together with the wide pyramidal axis, the resemblance to the *Homalonotus* aforesaid. The limb is concave all round, and the caudal fascia, rarely seen, is only slightly indented by the tail-axis.

In all the above characters,—the absence of glabella-lobes, the smooth and almost unfurrowed tail, with its broader and more pyramidal axis, &c., the species is distinguished from *A. tyrannus*. The labrum (fig. 6) is also different; longer, more indented on the sides, less raised in the centre, with very oblique lateral tubercles and shorter forks to the apex.

Localities. — LLANDEILO FLAGS. Treiorwerth, near Llanerchymedd, Anglesea. S. E. side of Arenig Mountain. Llangadoc, S. Wales. CARADOC or BALA ROCKS, everywhere in North Wales: Snowdon; Bettws-y-coed; Bala; Dinas Mowddwy; Meifod; Berwyn Mountains, &c.: fig. 5 is a fine specimen from near Waterloo Bridge in the Vale of Conway; and fig. 7 shows a specimen, distorted by cleavage, from Moel Benddu, Dinas Mowddwy, (both in Mus. P. Geol.). Figs. 3, 4, and 6 are from Horderly and the neighbourhood. It is extremely common in the CARADOC of Shropshire.

Foreign Localities.—None. It is, like *A. tyrannus,* a truly British species.

ASAPHUS (BASILICUS) MARSTONI, n. sp. Pl. XXIII, fig. 1.

A. parvus, vix uncialis, complanatus, ovato-oblongus, oculis magnis capite caudâque thorace brevioribus. Caput vix marginatum (angulis brevispinosis ?), oculis magnis remotis. Glabella anticè prominens, marginem ferè attingens, haud lobata? Cauda ut in A. Powisii, *nisi brevior, subtrigona, axe angustiori costis distinctis, lateribus latimarginatis, costisque utrinque 7 brevibus.*

This neat little species is evidently distinct from any other British form, nor do I think it has been figured in European works. It might have passed as a young *A. Powisii,* but for the great size of the eyes and the distinctness of the glabella; for the less relative size of the tail and the strength of its furrows would hardly be enough to distinguish it, as these are the usual characters of young specimens. I think the larger fragment belongs to it, but I shall describe the smaller one only.

Half an inch long, and three tenths broad at the base of the head—the widest part; oblong-ovate, blunt in front and rather pointed behind; the convex head as long as the thorax, and each of them much longer than the tail. Glabella large, oblong, reaching nearly to the front edge and almost overhanging it; about one third the width of the head. Cheeks strongly margined, bearing the large lunate eyes close to the glabella, and much nearer the hinder margin of the head than in *A. Powisii;* the facial suture beneath them not extending half-way out. Angles blunt? (or with short spines).

Thorax with a rather narrow axis, which tapers backwards, the rings somewhat nodose. Fulcrum of the nearly direct pleuræ placed beyond one third out, and strongly grooved. The tail is half ovate, somewhat triangular in outline, with a narrow conical axis reaching to the strongly concave margin, and sharply separated from the somewhat depressed sides. The axis has six or seven distinct rings (nodose, like the hinder thorax-rings) on the sides and middle, and the limb has about six abbreviated furrows, besides the upper strong furrow which reaches, much beyond the others, nearly to the margin. The concave margin has some concentric striæ, which may be, and probably are, pressed through the cast from the caudal fascia beneath.

Locality.—CARADOC. Lower shale, under the sandstones of Horderly, Shropshire. (discovered by Mr. A. Marston, and now in Mr. H. Wyatt Edgell's cabinet).

ASAPHUS (BASILICUS?) RADIATUS, *Salter*. Pl. XVIII, figs. 1—5.

OGYGIA RADIATA, *Salter*. Appendix to Ray edition of Burmeister, Org. Tril., p. 125, 1846 (and quoted by M'Coy).
ISOTELUS (BASILICUS) LATICOSTATUS, *M'Coy*, in part. Synopsis Woodwardian Fossils, pl. i E, fig. 18 (not fig. 18 *a*, or of Green), 1853.
ASAPHUS RADIATUS, *Salter*. App. Ramsay, Geol. N. Wales; Mem. Geol. Surv., vol. iii, p. 311, pl. xxiii, fig. 7, 1866.

A. subplanus, 7—8 *uncias longus?, obtusè ovatus, axe angusto, capite latissimo, glabellâ distinctâ. Thorax brevis dimidium caudæ efficiens. Cauda semicircularis lata, axe angusto abbreviato* 8-*annulato, vix per* ⅔ *caudæ producto, apice prominulo; limbo utrinque costis* 7—8 *abbreviatis tumidis radiato. Sulci æquales, recti, subsinuosi. Margo anticus valdè sinuosus. Superficies striis obliquis striata, perornata; fascia interna latissima, lineis remotis rugosa.*

This elegant and common species has not yet attracted much attention; the first specimen, found at Bala by myself in a pleasant summer's work with Prof. Sedgwick in 1844, was without much doubt referred to *Ogygia* in the MS. catalogue of British forms supplied to the Ray edition of Burmeister[1] by myself. A hasty field-note had previously assigned our fig. 2 to the *O. Buchii*, a species which does not occur in North Wales.

Prof. M'Coy supposed he could identify this, and another different species, with the cast published by Green. It is quite certain they are not the same, and, indeed, that fossil is a *Phacops* (see note, page 158). And though I do not admit his strong criticism as to the generic name I first gave it, for *Ogygia* has often simple furrows, it is more likely an *Asaphus*, of the group *Basilicus*, or of that to which *A. nobilis* of Bohemia belongs.

M'Coy describes the entire form as "obtusely oval; the cephalic shield rounded, about three times as wide as long, the thorax shorter than the head—only a little more than half the length of the tail—of eight slender segments; its axis rather less than two thirds the width of the pleuræ, which are nearly straight, slightly bent downwards and backwards; the tail nearly semicircular, length more than half the width, and one third" (it is nearly twice) "longer than the thorax." The rest of his description applies to the *A. laticostatus* (fig. 6), and not to our species.

I may add that the glabella is widest in front, and the head with a distinct margin; the neck-segment narrow, but strong. The pleuræ well grooved throughout. The tail has a curiously sinuated front margin, the outer portion at the angle almost forming a

[1] A most inaccurate edition it is, especially in the introductory portion, the German terms being misunderstood by the translator. Burmeister's descriptions are intelligible enough in the original.

lobe. The axis runs about two thirds down the tail, is narrow, not half the width of the side-lobes, of about eight distinct ribs and a terminal portion, all strongly covered by arched striæ, the tip quite prominent (*a* in fig. 1). The sides are radiated by eight equal strong furrows, which run direct to about two thirds to the flat margin, but are sinuated in their course. The five hinder ones scarcely reach half way. The whole tail (as shown in the splendid specimen, fig. 1, in the collection of Mr. Blunt, of Shrewsbury, who has had it for twenty years) is covered with a strong oblique striation. And the incurved fascia beneath is the widest and most coarsely ribbed that I know in any British species. It resembles that of the Bohemian fossils quoted above. *Asaphus nobilis*, Barr., however, has a larger cephalic shield and longer in proportion, and the tail has a longer and slenderer axis, more strongly ornamented. The side-furrows, which in ours are crowded at their origin, and so appear radiated, are in the Bohemian fossil more parallel.

There is an *Asaphus* (referred to *Ptychopyge*) described by M. Lawrow in the 'Verhandl. der Russ. Kais. Min. Gesellschaft,' 1858; pl. xiii, fig 2, which has radiating ribs (but fewer) and a similar sculpture.

Localities.—CARADOC only: Rhiwlas, near Bala (figs. 2, 3, Woodw. Mus.), Berwyn Mountains (Mr. Blunt's collection, fig. 1). Co. Louth, Ireland (figs. 4, 5, Mus. Pract. Geology).

ASAPHUS (B.) LATICOSTATUS. Pl. XVIII, fig. 6.

ISOTELUS (BASILICUS) LATICOSTATUS, *M'Coy.* Synopsis Woodw. Museum, pl. i E, fig. 18 *a*, 1851 (not of Green's Monogr. p. 45, which is a Phacops).

ASAPHUS LATICOSTATUS, *Salter.* Morris's Catal., 2nd ed., 1854, p. 100 (Builth locality only).

Asaph. (*B.*) *planus*, 5 *uncias longus? Cauda lentè convexa, semiovata, obtusa, margine angusto striato, haud concavo. Axis angustus longiconicus* $\frac{5}{6}$ *caudæ longus, sulcis axalibus exaratis bene notatus, quartamque partem latitudinis efficiens, annulis rectis* 10 *planatis, per medium interruptis; apice haud abrupto. Limbus lente convexus, costis* 10 *planis marginem ferè attingentibus, sulcisque profundis, ad apices recurvatis. Fascia angustissima.*

We only know the tail of this remarkable species, which Prof. M'Coy has unaccountably referred to the *Asaphus laticostatus* of Green. The American cast (No. 13 of the Monograph) shows that species to be a large *Dalmania*, which we have lately found in Britain.[1] Prof. Green himself rightly compared his species with the *Phacops* (*Dalm.*)

[1] In Mr. H. W. Edgell's cabinet. The specimen seems to be from Upper Llandovery rocks. Green's specimen has no locality, but he quotes it from the Helderberg (Ludlow) series.

Hausmanni, a Bohemian Upper Silurian fossil. There can be no reason, however, for instituting a new name, as that of *laticostatus* is not occupied for the present genus. *Asaphus radiatus*, with which Prof. M'Coy united it, has been described above, and is a totally different species, as our figures will show.

Tail obtusely half-oval, not semicircular, gently convex; length two and a quarter inches, width three and a quarter inches. The long-conical flat axis, with straight sides, occupies ten lines in breadth at top, and rather quickly and regularly tapers backwards for five sixths the length of the tail; it is not clearly marked out at the tip or at all prominent there. It is annulated by ten flat rings, which run direct across, and are interrupted in the middle (as in several species of *Phacops*), so as to leave a line of connecting points down the centre of the axis. The furrows which separate the axis from the limb are deep, though narrow, the limb itself gently convex to the very margin, and scored deeply by eleven furrows, which are at first direct or only slightly oblique, and radiate outwards, then at the apices enlarge and turn backwards rather abruptly. They reach nearly to the margin, and invade the striated border, which is in this species much narrower than usual in the genus. I suppose the caudal fascia has impressed itself upon the upper surface, as is frequently the case. But this character will serve well whereby to contrast the present with the next species. M'Coy's figure is not satisfactory, and we have corrected it very carefully from the original. His figure has the axis too narrow at top, and the furrows across it entire, not interrupted in the middle, nor does it shew well the character of the lateral furrows, especially the upper one.

This species differs at once from *A. radiatus* (fig. 1) in the narrower, longer form of tail and greater number of furrows. The furrows, too, are curved, not straight, and the striate border narrow, not broad, as in that species. It may be a *Ptychopyge*, but I think not.

Locality.—LLANDEILO FLAG. Maen Goran, Builth, Radnorshire (Woodwardian Museum).

[I must be pardoned for introducing here a figure or two of a species before described, and figured in Pl. XVI. The specimens are pressed in different directions, that from Builth (Woodcut, Fig. 32) being shortened by pressure, and that from the collection of Mr. Pardoe (Fig. 33) being narrowed by the same process—cleavage-action.]

The extreme similarity of caudal shields of this fine fossil (*Ogygia corndensis*, Murch.) to the fossil last described (*Asaph. laticostatus*) has rendered it necessary to show how the two may be distinguished, when found in company, as they are very likely to be often found, since they occur in neighbouring localities of the same rock.

The caudal shield much resembles that figured as *A. laticostatus*, but it differs from it in the greater width of the tail, and especially of the axis, which has continuous non-interrupted furrows across it; and also in the undulated, convex and not flat, ribs of the limb. The caudal fascia, too, is twice as broad. *O. corndensis* is, moreover, a larger species, and must have measured full five and a half inches in length when perfect.

FIG. 32.

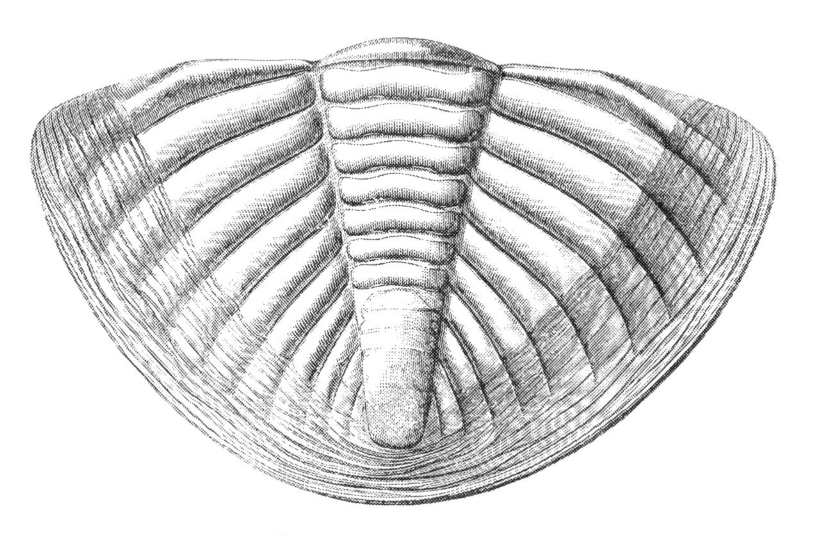

Ogygia corndensis, Murchison.

Very perfect caudal portion, from the Lower Llandeilo Flags, Builth, Radnorshire (Mus. P. Geology).

Tail two and a half inches long and four and a half inches wide, semicircular, gently convex all over, with the conical axis broader than in *A. laticostatus*, not one fourth of the whole width of the tail, and tapering gradually backwards to a very obscure apex, which reaches to five sixths of the whole length. It has about eight or ten rings, a little more curved than in the last species, not quite direct across, but not interrupted in the middle by any line of connected points. The axal furrows are shallow, and not very well defined in this fossil, while in the last they are deep and strong.

FIG. 33.

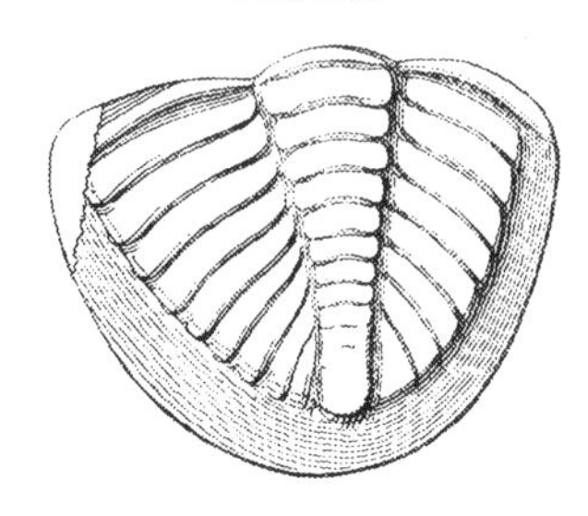

O. corndensis, Murch.

Small specimen from felspathic rock, probably a boulder. The late Mr. Pardoe's collection.

The sides show a broad gently convex surface, marked by eight gently undulating convex ribs, the furrows between which are shallow, and curved backward nearly to reach the margin, which is a little concave only just at the end of the tail, not on the sides. The furrows seem to radiate just as in the last species, but are more regularly curved and not bent back so suddenly as in that species. The caudal fascia is half as broad as the limb and coarsely striated throughout. The other figures in Pl. XVI will sufficiently shew the characters of the species.

Locality.—LLANDEILO FLAGS of Penrhiw, four and a half miles north-east of Builth. Collected by Miss Thomas, of Penkerrig, and presented to the Mus. P. Geol. several years back. The species is abundant at Gilwern, near Llandrindod; and also at Min Cop, Meadowtown, in the Corndon district, whence the species was named.

As reference has been made under the last two or three species to subgenera not yet *distinctly* recognised in Britain, I subjoin a figure of the subgenus *Ptychopyge*, Angelin, and also one of *Megalaspis*, to shew the extravagant form sometimes assumed by the latter group.

FIG. 34.

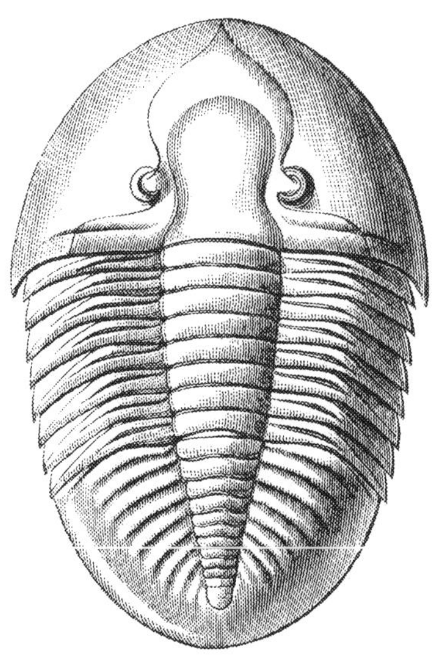

Asaphus (*Ptychopyge*) *latus*, Angelin.
'Pal. Suec.,' t. xxxi, Fogelsång, Scania.
Lower Silurian.

FIG. 35.

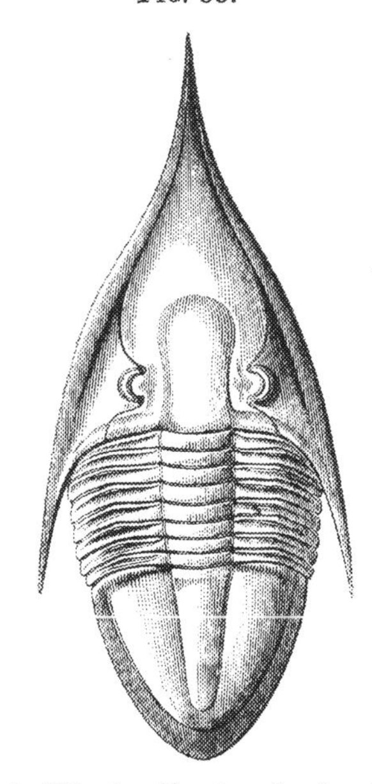

As. (*Megalaspis*) *extenuatus*, Angelin.
'Pal. Suec.,' t. xi, fig. 1, Ostrogothia. L. Silurian.

ISOTELUS, *De Kay.*, 1824.

ASAPHUS (ISOTELUS) GIGAS, *De Kay*. Pl. XXIV, figs. 1—5; Pl. XXV, fig. 1 (*var.*).

ASAPHUS PLATYCEPHALUS, *Stokes*. Trans. Geol. Soc., 1st series, vol. viii, p. 208, pl. xxvii, 1822.

ISOTELUS GIGAS and I. PLANUS, *De Kay*. Annals of Lyceum New York, vol. i, p. 176, pls. xii and xiii, fig. 1, 1824.

ASAPHUS GIGAS, *Dalman*. Palæadæ, p. 71, 1826.

ISOTELUS GIGAS, *Green*. Monograph Tril., 67. I. PLANUS; I. STEGOPS, *Ib.*, pp. 68—71. I. CYCLOPS; I. MEGALOPS, *Ibid.*, 1832.

BRONGNIARTIA ISOTELA, *Eaton*. Geol. Text-book, pl. ii, fig. 19, 1832.

ASAPHUS PLATYCEPHALUS, *Bronn.* Lethæa, vol. i, p. 115, pl. ix, fig. 8, 1835.
— GIGAS, *Emmrich.* Dissert. 32, 1839.
ISOTELUS GIGAS, *Milne-Edwards.* Crust., vol. iii, p. 298, 1840.
ASAPHUS PLATYCEPHALUS, *Buckland.* Bridgw. Treat., vol. 2, p. 76, 1840.
— — *Burmeister.* Org. Tril., p. 127, pl. ii, fig. 12, 1843; and Ray edition, p. 110, 1846.
ISOTELUS GIGAS; I. PLANUS, *Portlock.* Geol. Report, p. 295, pl. viii, fig. 7; pl. vii, figs. 2, 3 (except pl. viii, figs. 2, 3,). I. OVATUS; I. SCLEROPS. *Id.*, Pl. viii, fig. 5; pl. x, fig. 2. I. POWISII (not of 'Sil. Syst.') *Id.*, p. 297, pl. vi, fig. 1, 1843.
— MEGISTOS, *Locke.* Americ. Journ. Science, vol. xiii, p. 366, 1842; Trans. Assoc. Amer. Nat. and Geol., vol. i, pl. vi, 1843.
— GIGAS, *Hall.* Pal. New York, vol. i, p. 231, pl. lx, fig. 7; pl. lxi, figs. 3, 4; pl. lxii, figs. 1, 2; pl. lxiii, 1847.
— — *Billings.* Geol. Canada, p. 184, fig. 182. *A. platycephalus*, ib., fig. 183? 1863.
ASAPHUS (ISOTELUS) GIGAS, *Salter.* Mem. Geol. Surv., Decade xi, sec. 3, p. 1, pl. iii, 1864.

A. (*Isot.*) *ovato-oblongus, lævis, lateribus rectis; capite pygidioque ejusdem magnitudinis, subtrigonis et hyperbolicis. Capitis sulci axales minimè profundi. Oculi modici et pone medium caput. Sutura facialis intrâ marginem, cui parallela est; labrum ad basim angustum, in cornua longa parallela productum. Thorax axe pleurisque ejusdem latitudinis, fulcro ad tertias posito. Caudæ axis indistinctus conicus longusque; in latere quoque sulcus superior tantùm videtur, cæteri desunt; oculi modici pone medium caput.*

One sometimes persuades oneself of a thing by continually asserting it, and as this species has been continually quoted of late years as the *I. gigas*, it has come to be generally accepted as such. There are some differences from American specimens, however, which may possibly cause its separation when we have more abundant material. I have used General Portlock's original specimens, and, believing the species identical with the common American form, follow most writers in adopting De Kay's name. It must have been published at nearly the same time as Mr. Stokes' name *platycephalus*, though the paper of the latter, in the Geol. Trans., was read early in 1823. Moreover the fossil described by Stokes has a broader form and smaller eyes than the narrower one commonly known as *I. gigas*. But if these differences be regarded as due to sex, the *I. gigas* being the ♂, and *I. platycephalus* the ♀ form, there would be no difficulty in accepting all as of one species.

In the Decade XI of the Geol. Survey, I have fully described this fossil, and I do not know that I can do better than quote that description. I look upon all Professor Green's casts as one species. Professor Hall does the same; and Portlock's difficulties arose out of an attempt to identify his variably preserved specimens with the named casts of Professor Green. There is no occasion here to distinguish *I. gigas* from *I. Powisii;* they belong to different subgenera.

I would only repeat here that there is no warrant, in any fragments I have seen or heard of, for supposing the great *I. megistos* of American authors to be so large as the cast sold for lecture-purposes would indicate. Twenty-one inches is not too long for the largest *Paradoxides*, but is far too much for any species of *Asaphus*.

"General shape oval-oblong, with the sides rather straight; the head and tail nearly equal, and both subtriangular; the head pointed, the tail more obtuse at the tip; the surface is convex, a line taken from the snout to the apex of the tail being a regularly convex one, uninterrupted by neck-furrow, depression or convexity of the smooth and even body-rings, or by furrows on the axis of the tail. The axal furrows are very obscure in the head; they are neatly marked, but shallow, along the body, and only very faint along the tail; all the surface is smooth, and the sides are strongly deflected, but not steep The head has the shape of a broad and pointed Gothic arch, the breadth at base being to the length as three to two; the margin is very narrow and flat, not at all recurved; the facial suture, forming a broad ogive arch in front, runs for some distance close within and parallel to the front margin; and beneath the eyes,—which are large, placed near the glabella, and rather behind the middle of the head—the suture curves gently out and cuts the hinder margin midway; the head-angles are blunt-pointed, not rounded. On the under side of the cheek, near the angle, is a convex space containing an oval depression, which receives the apices of the front pleuræ in rolling up (fig. 6; see also fig. 5 for the cast of this depression on the matrix). The labrum (fig. 7) has a narrow base, then a strong constriction, and thence the sides are parallel; its apex is deeply furcate, the parallel forks occupying nearly half the entire length of the organ. Body-rings smooth, rounded at the apices, deflexed at the fulcrum, which is placed rather beyond one third, and with a broad strong groove. Tail subtrigonal, with straight sides and rounded blunt tip; the faint axis rapidly tapering, broad-conical, and reaching three quarters the length; sides quite smooth. In young specimens (says Professor Hall) the tail is more pointed, and exhibits eight faint articulations; in older specimens these increase in number, but the crust presents many traces of them when viewed from within, and they are often distinct (Hall, Pal. New York, *loc. cit.*, p. 231). Dr. Burmeister also calls attention to this character, which is, indeed, no more, as we shall see by and bye, than occurs in most of the smooth-shelled genera, the lobes being only obscured, not destroyed, by the even contour of the crust. *Illænus* shows this well, not only on the axis, but even in the inflated glabella. I have not seen internal lobes to the glabella in this species, but Dr. Burmeister figures them as strongly expressed."

Varieties.—If I am right in connecting the two forms above given—viz., *A. platycephalus*, Stokes, and *I. gigas*, there are two very distinct varieties of this species—the one, *A. platycephalus*, with "broad form and with small eyes; the other, of elongate form, and with variably large eyes, to which nearly all the above synonyms belong." I have quoted the best figures of these from the 'Geology of Canada,' (1863), by Sir W. Logan and E. Billings. But I have no difficulty in referring these differences to sex, the common form, *I. gigas*, being the ♂.

Localities.—CARADOC ROCKS. Desertcreat, &c., Tyrone, Mus. P. Geology. The species ranges all through the North American Continent, in Lower Silurian rocks.

ASAPHUS (ISOTELUS) AFFINIS, *M'Coy*. Pl. XXIV, figs. 13—14.

ISOTELUS AFFINIS, *M'Coy*. Synops. Foss. Woodw. Mus., pl. i F, fig. 3, 1851.
ASAPHUS — *Salter*. App., Ramsay Geol. N. Wales, Mem. Geol. Surv., vol. iii, p. 310, pl. viii, fig. 15 (pl. xii, fig. 4 ?), 1866.

A. (*Isot.*) *longè ovatus, capite rotundato, spiculis brevibus. Oculi submedii parvi. Sutura facialis ad frontem marginalis. Axis caudæ lævissimæ longus, et apice vix prominulo tenus prominens.*

There are possibly two species in the Upper Tremadoc Rocks, and I retain for one of them the name given by Prof. M'Coy to an imperfect specimen, without head and without apex to the tail, but which, on comparison, seems to agree (the figure does not) with our Tremadoc species. In case the *A. Homfrayi*, fully described by me in the volume of the 'Memoirs' above quoted, should prove to be identical with this, Prof. M'Coy's name must give way, as it is quite impossible to determine his specimen. Still, the great difference in the facial suture in front, rendering this species a sort of link with *Basilicus*, ought to distinguish it from the next, to which it has the closest resemblance. Were the two alive, we should probably find them totally different in colour. But palæontologists have no such advantages in pursuing their studies, and must make the most they can of form.

Oblong-oval, two and a half inches long, of which the head measures more than a third, and is rounded in front; the glabella is very obscurely marked out, and has no furrows: it is most convex forwards. The facial suture curves boldly out above the eye, and cuts the margin considerably outside of it. Below the eye, which is of moderate size, more than half-way up the cheek, and close to the glabella,—the suture again curves largely out. No distinct neck-furrow is to be seen; and this is usual in *Isotelus*.

Thorax shorter than the head, but as long as the caudal shield. The axis broader than the sides, and not strongly separated from them. Pleuræ flat as far as the fulcrum (which is at one-half in the middle rings), then curved rather than bent down. The facet broad and very well marked. Pleural groove obscure, except just beneath the fulcrum.

Caudal shield larger than a semicircle, but proportionately rather short, without any segmental furrows on the sides, except the uppermost. Axis narrow, tapering, well defined, and reaching three fourths the length of the shield; its end not prominent. Incurved fascia broad, striate.

The under side of the head has the broad epistome well divided by a strong vertical furrow, so characteristic of this subgenus; the labrum is subquadrate, broader above and narrower towards the truncate tip. It appears not to be at all bilobed (unless it be imperfect in our specimen; but if it be not, *A. affinis* does not belong to *Isotelus*, but must form the type of an entirely new subgenus). It resembles some of the Swedish species of *Cryptonymus*, but the course of the facial suture and the indistinct glabella will distinguish it from these; while from the species of *Isotelus* the well-marked-out axis of the caudal shield will easily separate this primordial species.

Locality.—Upper Tremadoc Slate. Pen-y-clogwyn, S. of Portmadoc, in flinty slate, much compressed by cleavage; at Garth, Penrhyn; and near Llanerch, by the roadside towards Treflys. Prof. Sedgwick's specimen came from the slopes above Tremadoc; and I have gathered this species at Tyddyn-dicwm, on the Tremadoc mountain-side.

[This and the following are the earliest true species of *Asaphus* known; if, as above said, the labrum do not betray an affinity with *Ogygia*.]

Asaphus (Isotelus) Homfrayi, *Salter*. Pl. XXIV, figs. 6—12.

Asaphus Homfrayi, *Salter*. App., Ramsay Geol. N. Wales, Mem. Geol. Surv., vol. iii, p. 311, pl. viii, figs. 11—14, 1866.

A. (Isot.) longè ovatus, lentè. convexus, capite ante subangulato, spiculis brevibus. Sutura facialis intramarginalis. Oculi submedii, parvi. Axis caudæ longus, ad apicem prominulus.

This is a more common species than the last, and is, indeed, not infrequent in the Tremadoc district. It is conspicuous too, being three inches long and one and a half broad. Compressed as our fine specimen from Mr. Homfray's cabinet is, it measures three and one third inches by three quarters of an inch. The Museum of Practical Geology has a fine series, presented by the same gentleman, to whom geological science is indebted for much hard work in the Tremadoc district.

The head is more than a third of the whole length, and longer than the thorax, which, in its turn, is longer than the caudal shield. The head is semi-oval, rather pointed in front, and has very short posterior spines; it is broadly depressed round the margin. The glabellar portion is scarcely marked out; the eyes are placed nearly half way up the head; they are small (two lines long), the facial sutures curving out boldly beneath them, and cutting the posterior margin more than half-way out from the axis. Above the eye they form a narrow ogive, and nearly follow the front margin. On the under-side of the head the vertical furrow on the epistome shows distinctly through the cast. The labrum is imperfect, but exhibits a strong marginal groove and two small lateral furrows.

The body-rings have the axis as broad as the sides, and moderately convex. The pleuræ are flat as far as the fulcrum, truncate at their ends, and have but a slight groove, which reaches only two thirds of the length. The fulcrum is at one third in front, and less than half-way out in the middle rings.

The caudal axis extends three fourths down the smooth tail, very indistinctly marked above, but in some specimens crossed by several faint rings, and is always prominent at the tip (see figs. 10, 11).

This has the characteristic facial suture of *Isotelus;* but if its labrum be like that of *A. affinis,* a point which is not yet known, it may belong to quite a distinct subgenus.

A. Homfrayi differs from its associate *A. affinis,* not merely in the facial suture, but in the much less pronounced axis of the tail; and from *A. Powisii,* with which it might be confounded by casual observers, the want of any tail-furrows, except obscure ones, on the axis, and the very slight convexity of the glabella, will easily separate it. Besides, it grew to only half the size of that Caradoc species.

Localities.—Upper Tremadoc Slate. Under Garth, near Tremadoc. Railway behind Penmorfa village. Tu-hwnt-yr-bwlch, near Portmadoc (cabinet of Messrs. Ash and Homfray, and Mus. Pract. Geol.).

§ Brachyaspis, *Salter,* 1866.

Asaphus (Isotelus) rectifrons, *Portlock.*[1] Pl. XXV, figs. 6—10.

Isotelus rectifrons, *Portlock.* Geol. Rep., p. 298, pl. ix, fig. 1, *a, b,* and pl. viii, figs. 2, 3, 7 (head), 1843.
— arcuatus, *Id.* Ib., pl. ix, figs. 2, 3 (tail).
— rectifrons, *Salter.* Mem. Geol. Surv., Decades No. xi, sec. 3, p. 4, 1864.
[Compare *A. lævigatus,* Angelin, Pal. Suecica, t. 29, fig. 1, 1852.]

A. (Brach.) latè ovatus, lævis, obtusus, cui caput caudaque transversæ, convexæ sunt. Caput breve, vix semicirculatum, angulis obtusis deflexis. Oculi arcuati, procul pone medium caput positi. Sutura facialis postica dimidium genæ efficit, anterior axi parallela est. Thoracis axis latus, vix á pleuris arcuatis distinctus, fulcro propinquo. Cauda brevis (long : lat : : 7 : 12), axe lato, 11 *annulis minimè profundis distinctus, haud prominente, sed in apicem prominulum desinente: latera, nisi sulco superno lato exarata, æqualia, limbumque angustum planiorem gerentia. Fascia modica, striis distantibus.*

I venture to consider this the type of a distinct subgenus, intermediate between *Isotelus* and *Cryptonymus;* the extremely short form of the head and tail, and the marginal suture in front, distinguishing it easily. *A. lævigatus,* Angelin, figured in the next page, belongs to the same group.

[1] General Portlock himself said that the species should form a new group, intermediate between *Asaphus* proper and *Isotelus.*

Head rather shorter than semicircular, gently convex; the angles rounded, but rather produced and incurved. The space between the eyes equal to the length of the head, the large, curved, semicircular corneæ being placed at one third from the base, about their own length distant from it; the eye-lobe horizontal, flattened above, and surrounded by the very convex lentiferous surface. The glabella quite obsolete, only marked by a deep punctation below the eye. The facial suture nearly straight above the eye, curving gently out, and cutting the front margin, along which it runs. Below the eye it runs outwards much beyond the eye about halfway along the cheek, and parallel to the posterior margin. On the underside of the head a narrow, convex, incurved margin shows the deep narrow pit above the angle for the reception of the pleuræ in rolling, and in front has a wide flat hypostome, divided by no vertical suture as in *Isotelus,* and showing a rather narrow base of attachment for the labrum, which is not yet known.

Of the body-rings we have only five, and these have a very broad axis, much wider than the pleuræ, and only divided from them by a shallow sulcus. They are much curved forward at their rounded ends. The fulcrum lies near the axis, about one fourth out.

The tail (*I. arcuatus,* Portlock) is wider than a semicircle. The upper angles are much bent down for the facet. The axis, marked out at its origin by two deep impressions, is at this part rather wider than the side-lobes. But from this point backwards it is not indicated, except by a slight prominence at the apex, which occurs at three-fourths the length of the tail. A broad shallow furrow beneath the fulcrum is all the marking that shows on the smooth convex sides. Caudal fascia concave, so narrow that its edge is not indented by the point of the axis; it is strongly lineated, the lines abutting sharply against the inner margin. *I. intermedius,* of Portlock, was formerly quoted by me wrongly as a synonym. But General Portlock erred in comparing this species with *A. tyrannus,* which it does not in the least resemble.

Fig. 36.

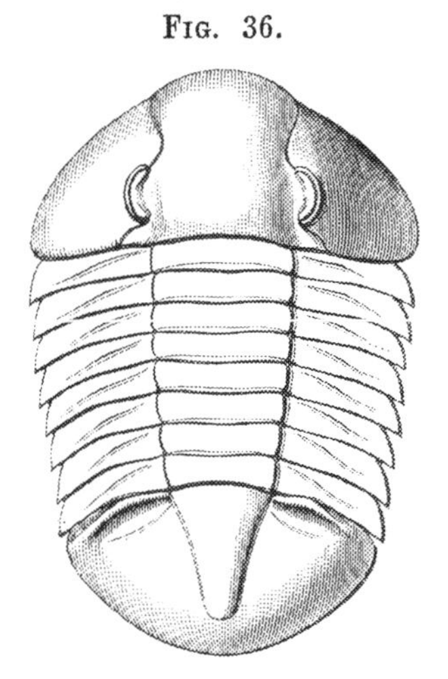

Asaphus Brachyaspis) *lævigatus,* Angelin, Lower Silurian, Sweden. From the 'Palæontologia Suecica,' pl. xxix, fig. 1. (The thorax is imaginary.)

But, if compared with the recently published *Asaphus lævigatus* of Angelin, the resemblance is so close that, were it not for considerable differences in the proportions of the head, and especially in the course of the facial suture beneath the eye, I should have combined that species with ours. They are, however, clearly distinct. We must include *A. lævigatus* in the same subgenus, and I give a figure of that species, which may serve to illustrate the subgenus *Brachyaspis.*

Other Species of the Subgenus.

To render our account complete, I must quote from the Decade XI, Geol. Survey, hoping in an Appendix to fill up these desiderata.

ASAPHUS (ISOTELUS), *n. sp.* *Salter.* Quart. Geol. Journ., Vol. VII, Pl. VIII, fig. 2.

A caudal portion of an undetermined species, figured by myself for Sir R. I. Murchison's paper on the Silurian rocks of the South of Scotland. Probably distinct from *I. gigas.*

Locality.—LLANDEILO FLAGS (?). Bogang or Bugan, Knockdolian, Ayrshire.

ASAPHUS (ISOTELUS ?), *sp.* *Salter.* Quart. Geol. Journ., Vol. IV, p. 205.

Quoted in Professor Nicol's paper on the Peebleshire Silurians. The specimen is now lost, but was stated to be allied to *A. gigas.*

Locality.—CARADOC. Wrac limestone, Peeblesshire.

CRYPTONYMUS, *Eichwald,* 1825.

The typical species of this section is the well-known *Asaphus expansus* of the Swedish and Russian works, of which a figure is given below. But though our British species does not show the lobes of the glabella, it has the hourglass-shape of that portion; the facial suture in a low ogive close to the contracted front margin; and the tail-axis with many obscure joints marked out within the crust. Lest, however, I may be mistaken in referring it to that subgenus, the figure will show what the type should be. I regard the British fossil as a link between *Cryptonymus* and *Isotelus.*

FIG. 37.

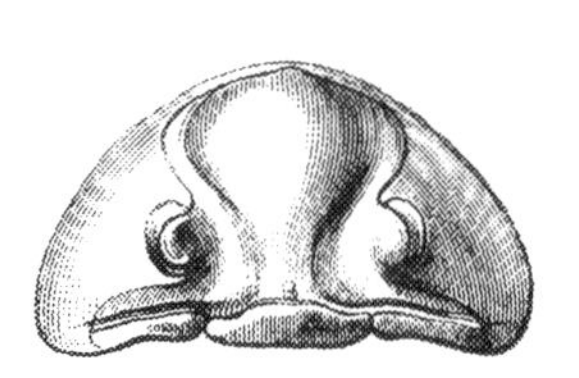

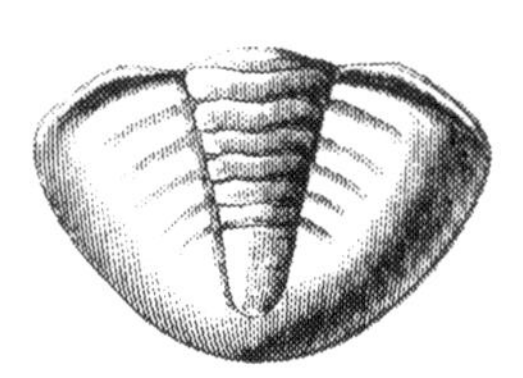

Asaphus (*Cryptonymus*) *expansus,* Russian variety. Lower Silurian, N. Russia. Mus. Brit.

FIG. 38.

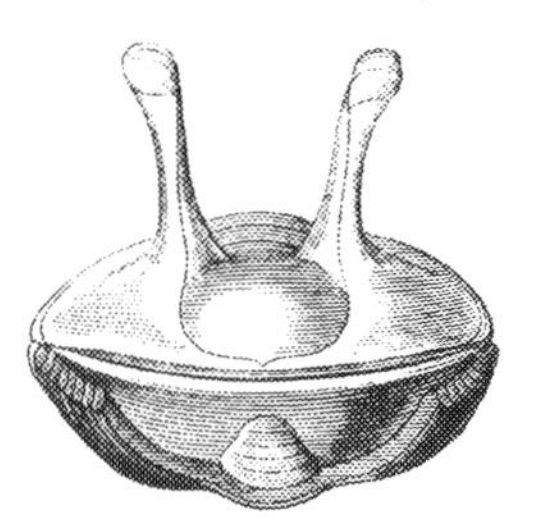

Asaphus (Crypt.) Kowalewskii, Lower Silurian. Pulkowka, Russia. (From Lawrow.)

ASAPHUS (CRYPTONYMUS) SCUTALIS, *Salter*. Pl. XXV, figs. 2 and 3.

ISOTELUS LÆVICEPS, *Portl.* Geol. Report, p. 299, t. ix, fig. 4 (not of *Dalman*, Paleadæ., t. iv, fig. 1), 1843.
ASAPHUS (ISOT.) LÆVICEPS? *Salter*, in Morris's Catal., 2nd edition, p. 100, including wrongly *I. intermedius*, Portl.; see *I. gigas*, 1854.
— (CRYPTON.) LÆVICEPS? *Ib.*, in Decade xi, Geol. Survey, under the description of *A. gigas*, sec. 3, p. 4, 1864.
— SCUTALIS, *Salter*, in the Catalogue of Fossils in the Mus. P. Geology, p. 5, 1864.

A. (Crypt.?) modicus, triuncis, latè ovatus, convexus, caudâ obtusâ, capite subangulato. Caput caudâ longius (long. $1\frac{1}{2}$ *unc., lat. 2 uncias), angulis obtusis. Glabella non lobata, lævissima, in medio capite constricta, et sulcis axalibus pone oculos profundis convergentibus, antè minimè profundis et divergentibus. Oculi ex medio capite prominentes, propiores. Suturæ faciales posticæ valdè arcuatæ, pæne angulum attingentes. Thorax axe lato convexo, pleuris brevibus. Cauda semicirculata ferè; axe conico prominente sub cortice annulato, per* $\frac{2}{3}$ *caudæ extenso. Latera caudæ lævia, sulco superno solum distincto.*

The *Asaphus læviceps* of Dalman's Palæadæ, T. IV, fig. 1, though in some general characters like this species, is quite another thing really, and is a species of the subgenus *Symphysurus* rather than of *Cryptonymus* or *Isotelus*, as Professor Burmeister pointed out. It is near to *A. palpebrosus*, Dalm., and Portlock himself pointed out some of the differences from his species.

I. intermedius of the last-named author, which he says he only separated provisionally from *I. læviceps*, is a crushed specimen of the *A. gigas*. Portlock seems not to have been clear about this form, for he has labelled the specimen here figured (fig. 3) as *Isotelus intermedius*, whereas the axis of the tail is well marked out, and the ribs visible.

I have figured the true *I. intermedius* of Portlock with ours at the top of the plate, that the difference in the eyes, thorax, &c., may be apparent. Our species is a marked one, and I am glad to have the opportunity of clearing it up. It has long been in the Mus. P. Geology as *A. scutalis* (MS.).

Our fossil seems to have been about three inches long. The general form is very broad-oval, blunt behind, but angulated in front of the semioval head, which is much longer than the tail, and a little longer than the thorax-rings. In our best specimen it is thirteen lines long by two inches wide, and of this width the hour-glass-shaped glabella occupies more than one third below, contracting to much less than a third midway, opposite the eyes, and again expanding to nearly half above, to form the forehead-lobe. No lateral lobes or furrows interrupt the very even contour of the glabella, which is divided from the equally convex cheeks by a rather sharp but not deep furrow, ending on the neck-margin in a distinct punctum, not strongly enough shown in the plate. In the extremely narrow and almost obsolete neck-segment there is so close an approach to *Isotelus*, that I think we may fairly look on the species as linking together the two subgenera.

The eyes are placed halfway up the head, and are much raised, not very large, approximate (measured from their outer edges they are as far apart as the length of the head). The cheeks are regularly convex, the angles rather acute than rounded, and above the angles is seen the pit for the reception of the pleuræ, so conspicuous in *Asaphus gigas*, Pl. XXIV, fig. 6. Portlock notices this as a tubercle, but a tubercle on a cast signifies a pit on the true surface.

The facial suture curves largely out below the eye to more than half the width of the cheek; above it keeps within the line of the eye to reach the margin, and then continues parallel and close to it all along the front, scarcely angular even at the point itself.

The body, of eight rings, and the tail, can only be described from a single specimen (fig. 3), and this presents some differences from the central figure (fig. 1), which belongs to *I. gigas*, though, like fig. 3, it has been labelled as *I. intermedius*, Portl., a species which must be entirely obliterated.

The axis of the thorax is as broad as the pleuræ, but more convex than in *I. gigas;* the fulcrum is rather close to the axis; the pleural groove short, faint, and not so strong as in that species. The tail of our specimen (fig. 3) is semicircular, with smooth, gently convex sides, and a narrow raised axis, extending slightly more than two thirds down the tail, and marked interiorly by ten or eleven rings, or rather by a double row of puncta (representing the glands described at p. 52), one on each side of the axis. The sides are furnished with one long and broad depression or upper furrow, but with none else, the sides being free from ribs, as in all *Cryptonymi* or *Isoteli*. The fulcrum is strongly marked, and the facet broad, long, and turned much backwards. Caudal fascia broad, strongly striated.

Locality.—CARADOC, Tyrone (Mus. P. Geology).

Fig. 39.

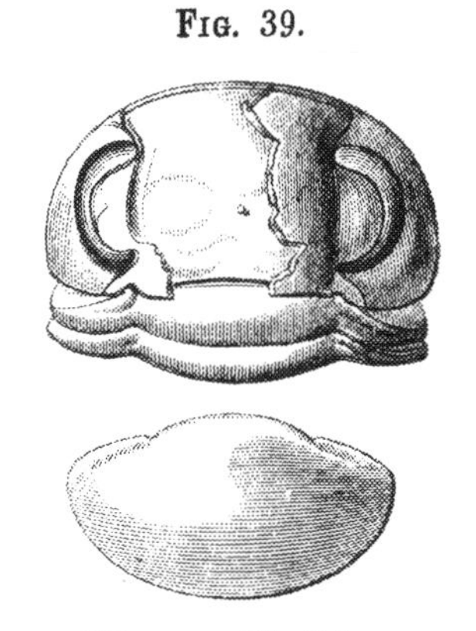

Nileus armadillo, Dalman.
Lower Silurian. Russia (Brit. Mus.).

The sub-genus *Nileus* has not yet, so far as I know, occurred in Britain, although it is so common in Northern Europe that it may reasonably be expected to be found in our own Lower Silurian rocks. I subjoin a figure of the common *N. armadillo*, Dalman, to show the near approach made to *Illænus* by this peculiar group of *Asaphidæ*. The internal lobes of the glabella are well seen in the cast; the specimen is a fine one, in the British Museum collection.

Stygina, *Salter*.

Among the many new and interesting forms of Trilobites described by Major-General Portlock in his work on Londonderry and Tyrone, a small species of *Asaphus* is recorded from the Lower Silurian of Tyrone, which he named *A. latifrons*, distinguishing it from some other species by the breadth of front included within the curve of the facial suture. The species is remarkable for the position of the eyes, which are placed so far backwards and inwards as to be close to the base of the small and narrow glabella. This peculiarity of habit is associated with some other characters which will remove the species from *Asaphus*. The flattened-oval form, long axis to the tail, and the head-spines, much resemble those of *Asaphus*, from which the nine ungrooved pleuræ effectually distinguish it. In the partial obliteration of the glabella, number of body-rings, and course of the facial suture, it is closely allied to *Illænus*, from which its habit differs so much; and there is enough of the under side preserved to show there was no rostral shield, which last is an essential character of *Illænus*. The technical generic character may be thus given:

Body ovate, flattened; head and caudal shield nearly equal; body of nine rings, and with narrow axis; eyes small, placed far backwards and inwards, near the base of the glabella, which is quite distinct above, and much contracted below. Facial suture marginal along a wide space in front, and below the eyes curved outwards and ending on the posterior margin; no rostral shield; labrum convex, entire; pleuræ without furrows; tail with a long and partially furrowed axis.

STYGINA LATIFRONS, *Portlock*. Pl. XVIII, figs. 7—10.

ASAPHUS LATIFRONS, *Portlock*. Geol. Rep. Londonderry and Tyrone, pl. vii, figs. 5, 6.
— MARGINATUS, *Portlock*. Ib., fig. 7, 1843.
STYGINA LATIFRONS, *Salter*. Rep. Brit. Assoc. Trans. Sect., p. 59, (read 1852), 1853.
— — *Id*. Siluria, 1st ed., 1854; and 2nd ed., p. 184, Foss. 26, fig. 2, 1859.
— — *Id*. Mem. Geol. Surv., Decade xi, sec. 2, p. 1, pl. ii, 1864.

S. sescuncialis, ovata, axe angusto; spicula capitis brevissima. Caput semiovatum, obtusum, glabellâ ad basim angustâ, oculis retrorsis, ad basim capitis pænè retractis. Cauda semiovata, obtusa, axe subannulato.

The general form is depressed and elliptical; the length about one inch and a half, the breadth one inch. The contour of the head, which is exactly as long as the caudal shield, and more than one third the whole length, is nearly a true semi-oval, evenly convex except on the median line behind (which is abruptly raised), and slopes on all sides to a concave border. The glabella, scarcely defined at all in front, though faintly indicated (more strongly so in young specimens), is of a pyriform shape. Posteriorly it is much contracted, and again suddenly expanded upon the neck-border. Its greatest width behind is not above one fifth that of the head. The eyes are small, convex, much curved, placed at less than their own length from the hinder margin, opposite the contracted part of the glabella, and rather further apart than the width of the thoracic axis. The facial suture runs nearly out nearly at right angles beneath the eye, and in front of it describes a large arc diverging from the eyes at an angle of 70°, and cutting the anterior border far outwards, in a line overhanging the fulcral points. The facial suture is strictly marginal in front, and the hypostome (fig. 4) appears to be quite continuous, without a rostral shield as in *Illænus*, or a vertical suture as in some *Asaphi*.

Two good specimens in Dr. Wyville Thomson's cabinet show the labrum, but its margin is broken off. It is wide at its attachment, considerably convex in the middle, more so than in *Asaphus*, and is marked with concentric lines on the sides. There is not enough to show that there was no marginal groove, or whether the tip was rounded and entire, as in *Illænus*, which is most probable.

Thorax of nine rings, not so long as the head, and with its axis only two thirds as wide as the pleuræ, convex. Pleuræ flat as far as the fulcrum, which is about the width of the axis remote from it. Thence the pleuræ are bent down and a little back, and facetted for rolling up. There is no groove whatever to the pleuræ, which in this respect resemble those of *Illænus*.

Tail semi-oval, blunt, not convex; the conical axis about half the width of the sides, and reaching fully two thirds the length of the tail. The axis has about eight faint furrows. The sides are gently convex at first, and then broadly concave, with a somewhat sharply defined margin; it is without any furrows, even the usual upper one is obsolete, or nearly so. The apex is very blunt, more so than the front of the head. The incurved striated portion is broad, and not indented by the point of the axis.

Locality.—Caradoc of Desertcreight, Tyrone. It will probably be found in the South of Scotland.

Stygina Murchisoniæ, *Murchison.* Pl. XVIII, fig. 11.

Ogygia Murchisoniæ, *Murchison.*	Silur. Syst., pl. xxv, fig. 3, 1837.
Stygina Murchisoniæ, *Salter.*	Rep. Brit. Assoc. Trans. Sect., p. 59 (read 1852), 1853.
— — *Id.*	Morris's Catal., 2nd ed., Foss. 10, fig. 4, p. 55, p. 115, 1854.
— — *Id.*	Siluria, 2nd ed., p. 55, Foss. 10, fig. 4; pl. iv, fig. 1, 1859.
— — *Id.*	Mem. Geol. Surv., Decade xi, sec. 2, p. 3, 1864.

S. convexa, valdè trilobata, capite longè semiovato cum spiculis productis. Cauda longè semiovata, axe prominulo lævi, non annulato.

In the black schists, which alternate with conglomerate and grit-bands at Mount Pleasant, Carmarthenshire, and which are most probably of Caradoc, not of Llandeilo age, a single specimen of this rather remarkable fossil occurred to Sir Roderick Murchison's hammer, and none other has since been found. The locality is worth searching. *Nacula* and *Orthoceras* are found with it.

The head is greatly longer in proportion than that of the species last described, forming a long semi-oval, produced behind into lengthy spines, and divided strongly into glabella and cheeks at the base. Above, our only specimen is indistinct, and it does not show the position of the eyes.

Of the thorax-rings but three or four joints are preserved, but these show the trilobation to be very strongly marked, and the axis is quite convex, while the pleuræ are considerably bent down from the fulcral point, and backwards too. They seem to be longer and narrower than those of *S. latifrons.*

The tail, reversed on the specimen (which was probably curled up when imbedded in the silt) is also of a long-oval contour, and its narrow axis is prominent to the very end; it shows no signs of transverse ribbing. The sides are faintly convex, then concave along the margin, which reveals, when the upper crust is removed, a moderately broad fascia, closely striated.

The species differs at a glance from *S. latifrons* in the much longer shape and more protracted head-spines. The tail, too, has a longer axis, without a trace of annular furrows, and the trilobation is throughout much more distinct.

Locality.—CARADOC? Black shales of Mount Pleasant, Carmarthen.

Presented to the Geol. Soc. Museum by Sir R. I. Murchison. It is named after Lady Murchison, and is as rare a species as any in Britain.

STYGINA —, sp. Woodcut 40.

I do not like to risk naming an apparently new *Stygina*, with much shorter tail than in the two preceding species. It is from the Caradoc slates of Pembrokeshire. The axis reaches only half way down the tail, and the general shape is very broad.

FIG. 40.

Stygina, sp.

Locality.—CARADOC. Sholes Hook, Haverfordwest. Mr. H. Wyatt Edgell's cabinet.

STYGINA? MUSHENI, n. sp. Pl. XXIX, fig. 1.

S. major, triuncialis, latè ovalis, subplanus, thorace caudâ obtusâ breviore. Axis corporis depressus, annulis arcuatis. Pleuræ lentè convexæ. Cauda semiovalis, obtusa (lat. 20 lin., long. 14 lin.), axe longo primùm conico, dein parallelo; apice haud prominulo, annulis obscuris. Limbus lente convexus, margine angusto concavo.

A larger species than either of the two preceding, and unfortunately not perfect enough to determine the genus. I have figured it with the *Illæni* only because there was most room on the plate. But it can hardly belong to any known British genera except *Asaphus* or *Stygina*, and I refer it to the latter for choice.

The whole fossil must have been above three inches long, for without the head it measures fully two inches. The thorax, with a gently convex surface, is shorter than the tail, and has a much wider axis, of eight arched rings, and with the axal furrows well marked out. The pleuræ are so much lost and obscured by rubbing, that we can only see they were unfurrowed and much arched, not bent down greatly. This is like *Stygina* or *Illænus*, not at all like *Asaphus*.

The tail is more perfect, and shows a wide semioval plate, blunt behind, gently convex, except the narrow concave border. The axis is not convex, and is very narrow, broadest and conical at the base, then parallel-sided, and extending to the edge of the concave border, but there indistinct, and, as it were, connate with the border. This is partly the case with *S. latifrons*. A few obscure annuli show at the upper part. The sides of the tail are smooth, and show no trace of furrows, not even the top one.

Locality.—In a gray calcareous flagstone boulder (from the CARADOC probably) in the drift of the Severn, near Buildwas. Cabinet of the late James Mushen, of Birmingham, whose persevering labours collected so fine a series of Silurian fossils.

PSILOCEPHALUS, *new genus*, 1866.

This has been long a MS. name, but only not yet made public owing to the great delay in the publication of vol. iii of the 'Memoirs of the Geol. Survey of North Wales.' The genus is an inconspicuous one, and at first sight it looks as if it might be placed with either *Asaphus* or *Illænus.* But the shape of the head, all but lobeless, as well as the forward position of the small eye, easily distinguish it from *Nileus,* the group of *Asaphi* which it most nearly resembles, while the eight grooved body-rings effectually exclude it from *Illænus.* In the absence of the hypostome and labrum—essential characters in the *Asaphidæ,* it is impossible to characterise it fully. But I feel assured it is a distinct genus.

Oval, convex, especially the head, which may be described as inflated, with a faintly marked out parallel-sided glabella, undefined in front; no marginal furrow; eyes very forward and small; head-angles obtuse; body of eight rings, with grooved and facetted pleuræ; it has an ungrooved tail, with a rather long distinct axis; the metamorphosis is only partly known.

Psilocephalus may be considered as forming the passage from *Illænus* to *Asaphus,* by means of the abnormal subgenus *Nileus* of Dalman.

PSILOCEPHALUS INNOTATUS, *Salter.* Pl. XX, figs. 13—19.

PSILOCEPHALUS INNOTATUS, *Salter.* App., Ramsay, Geol. N. Wales; Mem. Geol. Surv., vol. iii, p. 315, pl. vi, figs. 9—12, 1866.

P. latè-ovatus, lævis, sescuncialis; cujus caput semiovatum caudâ semicirculatâ majus; sunt 8 *pleuræ, fulcro tenus (ad tertias posito), planæ dein decurvæ. Cauda lævissima, axe prominulo.*

This neat and rather conspicuous form is the most abundant fossil in the Lower Tremadoc beds, and usually found in company with the much rarer *Niobe Homfrayi,* described at p. 143. We owe our best specimens to David Homfray, Esq., of Portmadoc.

The general shape is a rather broad blunt oval. It is scarcely an inch and a half, seldom more than an inch, long, by about three quarters of an inch broad. The form is very convex, especially in the head, which is but slightly lobed; the tail more strongly so, and the thorax strongly tri-lobed throughout.

The head may be called hemispherical; it is considerably longer than the tail, semi-oval, blunt, very convex and smooth, not margined at all in front. The glabella is confused with the cheeks, and scarcely marked out except near the base, as in *Illænus;* it is about as broad as the cheeks. Neck-furrow quite absent. Fixed cheeks narrow, and the free cheeks a small segment of a circle; the facial suture taking its rise near the obtuse outer angles, running obliquely forwards to the front margin, which it cuts rather within than immediately over the eye. The eye is very forward, fully two thirds up the head, and small. The body is strongly trilobed, a little shorter than the head; the axis narrower than the sides, tapering backwards slowly, and very convex. The pleuræ have square ends, are blunt, a good deal curved down from the fulcrum, which is at one fourth in front and thence extending further and further out till it reaches one third in the hinder rings, an unusual arrangement. The pleuræ are flat or but slightly concave as far as the fulcrum, thence sharply facetted and decurved, and striated lengthwise. The pleural groove shows strongly, but only towards its lozenge-shaped outward termination.

The semicircular, somewhat convex tail has the axis rather long, blunt, sub-cylindrical, reaching three quarters down the tail; it is smooth and has a blunt prominent tip. The sides are convex, smooth, with no flat or hollow margin along the striate border, and no furrows or ribs whatever.

Metamorphosis; one young specimen (fig. 17) shows only five rings.

Locality.—Lower Tremadoc beds of Penmorfa, Tremadoc, and the neighbourhood of Borth and Moel-y-gest; *e.g.*, Borthwood, Tyn-y-llan, Tyddyn-llwyd farm, &c. Mr. D. Homfray's fine specimens are figured, but the species is now in many other cabinets.

Psilocephalus inflatus, *Salter*. Woodcut 41.

Psilocephalus inflatus, *Salter*. App., Ramsay, Geol. N. Wales; Mem. Geol. Surv., vol. iii, p. 316, woodcut fig. 8, 1866.

P. prorsus trilobatus, capite inflato, et glabellâ conspicuè rotundâ. Axis corporis prominulus, caudæ brevissimæ distinctissimus, et valdè a lateri sejunctus.

This small species has always appeared to me to differ from the preceding. The head is very convex, the glabella rounded, and separated by its convexity and by distinct furrows from the head. Eye?

Fig. 41.

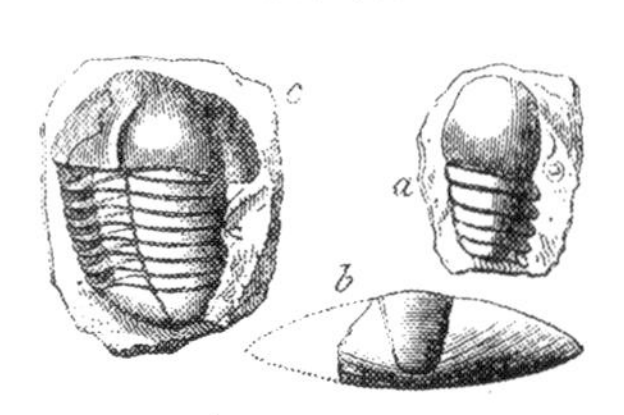

Psilocephalus inflatus, Salter.
Lower Tremadoc, N. Wales.
Mr. D. Homfray's cabinet.

Body with a moderately broad convex axis, also well distinguished from the somewhat convex pleuræ, the hinder of which are curved. I can only find seven body-rings. The tail, if it belongs to the same species (and I believe it does) is very short, and has the axis not only convex but short-conical, and marked out clearly all round.

Locality.—Lower Tremadoc, lowest beds, north-west of Penmorfa, on the Caernarvon Road (*a*, *b*). Upper beds, Trwyn cae Iago, Portmadoc Harbour (*c*). Borthwood.

ASAPHIDÆ—*continued:* Ogygia, &c.

As must sometimes happen, in writing a continuous monograph, new materials, connected with plates and descriptions already completed, turn up just as the work has passed through the press; and if such material refer to plates long since issued, it would be of course desirable to leave the new matter to the end of the work; but if not too far removed, it is as well to insert it as nearly as possible in its place.

Before proceeding, therefore, to the description of *Illænus*, to which group the present part of the Monograph is restricted, we give a plate (numbered Pl. XXV*) of *Ogygia peltata*, containing the new and more perfect specimens to which reference has been made in the "Corrigenda" of the volume for 1864. A new species has also occurred in the same locality, and from the same formation (the Arenig or Skiddaw Rock, a set of beds but little known). I had better describe the two forms of *Ogygia* together; I take also the opportunity of inserting a new *Barrandia* at the same time.

Ogygia peltata, *Salter* (Pl. XVII, figs. 8—10). Pl. XXV*, figs. 1—4.

In page 133, line 1, it was stated that on Pl. XVII, figs. 9 and 10 are wrongly marked as *O. peltata;* and in page 134, our fig. 10 was described as *O. scutatrix*, a species which is only yet known from the Tremadoc Rocks.

The mistake arose, as stated in the Corrigenda, from laying too much stress on the broad form and broad axis of the female form ('*forme large*' of Barrande), which in *Ogygia peltata* resemble those of *O. scutatrix:* and, following the same idea, the head of the broad form, with its wide labrum, was also described as *O. scutatrix* (p. 133, line 22, &c.).

These mistakes are now rectified by the accompanying plate (Pl. XXV*), which exhibits fine specimens of the broad form (figs. 2, 4); the shape of the head, with its oblong parallel-sided glabella, in fig. 1; and the labrum, of full size (fig. 3). The eyes, it will be observed, approach the glabella closely, as in Pl. XVII, fig. 10, and are not placed wide apart, as in fig. 8 of that plate, which, copied from the best reconstruction I could formerly give of the species, gives an incorrect idea of all but the general proportions. The fulcrum-points of the axis, &c., are also imperfect, so we had better erase the figure entirely. It is not very pleasant to admit so many errors, but it is a shorter plan than trying to defend or excuse them. The description in page 135 is more correct; but the glabella is not so wide as the cheeks, nor the axis of the thorax as the pleuræ; and these latter have the fulcrum placed at less than one half out instead of two-thirds. The pleural groove is nearly straight, only a little sigmoid; and the tips are truncate, not falcate. The groove ends abruptly a little way within the tip. In the tail eight furrows cross the axis, the last three not quite complete. The interlining furrows are certainly not so long nor so distinct as in our former description and figure.

The following is the new form which occurs with *O. peltata* at St. David's:

OGYGIA BULLINA, n. sp. Pl. XXV*, fig. 5.

O. plana, modica, biuncialis, capite semicirculato, fronte latiori, angulis—? Glabella pyriformis, ad basin angustata, lobisque basalibus inflatis; genæ absunt. Oculi prope glabellam, in medio capite positi. Thorax axe angustissimo, pleuris rectis, fulcro ultra medium posito. Cauda? Labrum?

A neat species, which was sent by my correspondent Henry Hicks, Esq., of St. David's, from the beds at Whitesand Bay.

The form is broad-oval. We have but the head, without free cheeks, and eight or nine of the body-rings, but there is enough to show that the fossil is quite new. The head is broader than a semicircle, and has a wide, coarsely striated front (the squamate striæ are not visible in our figure); a pyriform glabella, inflated in front; strong basal obliquely oval lobes, projecting beyond the neck of the glabella, and separated from it by rather strong lines. Neck-furrow of the glabella tolerably strong, and much broader than that beneath the cheeks. Eye-lines above and below the eye widely diverging; the eye short, and placed about halfway up the head, but even closer to the glabella than in *O. peltata*, which species accompanies it. Beneath the eye the suture curves so largely out that it must nearly reach the angle of the head. We have not the free cheeks, and of course do not know anything of the spines; I suppose these must have been short.

Thorax with a very narrow and somewhat prominent axis, which is about half as wide only as the flat pleuræ, and these are nearly direct; the slight fulcrum, placed beyond the half of the pleura, being scarcely an angular bend at all. The facet, however, is very distinct and sharp-edged. Fine curved transverse striæ cover closely the glabella, and are traceable on the axis of the thorax. The much coarser striæ on the fascia beneath the pleuræ range up to the fulcrum, but not within it.

Locality.—ARENIG ROCKS. North-east angle of Whitesand Bay, St. David's (Woodwardian Museum, Cambridge). The species should be sought for in the same beds on Ramsey Island.

My lamented friend Mr. H. Wyatt-Edgell found a new *Barrandia* in the black shales of Abereiddy Bay. (For this genus, see p. 137, &c.) We are only able to give a woodcut; and it is a melancholy pleasure to introduce a description in this work from his pen.

"BARRANDIA (HOMALOPTEON) LONGIFRONS, *Edgell.* Woodcut, fig. 42.

"*B. (Homalop.) magna, 4 uncias ferè longa; cui glabella producta claviformis, oculi propinqui (sulcus cervicalis abest?). In thorace axis latitudo dimidium latitudinis pleurarum, quarum sulci pæne aciem attingunt, efficit. Cauda semiovata axem conicum gerit, dimidium longitudinis illius efficientem; latus quodque sulco unico furcato distinctum.*

"A very large species, with a projecting club-shaped glabella; the axis of the thorax about half the width of the pleuræ; and the tail semicircular or rather semi-oval, each side bearing one strongly branched furrow only. The long clavate glabella, with undulated sides corresponding to the glabella-furrows, projects about one third of its whole length beyond the anterior margin of the cheeks, which are at their base one third wider than the base of the glabella. The rather large eyes are placed considerably forward and close in to the glabella. The fixed cheek is very broad, and the facial suture curves widely out beneath the eye, so that the free cheek is much narrowed, its outline arched almost in a semicircle; it has a narrow but very distinct margin; it shows, so far as our only specimen can be trusted, no neck-furrow (the front thorax-ring has been pushed under the head and appeared to us at first sight as a branched neck-furrow), the cheek ends posteriorly in a small needle-shaped head-spine.

FIG. 42.

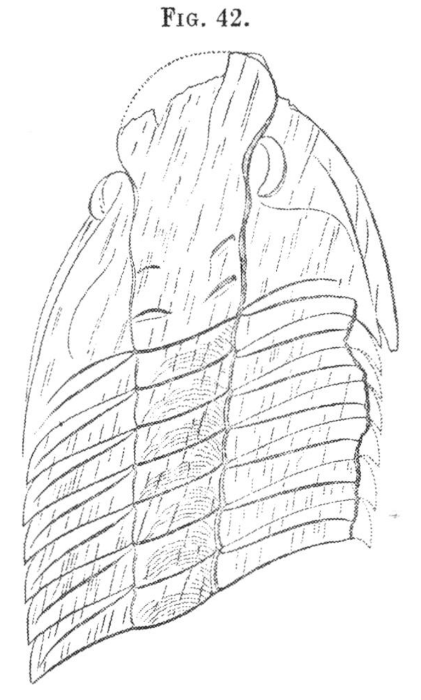

Barrandia (Homalopteon) longifrons, Edgell. Llandeilo Slates, Pembrokeshire. Mr. H. Wyatt-Edgell's Cabinet.

"There are, apparently, seven thorax-rings, as in some other species of this genus. The pleural furrow is well defined, and reaches almost to the tip of each of the recurved flat pleuræ.

"The tail, fig. 43 (natural size and enlarged), found in the same locality, is supposed to belong to this species, as only one other *Barrandia* (*B. Cordai*, M'Coy, a small species) is known to occur there. The length of this caudal portion is almost equal to its breadth; the axis is conical, the base being about half the length. The side has two furrows, or perhaps a strongly branched one (the branch arising as usual from the very base, and indicating the suture of the first tail-segment): these nearly reach the outer margin; but no corresponding furrows occur on the smooth conical axis, which is half the length of the whole tail, and ends apparently in a sharp tip, from

FIG. 43.

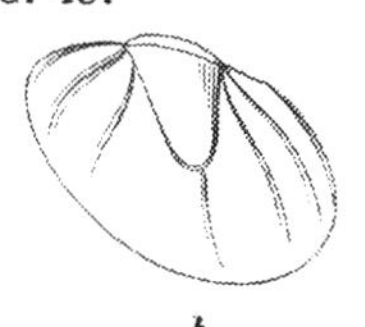

a b

Tail of *B.* (*Homal.*) *longifrons*? Llandeilo Slates, Pembrokeshire. From the same Cabinet. *a*, Natural size; *b*, enlarged.

whence a raised line runs to the margin, a common circumstance where the axis is abbreviated.

"Length of head and thorax taken together three inches; length of the supposed tail (of a younger individual) five lines.

"*Locality.*—LLANDEILO SLATES of Abereiddy Bay, Pembrokeshire; in my collection."—*H. Wyatt-Edgell.*

We may now proceed with the most abnormal of all the *Asaphidæ*—the group of *Illænus*, so very characteristic of Lower Silurian—the "*Faune Seconde*" of Barrande. It is true that in one of its subgenera it rises into the Upper Silurian; but the mass of the species, and especially the abundance of individuals of this genus, mark Llandeilo and Caradoc rocks (Lower and Middle Bala of Sedgwick) most effectually. [Lower, Middle, and Upper Bala rocks of Sedgwick are equivalent to Llandeilo, Caradoc, and Llandovery.]

ILLÆNUS, *Dalman*, 1826.

Of all the higher Trilobites, this is the most abnormal in shape;—the segments of the large head, and of the equally bulky tail, being so condensed and obliterated as to leave those portions like the rudimentary extremities of an *Agnostus;* while the structure of the pleuræ, the form of the head, eyes, labrum, and rostral shield,—the thick sculptured crust, and the perfect mobility of the thorax-rings, all mark the genus as one of high rank among the *Asaphidæ*, and as nearly allied to the *Proetidæ*, the highest of the smooth-eyed groups.

All the species of the genus have a common facies:—the similar head and tail, unmarked by lobes or annulations; the glabella, cheeks, and margin being usually confounded in one hemispheric mass; and the axis, in most instances, scarcely indicated outside the caudal shield. The creature could roll itself up into a perfect ball, and, no doubt, chiefly suggested to Dr. M'Leay the analogy with *Bopyrus* and other *Isopoda;* the resemblance to the Armadillo Woodlice is not less striking. Yet the *Isopoda* have crustaceous feet well articulated, and we have yet to learn that Trilobites had any crustaceous feet at all.[1]

[1] I find that every one in our own country has overlooked a very important notice, by Dr. Volborth of St. Petersburg ('Verhandl. Min. Gesellsch. Petersburg,' 1857-8, p. 168), of the discovery, by Dr. Pander, of some appendages, which may be membranous feet, attached to definite parts of the outer portion of the pleuræ beneath. I do not know that Dr. Pander has published this in any definite form, but Dr. Volborth has fully described and figured these points of supposed attachment for swimming feet ('Mém. Acad. Imp. Petersb.,' 1863, tom. vi, No. 2). Barrande observed them in 1855, in *Ogygia*, and described them in 1858. Mr. E. Billings, Palæontologist to the Canadian Survey, has also found in *Asaphus gigas* a trace or two of curious processes below the crust, but close to the axis. All these data have great value; and those who have abundance of specimens of *Asaphus expansus*, from the fine Lower Silurian mudstone of St. Petersburg, will do well to cut and polish specimens, observing all traces of fragments of the general lower membrane, as well as of distinct processes (for details see 'Mém. Acad. Imp. Petersb.,' 1863).

Illænus is as compact a genus as any in the whole Trilobite Order. It comprehends almost as many distinct, if less varied groups, as *Phacops* itself; and all of these have the common facies of an inflated head and caudal shield, without external lobes or rings, and that peculiar hemispheric contour to each, which is so marked a character of the whole genus. Sometimes, the angles of the head are shortly spinous (*Dysplanus*), but usually they are rounded off and unarmed (*Illænus, Bumastus*); sometimes the eye is subcentral (*Il. centrotus*), but more commonly it is placed behind the middle of the head, and near the posterior margin (*Il. crassicauda, Il. Bowmanni, &c.*). A few have the axal furrows reaching quite up the head (*Il. distinctus*, Barr.), but in the great majority they reach less than two-thirds this distance—and are often less than half-way. In a very few the eyes are large (*Il. ocularis*), but usually they are small. The labrum varies in shape in the different subgenera (see figs. 44 *a, b*).

FIG. 44.

a. b.

a. Labrum of *Illænus centrotus* (after Volborth.)
b. Labrum of *Illænus crassicauda* (ib.).

Lastly, the number of body-rings varies.[1] Two sections, perhaps distinct genera, *Panderia* and *Octillænus*, have but 8 rings. *Dysplanus*, including species with both spinous and rounded head-angles, has only 9 rings. *Illænus* proper has 10 body-rings, and this includes many well-known species; while *Bumastus*, Murchison, which, like *Illænus*, has 10 rings, has the axis very wide, and so little distinguished from the pleuræ as to give a unique aspect to the fossil.

I agree with Barrande,[1] that it is wiser, in the present state of our knowledge, to include all these subgenera under one common name; dividing them into groups, which may, by and bye, if the characters prove constant, be called genera. All of them appear to agree in the following characters:

Head hemispherical, or rather quarter-spherical, with rounded contour, no marginal furrows, and a very slight indication of neck-furrow. Glabella lobeless externally; within it has often 4 pairs of lobes, and is separated by incomplete axal furrows from the tumid cheeks. Eyes lateral, remote. Epistome striate, with a well-defined transverse rostral shield, and bearing a somewhat triangular tumid labrum, which has a strong border and expanded base (see fig. 44 and Pl. XXVII, fig. 7).

FIG. 45.

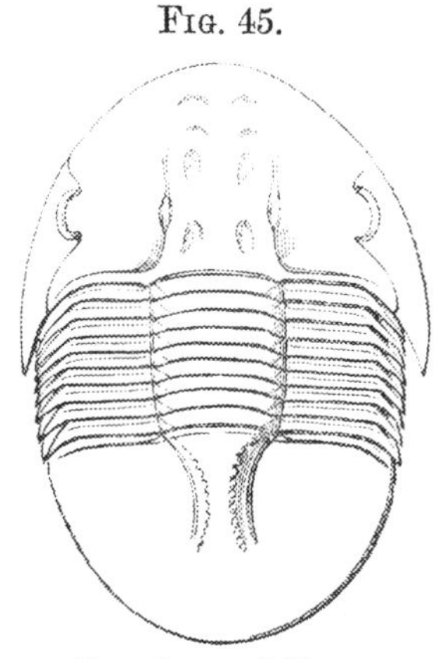

General type of *Illænus*.
Il. centrotus, Dalm. North Russia.
(After Volborth).

Body-segments 8—10; usually 10, without grooves to the pleuræ. Tail large, hemispheric, with a short incomplete axis, showing numerous rings, but only within the crust.

The genus, world-wide in its distribution, is neatly distinguished from all others by habit, as well as a combination of characters. These are

[1] Barrande has enumerated 11 genera in which the number of rings is variable.

maintained, though under considerable variations, throughout the North and Central European types and those of Canada and North America. The greatest variation is found among the oldest forms, viz., those from the Arenig and Llandeilo rocks, where the genus first appeared. If we look to our plates, we shall find that the ordinary species of *Illænus* have the furrows which mark out the glabella very short externally, and the eye placed at a moderate distance or somewhat remotely from the glabella, and behind the middle of the cheek. But in the oldest group of Silurian rocks known to contain *Illænus*, the eyes are placed so far forward and outward, that in one form, *Ectillænus* (Pl. XXVI, fig. 8), they seem at first to be absent altogether; and in *Hydrolænus*, a Canadian subgenus (*Illænus conifrons*, Billings), the glabella-furrows are so long as to reach the front, the glabella so gibbous as to overhang it, and the eyes are extremely forward. These two subgenera lead directly to *Illænopsis* (Pl. XX, fig. 1), which is from the Arenig rocks (it has been quoted from May Hill Sandstone !).

If we divide Illænus into the undermentioned sections, following the lead of previous authors, it is not because we think the divisions quite natural, but because they are convenient. Some of the subgenera will be hereafter elevated to generic rank when we know more of the lobes of the head, not often visible even in internal casts.

§ 1.—Trilobed: labrum elongate.

1. *Octillænus*, Salter, 1866. Eight rings; a well-defined glabella, and the upper pair of pleuræ produced. Lower Silurian, Central Europe.

Illænus Hisingeri, Barr.

2. *Panderia*, Volborth, 1863. Eight body-rings; short glabella-furrows, and ordinary ungrooved pleuræ. Lower Silurian, North Europe.

Il. triquetra, Volb. *Il. Lewisii*, Salter.

3. *Dysplanus*, Burmeister, 1843. Nine rings; ordinary pleuræ. Lower and Middle Silurian.

Il. centrotus, Dalm. *Il. Bowmanni*, Salter.

4. *Illænus* proper, Dalman, 1826. Ten body-rings; ordinary pleuræ; eyes subcentral. [The greater number of species belong to this subgenus.] Lower Silurian, universal.

Il. crassicauda, Wahl. *Il. Murchisoni*, Salter.
Il. Davisii, Salter. *Il. Rosenbergii*, Eichw.

5. *Ectillænus*, Salter, 1866. Ten body-rings; ordinary pleuræ; eyes remote and very forward; glabella-furrows short. Lower Silurian, only British as yet.

Il. perovalis, Murch.

6. *Hydrolænus*, Salter, 1866. Eyes remote, forward; glabella-furrows reaching the front; head most gibbous. Lower Silurian, Canadian.

Il. conifrons, Billings.

7. *Illænopsis,* Salter, 1865. Eyes remote, forward; glabella-furrows reaching the front; head not gibbous, only convex; pleuræ grooved. [Probably a distinct genus.] Lowest Silurian, British.

Il. (*Illænopsis*) *Thomsoni,* Salter. [Not *Illænus Thomsoni.*]

§ 2.—Body scarcely trilobed: labrum short.

8. *Bumastus,* Murchison, 1837. Eyes large, remote, backward; ten body-rings. Lower and Middle Silurian, but chiefly Upper; universal.

Il. Barriensis, Murch. *Il. insignis,* Hall.
Il. carinatus, Salter. *Il. Trentonensis,* Hall.

Subgenus 1.—Of Octillænus we have no British types. It is a Lower Silurian Bohemian form. We give a woodcut (fig. 46), as it is so peculiar a form of the group, from Barrande's figure. The enlargement of the first pair of pleuræ is a very unusual character in the *Asaphoid* or smooth Trilobites.

Fig. 46.

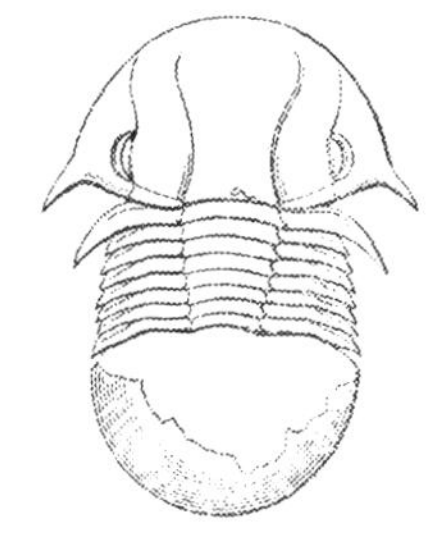

Il. (*Octil.*) *Hisingeri,* Barr. Lower Silurian, Bohemia.

Subgenus 2.—Panderia, *Volborth,* 1863.

Illænus (Panderia) Lewisii, n. sp. Pl. XXVI, figs. 2, 2*a*, *b* (nat. size and magnified.

Il. (*Pand.*) *minimus* (*forsan junior?*), $\frac{7}{12}$ *unciæ longus, vix* $\frac{5}{12}$ *latus, ovatus, capite* (*glabellâ præcipuè*) *gibbo, thoracem longè superante, et caudâ planatâ bis ferè longiori. Caput subtrigonum, fronte paullo producta, angulis obtusis marginatis. Glabella post verticem gibba,* $\frac{2}{3}$ *latitudinis capitis efficiens, oculos attingens. Sulci axales brevissimi, ad basin glabellæ solum conspicui. Oculi magni, glabellam appressi, dimidium diametri suorum a margine distantes. Thorax axe modico, anticè latiori, pleuras subplanas superante. Cauda brevis semiovata planata, axe longo, per totam longitudinem conspicuo.*

Had not Dr. Volborth figured two species of this very distinct subgenus, one of which is given on the next page, I should have taken this for the *young* of some undescribed species. The large size of the head and eyes relatively to that of the tail, and the more complete marking out of the axis in the caudal shield, are characters belonging usually to young specimens. But there can be little doubt we have here nearly, if not quite, the adult form of a subgenus not heretofore described from Britain, and apparently rare even in Northern Europe. We owe our single specimen to the care of Mr. Lightbody, who obtained it for this work from the cabinet of the Rev. D. P. Lewis, of Guilsfield, near Welchpool. It is a valuable addition to British fossils.

The width of our single rolled-up specimen at the base of the head is 5 lines, and, the length of the large head being fully 4 lines, we may reckon the whole specimen as 7 or 8 lines long. The form, unrolled, would be ovate, the gibbous head being somewhat trigonal, with the front a little produced, the sides flattened, and the hinder angles much rounded. The head is very gibbous, a section of it across the eyes, as shown in fig. 2, being a broad parabola. The principal gibbosity lies in this line, about the base of the great eyes. These are closely appressed to the glabella, so as to abut at their lower end on the short axal furrow, which is conspicuous here, but is not carried further up. And the eyes are placed about half their length distant from the posterior margin of the head.

The course of the facial suture, below the eyes, is abruptly outward; above them it runs direct to the front margin, in a line continuous with that of the short axal furrows before noticed. The cheeks are much flattened, and decline rapidly, and our figure 2*a* rather too strongly expresses a marginal furrow within the angle—a character common enough in other genera, but extremely rare in *Illænus*.

The thorax, of only 8 rings, has a broad axis, which tapers backwards, and is gently convex. The pleuræ are not so wide as the axis; they are flat as far as the fulcrum, which is placed at one half in the hinder rings, and is pretty close to the axis in the front ring; thence they bend a very little downward and backward to the oblique tips.

The tail is less than a semioval, and has the front edge straight. The axis is equal in breadth to the sides, and reaches down three fourths the length of the tail; it is rounded at the tip, flattened above, and with a distinct furrow all round it. The tip is very little prominent, but clearly defined by the prominence, as the sides of the axis also are by the axal furrows. The limb is flat, as, indeed, the whole tail is; and only the edge turns gently downward.

The relatives of this little species are clearly to be found in the *Panderia*[1] *triquetra* and *P. minima* described by Dr. Volborth in his paper in the 'Transactions of the Imperial Academy of St. Petersburg,' vol. vi, No. 2, 1863, pl. iii, figs. 13—19. One of these is given in our woodcut, fig. 47. Though I think *Panderia* only a subgenus, it is a very good and distinct one; and the eight body-rings, with their tapering axis, the flattened caudal shield with its long distinct axal lobe, and the large relative size of the head and eyes, mark it as an embryonic form (not using this word in a strict sense) of a comprehensive genus. The figure of *P. minima* is especially like ours in the gibbosity of the vertex, direct facial suture in front, and approximated eyes; and it is about the same size. Still, it differs in nearly all its proportions from *P. Lewisii.*

Fig. 47.

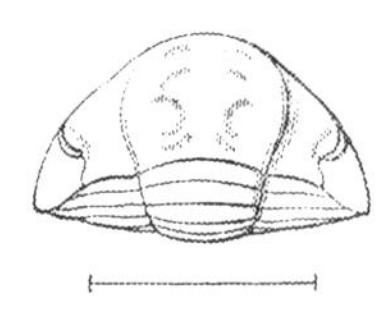

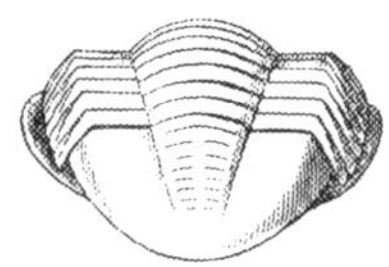

Ill. (*Panderia*) *triquetrus*, Volb.
L. Silurian, N. Russia.

Locality.—Caradoc or Bala limestone of Moelydd, Oswestry,—near Llanymynech quarry (where "Fossils" is printed on the Ordnance Map). Cabinet of the Rev. D. P. Lewis).

[1] Named in honour of the veteran naturalist Dr. Pander, so well known for his works on the fossils of Russia. See foot-note, p. 180, for a reference to his discovery of the membranous feet of Trilobites.

Subgenus 3.—DYSPLANUS, *Burmeister*, 1843.

ILLÆNUS (DYSPLANUS) BOWMANNI, *Salter*. Pl. XXVIII, figs. 6—13; Pl. XXX, fig. 6.

ILLÆNUS CENTROTUS, *Portlock*. Geol. Rep., pl. x, figs. 3—6 (not fig. 9), 1843.
IL. BOWMANNI, *Salter*. Mem. Geol. Surv., vol. ii, pt. 1, pl. viii, figs. 1—3, 1848; Decade 2, art. 2, p. 3, 1849; in Morris's Catalogue, 2nd ed., p. 110; Appendix A to the Woodwardian Synopsis p. 4, 1852. Siluria, 2nd ed., passim, 1859; also Mem. Geol. Surv., vol. iii, p. 317, pl. xviii, fig. 8, 1866; and Catalogue Mus. Pract. Geol., pp. 5, 19, 1865.
IL. CENTROTUS, *M'Coy*. Synopsis Sil. Foss. Ireland, p. 54, 1846; *Dysplanus*, Synopsis Woodw. Fossils, pl. i E, fig. 19, 1852.
IL. LATUS, id. Ibid., fig. 17.

Il. (*D.*) 3-4 *uncias longus, convexus, lævis nisi cauda lineis nonnullis* Λ-*formibus ornatâ; capite valde convexo caudam obtusam longitudine superante, thorace brevi. Caput semi-ovatum, convexissimum, sulcis axalibus brevibus, superne recurvis sæpè claviformibus, per tertias longitudinis capitis extensis. Genæ parvæ declives. Oculi minimi, distantes, basi capitis proximi. Anguli obtusi, nec spicula gerentes. Thoracis longitudo dimidium capitis totius efficit, annulis* 9; *axis ejus convexus, et pleuris antè latior est; at posticè, in pullis præcipuè, angustior. Pleuræ fulcro tenus planæ, dein recurvæ paulloque decurvæ. Fulcra anticè in tertias, postice in dimidium latitudinis pleurarum posita. Cauda lentè convexa, semiovata, obtusa: cujus longitudo* $\frac{2}{3}$ *latitudinis suæ efficit; axe magno supernè tantum inter foveas latas conspicuo, sulcis obscuris. Latera convexa, sulco superno lato, et interdum* (*in pullo*) *sulcis binis obscuris.*

A large species, frequently four inches long; oblong, with blunt and very convex extremities: the head not much larger than the tail, the thorax scarcely more than half the length of the head, and but little more compared with the tail; the eyes small and remote; the axis of the thorax wider than the pleuræ.

Head regularly convex, semi-oval; glabella wider than the sides, its furrows slightly converging as they reach one third up the head. Cheeks gently declining to the remote eyes, which are distant from the axal furrows about half the width of the glabella; and thence bend steeply down to the incurved margin. The eye is very short and small, and placed so far backwards as to be only its own length distant from the straight posterior margin. Free cheeks very small and narrow, with blunt rounded angles (not acute and produced as in *Il. centrotus*). The facial suture is nearly direct, *i. e.* vertical above and below the eye. The neck-furrow is distinct in casts, but beneath the cheek only; a deep punctum at the base of the axal furrows defines its position under the glabella. The section of the head across the eyes is an arc of a circle about one third of the whole circumference, and from front to back the curvature is similar, but more abrupt in front, where the forehead overhangs.

Thorax about half the length of the head, and about three-fourths the length of the tail; of 9 rings. The axis gently convex, broader than the lateral lobes, and tapering but little backwards; marked out by strong but not deep axal furrows from the pleuræ. These are flat as far as the fulcrum, which is at one third anteriorly, and at one half posteriorly. From this point the pleuræ are bent gently down and a little backwards; the front ones more so than the hinder ones, which last are nearly direct.

Tail regularly and gently convex, half a broad oval; the length being two thirds the breadth. The upper angles are strongly truncated; the axis, which is more than one third the breadth, being only marked out by broad indentations in the upper part, and having no true axal furrows. The tail-margin is neither obtuse nor recurved, but regularly and evenly declines to the smooth edge. Caudal fascia concave,[1] rather narrow, especially towards the upper angles, broader below, and not indented by the axis internally.

Variations.—In young specimens the glabella is narrower in proportion (Pl. XXVIII, fig. 8; and see also Portlock, Pl. 10). The thorax too is longer in proportion and more quickly tapering behind; the tail is flatter, and so short as to be less than a semicircle (see also fig. 9). Such variations in the young state are common among the species of this genus. Rarely, as in Pl. XXVIII, fig. 11, the glabella-furrows are longer and more connivent, and more recurved above; fig. 6, again, same plate, represents a May Hill Sandstone specimen, in which these furrows are shorter than usual. Some specimens have them straighter than others, and they vary a little in depth.

Illænus latus, of M'Coy, is, I think, only a compressed specimen. We have several such forms from Pembrokeshire and elsewhere. It is figured in p. 215, woodcut 54.

Our fossil has been referred both by Gen. Portlock and Prof. M'Coy to the Swedish form, *Il. centrotus.* I cannot clearly make out why; for the distinction of rounded (instead of spinose) angles was clearly pointed out by myself, in the Appendix to the Woodwardian Synopsis; and I communicated freely with Prof. M'Coy. I have re-examined the foreign specimens, and am convinced the species are only allied, not identical. The position of the eye at once separates our fossil from *Il. Davisii;* these two hardly ever occur together, *Il. Davisii* requiring, to all appearance, a purer element than the more common *Il. Bowmanni*, which could live on mud, sand, gravel, or any kind of sea-bottom. It, however, seems to have preferred a calcareous sea-bed when it could get it.

Localities.—LLANDEILO FLAG, or *Lower Bala*, rare. Knockdolian, Ayrshire (*Woodw. Mus.*)? Everywhere in CARADOC or BALA Rocks, N. and S. Wales, Westmoreland; S. Scotland; North, East, West, and South Ireland. A few localities may be given: Bala, abundant, especially east of the lake (*I. Davisii*, west and north); Dinas Mowddwy; Llanfyllin; Llanfairynghornwy, Anglesea; N. W. of Llandeilo; Shoal's Hook, &c., in

[1] That is, reflected closely for its anterior half upon the upper crust; most of the species have it convex, *i. e.*, *concave* upon the cast. Probably the difference is due to greater flexibility in the membranous fold.

Pembrokeshire. In Ireland:—Tyrone; Portrane, Dublin; Chair of Kildare; Wexford; &c. LLANDOVERY Rocks, Girvan, Ayrshire, abundant; Builth; Llandovery; Haverfordwest (with *Ill. Thomsoni*), and other localities in S. Wales. Kilbride, Co. Galway (Sir R. Griffith). MAY HILL SANDSTONE, Shropshire, rarely, viz.—Norbury, Pl. XXVIII, f. 6; Chirbury. Also in the Purple ('Tarannon') shales of the Onny River, Shropshire (Cabinet of the late H. Wyatt-Edgell, Esq.). Our figures are chiefly from the Museum Pract. Geology. The species never occurs in Wenlock rocks, and but rarely so high as May Hill Sandstone.

Mr. Wyatt-Edgell's cabinet contains a caudal shield, clearly intermediate in form between *Il. Bowmanni* and our next species, which must be regarded therefore as a subspecies only. We shall term it *Il. æmulus*:—

Subspecies I.

ILLÆNUS (DYSPLANUS) ÆMULUS, n. sp. Pl. XXVIII, fig. 5.

Il. (D.) modicus, forsan triuncis, depressus, superficie imbricatâ, caudâ (cætera non adhuc inventa sunt) latâ. Long. caudæ 16 *lin., lat.* 2 *unc. Cauda semicirculata, antè subrecta (angulis truncatis exceptis), posticè æquabiliter rotundata. Axis dimidium caudæ vix efficiens, brevis; sulcis axalibus latis convergentibus, ad marginem superiorem profundis latisque. Annuli in axe antico* 5-6, *inconspicui, arcuati. Latera sulcis* 3-4 *brevibus et radiantibus distincta, quorum supernum latus profundius. Fulcra ab axe dimidium latitudinis ejus distantia. Anguli truncati, facie externâ curvâ. Margo valdè deflexus. Fascia lata.*

A single caudal shield or two only of this remarkably fine fossil have yet occurred. They are from that as yet little known formation, the May Hill Sandstone, and are in the collection of Silurian fossils in Jermyn Street. Collectors should search for the head and body-rings, which probably would indicate a short obtuse species, depressed above, but convex on the borders. It can only at present be regarded as a subspecies of the preceding.

Tail 2 inches wide, and 1 inch 4 lines long; semicircular, convex on the steep sides and depressed above. The anterior border is rather straight. The axis is broad above, defined there by wide axal depressions, and not quite equal in width to half that of the front margin; from thence greatly converging shallow axal furrows reach one third down the tail. The axis is, however, really extended further, and is faintly indicated as far as nearly half-down the tail; it forms nearly an equilateral triangle—our figure does not show this quite correctly. Faint arched rings, five or six in number, ornament the upper part, and seem to run almost continuously into the strong squamate Λ-shaped plicæ which ornament the surface. These are stronger, sharper, and more remote than in any species with which I am familiar.

The sides show a remote fulcrum, placed at about half the width of the axis away from it, and thence a suddenly recurved facet, which is convex outwards, while in most species it is nearly flat. The angles are thus truncated considerably; and this gives this rare species much the look of the ordinary *Il. Bowmanni*, figured in the same Plate, fig. 7.

The sides are moreover radiated by 4 short furrows; of which, as usual, the uppermost, subtending the fulcrum, is broad and strong; the second fainter, and the rest obscure. These furrows do not reach above half across the sides, which are gently convex only above, and then slightly decurved. Our figured specimen is a perfect internal cast in arenaceous limestone, and therefore, not being broken, conceals the fascia entirely. But a Presteign specimen, in the 'Mus. P. Geology,' shows a broadish *fascia* with remote striæ.

The ornament of the surface is very remarkable, as above noticed. Strong, arched, or rather bent striæ, which run up obliquely from the sides towards the axis, inosculating as they go in rather an unusual manner. They keep clear of the deflexed margin, and do not cover the upper half of the sides, but reach the lower half of the short axis, and seem to be continuous with its obscure arched furrows, indicating the many axal rings. Where they meet from either side they take a Λ-shaped form, something like the ornament on *Lucina divaricata* and many species of *Pecten*.

Localities.—MAY HILL SANDSTONE of Upper Snead, near Chirbury, Shropshire. Also Presteign, Radnorshire. (Coll. by Mr. J. E. Davis: both specimens are in the Mus. Pract. Geol.)

ILLÆNUS (DYSPLANUS) THOMSONI, *Salter*. Pl. XXVIII, figs. 2—4; Pl. XXX, figs. 8—10.

ILLÆNUS (DYSPL.) THOMSONI, *Salter*. Quart. Geol. Journ., vol. vii, p. 171, pl. ix, fig. 3, 1851.
— — — *Id.* Siluria, 3rd ed., Appendix, 1867.
— — — *Id.* Mem. Geol. Surv., vol. iii, p. 360, as *Illænopsis*, in p. 231, 1866.

Il. (D.) maximus, 6 uncias longus, 3¼ latus, lævis; capite lentè convexo, caudam semi-ellipticam vix superante, thoracem valde trilobum hac superante. Oculi modici, haud distantes, longitudinem eorum à margine postico distantes. Anguli rotundati? Genæ parvæ. Thoracis longitudo ⅔ capitis efficit, annulis 9;[1] *axis ejus valdè convexus pleuris latior est. Pleuræ fulcro tenus (hóc postice ad dimidium posito) planæ, dein abruptè declives et reflexæ. Cauda lentè convexa, axe magno tantum inter foveas conspicuo, sulcis omnino obsoletis. Latera abruptè latèque truncata. Fascia latissima concava, striis raris conspicuis.*

There is a close general resemblance in this bulky Llandovery species to the more common *Il. Bowmanni*. I have therefore contrasted the diagnoses minutely, and beg the

[1] I think I am not mistaken in this number. We have two perfect specimens, but in both the rings have somewhat slipped over each other. The relation of the species to *Il. Bowmanni* is so close in many respects, that it can hardly be other than a *Dysplanus*.

patience of classical readers for their unwieldy length. *Il. Bowmanni* has the greater range, for our fossil is confined to the Llandovery and May Hill rocks; but in the former the two species frequently occur together, and their differences then become manifest enough. The larger form has much less remote and less backward eyes, a stronger trilobation, and a more tapering shape in the axis of the thorax; the pleuræ are more abruptly bent backward; and the tail longer in proportion to the width.

It needs close examination to detect these proportional differences in the *Illæni*; but the habit and aspect would be alone sufficient, for *Il. Thomsoni* is very much less convex, the thorax much longer, &c. These differences will appear more fully in the course of description.

A large oblong species, not less than six inches by three inches and a quarter! Rounded, but not obtuse, at either end; regularly convex, not at all gibbous; and with the thorax strongly trilobed, the semioval head faintly so, and the tail scarcely at all trilobate. The head appears to be scarcely at all longer than the tail, but both are longer by one third than the thorax. The fine specimen in Pl. XXX, figured from Prof. Wyville Thomson's Cabinet, shows these proportions.

Head only gently convex, and divided pretty equally into glabella and cheeks by long straight axal furrows, faint on the outside, but strong in the cast, with a pair of ovate-lanceolate glands.[1] The cheeks outside the facial suture are narrow and abruptly depressed, but are level from the glabella to the eyes, the latter being large and prominent, and placed at about two thirds the glabella's width away from it; they are fully their own length from the hinder margin, and being larger than in *Il. Bowmanni,* of course appear further forwards than in that species. The sculpture of the head, so far as seen, consists only of short broken arched lines and a few puncta, except over the front, where larger and more remote lines occur at rare intervals. We do not know the actual front margin.

Thorax of nine broad rings, greatly arched forwards on the axis, which even posteriorly is wider than the pleuræ, and greatly so in front; it is divided from these by strong axal furrows. The pleuræ rise a little as they leave the axis, and are gently convex rather than flat[2] towards the fulcrum, placed at less than half out in the hinder rings, and at one third in front. Thence they bend strongly downward and backward, and are a good deal thickened, as shown in the cast (Pl. XXVIII, fig. 2). The tips are blunt-pointed, not at all truncate as in fig. 7 (*Il. Bowmanni*).

Tail long, semioval, except for the strong re-entering curves and angles of the

[1] I can only suppose these oval spaces, frequently punctate-granulate, to represent one pair of the curious gland-like markings which occur on so many segments of the head, body, and caudal axis in various genera (p. 50, &c.). Prof. W. Thomson does not believe them glands; what are they?

[2] *Illænus* has apparently ungrooved pleuræ in all the species. These specimens show the meaning of this; for the true position of the groove, which, distinct or not, exists in all Trilobites, is here quite at the hinder edge of the pleura, and is visible internally in the two front rings.

front edge. It is as broad as long, slightly and very evenly convex all over, the broad axal lobe being marked out in front by shallow depressions only, but no trace of axal furrows, however short. The sides, too, show scarce a trace of the usual strong upper groove—so conspicuous in most species. The greatest convexity is about the middle of the caudal shield, and the edge is neither obtuse nor recurved. The front margin of the tail is unusually sinuous. The front edge of the axis is greatly arched forwards; the fulcral point rectangular, and the outer angles widely truncate, so that from the fulcrum the edge is almost vertically cut off. The *fascia*, narrow and convex at the outer angles, becomes very wide and quite *concave* (*i. e. convex* in the cast) round the posterior border, and is coarsely striate.

Localities.—LLANDOVERY ROCKS. Abundant in the light-coloured sandstones of Mullock, Girvan Water, Ayrshire (Mus. Pract. Geology and of Prof. Wyville Thomson). Also the species is common at Haverfordwest, Pembrokeshire (Museum of the late H. Wyatt-Edgell, and of Mr. Lightbody). MAY HILL SANDSTONE, Presteign: Builth: and Pen-y-lan, Llandovery; S. Wales. TARANNON SHALE: Onny River, Shropshire (Mus. Pract. Geol.).

Subspecies II.

ILLÆNUS NEXILIS. Pl. XXX, figs. 4, 5.

This fossil appears to be intermediate between the ordinary *Il. Bowmanni* and our next species *Il. Thomsoni*, which is evidently distinct. I cannot be so sure about the present one, which, while intermediate in form, possesses some characters not common to either species. It has the aspect of *Il. Thomsoni*, the even contour of tail, and arched convex axis of the thorax, but not the abruptly reflexed pleuræ or long truncate angles to the tail of that species. The position of the eye and the convexity of the head are like those of *Il. Bowmanni;* but it differs from that species—by having no trace of the narrow neck-furrow so conspicuous in the Caradoc fossil; and by having a *thickened* line of apophyses below the axal furrows of the longer thorax, which thickened ridge produces a broad and very definite axal furrow in the cast; this is absent in *Ill. Bowmanni.* Again, the tail is longer; the front margin of its axal portion arched, not truncate; the upper lateral furrows all but obsolete; the fulcrum closer in; and the facet much more oblique, thus truncating the angle very slightly. And the *fascia,* which first drew my attention to this marked subspecies, differs entirely from that of *Il. Bowmanni,* and is not quite like *Il. Thomsoni.* I have only seen three specimens, all from the same locality, one of which is a caudal shield, showing the considerable size of the species.

Elliptical, 4 inches long, of which the large semioval tail forms nearly one half, while the very convex (not gibbous) head is about one third longer than the strongly trilobed thorax of nine rings.

The head forms a regular quarter of a sphere, and is very equally and highly convex; it is marked for one third up by axal furrows, which first converge a little, and then turn outward above the oval gland. Eye distant from the glabella about half the width of the latter, of moderate size, and placed at less than its own length apart from the neck margin. Cheeks broad, with rounded angles—the facial line divergent above the eye. No neck-furrow, even in the cast.

Thorax with its axis nearly parallel-sided, a good deal wider than the pleuræ, and very convex, separated (in the cast, at least) by a broad, flat-bottomed furrow; the pleuræ convex thence as far as the fulcrum, which is placed at one third out in all the rings; the pleuræ bent down and a good deal backwards from the fulcral point; the ends pointed, and curved backward (they are straight in *Il. Bowmanni*).

The tail a long half-oval, broader than long, and not obtuse at the tip; the front margin uneven, divided into a broad axal lobe much arched forward, which is distinctly marked out by axal pits, but no axal furrows. Thence to the fulcrum the margin is straight, and beyond it a wide oblique truncation, with the edge of the facet vertical. No axal furrows or depressions interrupt the smooth contour, except at the extreme front margin.

Our figured specimen shows (the cast being broken) a very broad and remarkably bent fascia beneath, extending near half-way up the tail, and divided into two very distinct portions. The outer part is narrower and only slightly concave round the hinder margin (Fig. 48 *a*); but at the upper angles (in Fig. 48 *b*) this part becomes first flat, and then strongly concave; and thence bends upwards, with a sharp angle (making a deep furrow on the cast) between it and the inner half of the fascia. This *inner* portion is then reflected closely beneath the upper surface, and follows it. The squamous lines which cover the inner half are wide apart; those on the marginal portion closer together. Round the extreme edge the fascia forms a convex fillet, so that the section of the whole fascia, very perfect in this specimen, would be somewhat like our figure 48. The variations of the caudal fascia deserve close study; but the details of its shape are better understood by a figure than description; a circumstance not at all uncommon in natural-history illustration.

Fig. 48.

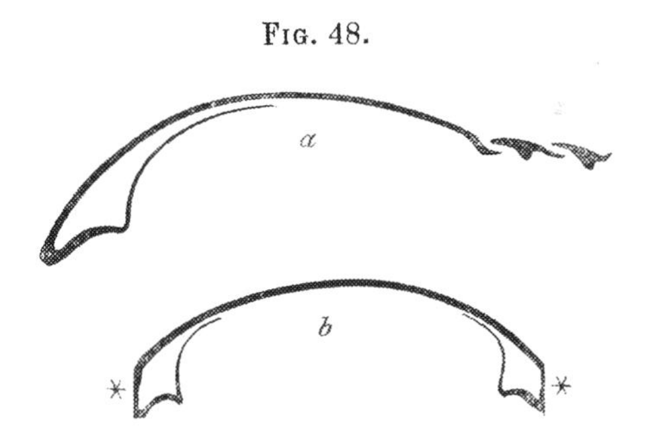

Section of the tail, with its incurved fascia: *a*, longitudinal section; *b*, transverse section across the facets (* *).

Locality.—Llandovery schists of Mullock, Girvan, Ayrshire (Prof. Thomson's and Mr. Wyatt-Edgell's cabinets).

Subgenus 4.—ILLÆNUS proper.

ILLÆNUS (IL.) BAILYI, n. sp. Pl. XXVIII, fig. 14.

Compare with ILLÆNUS CRASSICAUDA, *Wahlenberg*. Nova Acta Reg. Soc. Sci. Upsal., vol. viii, p. 27, tab. ii, figs. 5, 6, 1821—not of Portlock nor M'Coy.
— also with ILLÆNUS DALMANI, *Volborth*. Mém. Acad. Imp. Sc. Petersb., vol. vi, No. 2, p. 13, tab. ii, figs. 7—13. *Il. crassicauda*, Dalm. et Auctorum, fide Volborth, 1863.

Ill. (*Il.*) *magnus*, 4 *uncias longus*, $2\frac{3}{4}$ *latus, ellipticus, obtusissimus, profundè trilobus; caput gibbum multùm caudâ planiore majus, oculis distantibus retrorsis; angulisque rotundatis; thorax caudâ brevior. Caput transversum valdè convexum, vix gibbum, glabellâ angustâ, sulcis axalibus brevibus tertias capitis efficientibus. Genæ abruptè declives. Oculi modici eminentes, longè à glabellâ positi, et dimidium longitudinis eorum a margine postico distantes. Axis corporis convexus, parallelus, pleuris vix latior: his fulcro tenus* (*ad dimidium posito*) *planis, dein valdè deflexis, paullò* (*posticis præcipuè*) *recurvis. Cauda semicirculata* (*angulis obliquè truncatis*), *subplana seu lentè convexa, axe lato, inter foveas latas tantùm conspicuo, sulcis axalibus nullis: margo acutus, nec re-curvus.*

I have purposely confined the diagnosis and description to our single British specimen. It may, indeed, possibly prove identical with the common Scandinavian form above quoted. But it is much larger than *Il. crassicauda*, at least if the specimens which reach England are fair samples of that species. (I have here figured *Il. crassicauda* and its ally *Il. Dalmani*, for comparison.) And while in the position and size of the

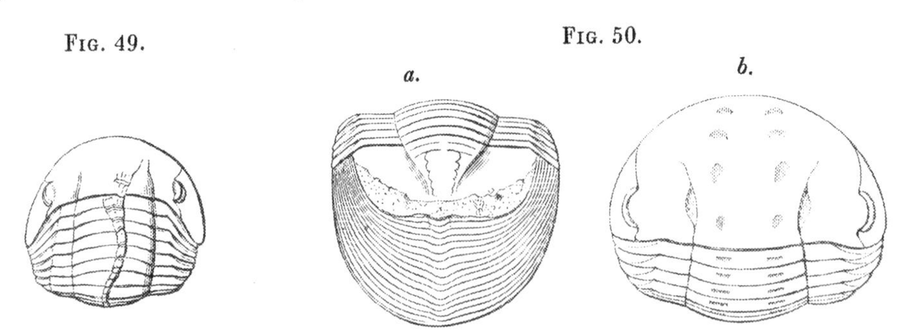

FIG. 49. Internal cast of *Illænus crassicauda, Wahl.* showing alimentary canal (full of ingested matter ?), from Volborth.[1]

FIG. 50. *Illænus Dalmani*, Volb. (After Volborth), Lower Silurian, North Russia.
a. The caudal fascia, seen when the surface of the cast is broken away.
b. Interior of cast, showing lobes (attachments of muscles ?).

[1] Volborth "On the Smooth-tailed Russian Trilobites," 'Mém. Imp. Acad. Sc. Petersburg,' 7th ser., tom. vi, No. 2, 1863. From this important and philosophical memoir our woodcuts illustrative of the structure are all taken. Dr. Volborth discusses all the accessible points of the anatomy of Illænus. He regards the internal tube (fig. 49) as the dorsal heart, comparing it with that of *Apus cancriformis*. I trust I am not presumptuous in supposing that the viscus represented by Dr. Volborth as a heart or

eye, the convexity of the head between the eyes, and the proportionally longer tail-piece, our fossil differs from the ordinary *Il. crassicauda.* A nearer approach is made in these respects to Dr. Volborth's new variety or species *Il.* (*crass.*) *Dalmani*, Volb.; but I will not unite it with that form, which has larger eyes, nearer the glabella—a longer head, and squarer head-angles. The species is so near to these two Russian forms, that I give a woodcut of both, more especially for the purpose of showing the structural points gained for us by the acumen of Dr. Volborth, in the work cited in the foot-note.

We have but one specimen, rolled up, of this fine species, for the use of which I am indebted to the Irish Geol. Survey. At first sight it was naturally taken for *Il. Bowmanni*, the common Lower Silurian form; and it occurs in beds of the same age. To that species, too, it bears much resemblance in the shape of the head, position of the eyes, and the general aspect.

But it is a true *Illænus*, with 10 rings—not a *Dysplanus;* and its alliance is with the Swedish forms of *Il. crassicauda*, Wahl., and *Il. Dalmani* of Volborth. *Il. crassicauda* has been often quoted from Britain, but *always erroneously;* whether this may be an extreme variety of that northern species, uniting the above-mentioned forms, I am not quite sure; but I shall try to point out its proportional differences, which are nearly all the characters we have to rely on; at least, in the great majority of the species of this very uniform genus.

Our coiled-up specimen, with the crust preserved, measures rather more than three inches in breadth, and when uncoiled must have been four inches long. Of this length the large semicircular head, more than a quarter of a sphere in convexity,[1] is two and three quarter inches long, obtuse in form, but regularly convex, and three inches broad. The glabella occupies less than a third of this breadth, and is defined only by short axal furrows, strongest behind, which do not reach one third up the head. The prominent eye is placed as far from the glabella as the width of the latter, and very close to the hinder margin—not its own length from it. The facial suture beneath it turns sharply outward. In front of the eye its course is not known. The cheeks slope gradually down, not abruptly, and are greatly rounded off at the angles—more so than in kindred species. The chief convexity of the head is more than midway up—not near the vertex as in *Il. crassicauda*, which, seen from above, has a shorter head and less rounded angles (Volborth, l. c., pl. ii).

The thorax is much shorter than the head, and rather shorter than the tail, the proportions of the three being nearly as 4 : $2\frac{1}{3}$: 3. This may probably vary in some individuals, and is sure to be different in younger specimens, which would have the tail shorter.

dorsal vessel (see woodcut 49) must be the cast of an annulated or saccate intestine. It seems to me all but impossible that the contents of the heart should be preserved in a solid state; but the *ingesta* may easily be so, if I am right in supposing the food to be silty matter.

[1] A precise English term conveying this idea—parallel to the terms spherical, hemispherical—is wanted. Will any Greek scholar supply it? *Tetartospherical* is a long word to use, and I hate *verba sesquipedalia;* still, we require a term for a shape very common in the carapaces of Crustacea.

The axis of the thorax is regularly convex; it is divided from the pleuræ, which are narrower by one fourth than the axis, by straight and parallel furrows, sharp but not deep, and continued a very little way into the tail. The fulcral points are placed rather more than one third out, and at nearly equal distance in all the rings; and the pleuræ are thence curved, rather than bent downward, and turn but little backward except in the front rings. The tips seem to be squarish or obtuse, and are not recurved.

Tail semicircular, not much sinuated on the front border, the broad axis being scarcely arched forward; and the facet beyond the fulcrum, following the line of the pleuræ, forms a very obtuse angle with the straight portion (not abruptly cutting off the angle, as in many species,—*Il. Thomsoni*, for instance). The general convexity is slight and very regular, the border not being abruptly decurved; and the axis is obscure, except for the broad obtuse depressions on the front border; it does not form any axal lobe.

Comparing this form with the typical *Il. crassicauda*, as given by Volborth, we find the head less gibbous, and more generally convex; the eyes smaller, and placed further out; the fulcra of the pleuræ at nearly equal distance from the axis in all the rings; and the tail proportionately longer, with the axis indistinct, except quite in front.

Some of these differences become resemblances when we compare this form with *Il. Dalmani*, which Volborth has critically distinguished. But the head of that species is less convex than ours, and longer too—semi-elliptical rather than semicircular. The transverse section of a rolled specimen of that species is more obtuse than in ours, and less pointed at the ends. The cheeks bend more steeply down. The tail is longer than in the British specimens, and has the axis well marked out, &c.

I cannot think, therefore, I shall be wrong in distinguishing our fossil by the name of W. H. Baily, Esq., whose useful work as Palæontologist to the Irish Survey deserves full recognition. Mr. Baily has sent me several undescribed forms, which will be referred to under their proper genera.

Locality.—Caradoc or Bala. Dunabrattin Head, Waterford. (Mus. Irish Geol. Survey, Dublin.)

Illænus (Ill.) Davisii, *Salter*. Pl. XXIX, figs. 10—16.

Illænus crassicauda, *Sharpe*. Quart. Geol. Journ., vol. iv, p. 149, 1848.
— Davisii, *Salter*. Decades Geol. Survey, No. 2, pl. ii, 1849.
— — *M'Coy*. Synopsis Woodw. Fossils, p. 171, pl. i G, fig. 36, 1852.
— — *Salter*. Ibid., Appendix, p. iv, 1852.
— — *Id*. Morris's Catal., 2nd edit., p. 110, 1854.
— — *Id*. Siluria, 1st edit., woodcut 29, fig. 2, 1854; 2nd edit., p. 223, Foss. 44, fig. 2, 1859.
— — *Id*. Catalogue Mus. Geol. Survey, pp. 5, 19, 1865.
— — *Id*. Memoirs Geol. Surv., vol. iii, pl. xviii, fig. 9, p. 317, 1865.

Il. (*Il.*) *minor*, $2\frac{1}{2}$ *uncias longus, ellipticus, obtusus, convexus, benè trilobus; capite et pygidio ejusdem magnitudinis ferè, oculis parvis approximatis. Caput semiovatum, convexum nec gibbum, glabellâ angustâ sulcis axalibus brevibus convergentibus. Genæ declives, angulis quadratis. Oculi parvi, elongati, glabellæ propinqui, â basi capitis diametrum eorum distantes. Thorax minùs convexus, fusiformis, sulcis axalibus minimè profundis. Pleuræ fulcro axi propinquæ, rectæ, angustæ. Cauda semicirculata ferè, margine antico vix sinuato, angulis vix truncatis; axe lato, inter sulcos latos* (*in pullo conspicuos*) *paullulum elevato; margine deflexo. Fascia lata, striis crebris.*

One of the many peculiar fossils which abound in the more argillaceous parts of the Bala Limestone, but which give place to others directly the matrix becomes sandy. A list of such species is given in the third volume of the Memoirs of the Survey, quoted above.[1] This difference led Prof. Sedgwick and myself, in 1844, to believe there were two bands of limestone; but the researches of the Geological Survey have shown that there is but one principal band. Indeed, the calcareous bed called the Bala Limestone is continuous, or at least intermittent, over something like 4000 square miles. From Kildare, in Ireland, to the Grug limestone[2] at Llandeilo, in South Wales, in one direction; thence to Bala, in North Wales; at Horderly, in Shropshire; as the Coniston band in Westmoreland; in Ayr, Peebles, and probably through all the Western Highlands, this remarkable band of limestone ranges, and keeps the same fossils throughout, with mere local variations, such as I have just noticed: and, perhaps, there is no species of Trilobite so characteristic of this band (it is not the most abundant) as *Illænus Davisii*. *Il. Bowmanni* nearly everywhere accompanies it, but that species has a wider range, and seems to have been necessarily less confined to one kind of sea-bottom. The reader will pardon this digression for the sake of the facts; the Bala Limestone being the key to the geology of the Lower Silurian (Middle Bala group, Sedgwick).

Il. Davisii is a small species, seldom more than two or two and a half inches long, convex, but not gibbous, well trilobed, but not with deep axal furrows, and with the margin rather suddenly bent down. The eyes are forward and approximate, as com-

[1] I may, perhaps, not have so good an opportunity of showing this change which takes place in one and the same bed of limestone, at a distance of certainly not two miles. Where the Bala Limestone has a muddy matrix the Trilobites are as in the first column: where it is sandy, the other group is conspicuous.

Argillaceous Bala Limestone.	*Arenaceous Bala Limestone.*
Trinucleus seticornis.	Trinucleus concentricus.
Illænus Davisii.	Illænus Bowmanni.
Cheirurus bimucronatus.	Phacops apiculatus.
Asaphus radiatus.	Asaphus Powisii.
Agnostus trinodus.	Calymene senaria
Ampyx tumidus.	(Beyrichia complicata).

The shells follow the same rule, but less strictly.—J. W. S., 1853; 'Mem. Geol. Surv.,' vol. iii, p. 273.

[2] This Grug limestone is often confounded with the Llandeilo limestone: it is brought close to it by faults, but has quite different fossils.

pared with *Il. Bowmanni* and its allies; but this is a true *Illænus*, and does not need comparison with that species. I give the description, revised, from 'Decade 2.'

General form oval; length to width, as 17 : 10; the semioval head and semicircular tail nearly equal in length; the thorax about two thirds the length of either, and strongly, but not deeply, trilobate. The axal furrows are carried less than half way up the head, and but a little way down the tail (except in young specimens, fig. 16).

Head more depressed than a quarter of a sphere, and regularly convex, not gibbous behind (our fig. 13 is too much depressed, being crushed a little). It is divided into three nearly equal parts by the short and slightly converging axal furrows, which turn out again, and then cease at about the level of the top of the eye. The latter is of moderate size, gently lunate and narrow, bounded beneath by a slight furrow, and placed fully its own length from the posterior margin, which shows no trace of a neck-furrow, within or without the crust. The facial suture is divergent above the eye, and slightly so below it, so as to cut the margin beneath the most prominent curve of that organ. The rostral shield (fig. 14) is shuttle-shaped, more than twice as wide as long, and produced into an angle below, where the labrum, which we do not yet know, would fit to it.

Thorax of ten narrow segments, the axis well marked, gently convex, and subfusiform; wider than the pleuræ in the forward segments, in the last only equal to them. The fulcrum is very near the axis in the first segment, and in the last placed scarcely more than a third along the pleuræ, which have a distinct facet and oblique ends. The front pleuræ bend down and a little back; the hinder ones are straight, only bent downward.

Tail semicircular and moderately convex, chiefly so toward the margin, which descends abruptly, but rather flattened along the anterior two thirds. The axis is indistinctly marked out by two deep impressions, which sometimes form short, rapidly converging furrows; the upper corners are bent sharply down beyond the fulcrum, in order to pass freely under the thorax-rings in rolling, but are not truncated as in many species; so that the outline of the tail is tolerably semicircular, and straighter in front than in many forms of the genus. The fascia is of even width all round, and not very broad; it is rather finely striate. The tail, too, has oblique ornamental lines round the margin.

In the tail of young specimens the axis is marked out nearly all round (fig. 16), and extends three-fifths down the tail, which is also flatter.

Var. β, Involutus.—'Decade 2, Geol. Survey,' pl. ii, fig. 8.

The axis in some specimens is so much narrower, and the tail-margin so much more incurved, that the specimens possessing these characters might well pass for examples of a new species. The axis is truly very narrow, and the fulcrum more remote, as usual in all such cases. I find the same form in Dr. Wyville Thomson's Cabinet, from the Ayrshire district, and note it under a varietal name, not much doubting that, when we know the perfect form, we shall find this variety a true species, or, at least, one of those constant forms which botanists call *sub-species*,—a very useful term.

Comparing other species of the same sub-genus with ours, we find that *Il. crassicauda* is not only a much more convex form, but has the head rather gibbous behind; the eyes more remote and further back; the tail scarcely longer than the thorax, the latter having the fulcrum more remote in the front rings, and placed halfway out in the hinder ones; the incurved under portion or fascia (for example see fig. 50) far broader and less concave (*i. e.* less reflected), and the surface sculptured by strong sharp lines; these lines are certainly not conspicuous in *Il. Davisii,* though the latter is not quite smooth.

From *Il. Portlockii* its much greater convexity and the approximate eyes readily distinguish it. The semicircular, not long tail, and oblique pleural tips distinguish it from the species we have named *Il. Rosenbergii. Il. Murchisoni* does not need comparison; and the position of the eyes will at once separate our neat regularly convex fossil from the *Il. Bailyi.* I hardly know any species that has better characters of habit; and it is rather strange it should ever have been confounded with the Swedish forms.

Localities.—Caradoc or Bala limestone and slate, west and north of Bala Lake, at Rhiwlas chiefly; also Pont-y-Glyn, Diffwys, Corwen, and other intermediate places; Llanwddyn, Montgomeryshire. (Mus. P. Geol. and Woodwardian Mus.)

In Scotland,—the Wrae limestone; and at Biggar, Peeblesshire. (Mus. P. Geol.)

Illænus Portlockii, *Salter.* Pl. XXVI, figs. 3, 4.

Illænus crassicauda, *Portlock.* Geol. Rep. Tyrone, &c., pl. x, figs. 7, 8, 1843.
— Portlockii, *Salter.* Decades Geol. Surv., 2, pl. ii, p. 3, 1849.
— — *Id.* Siluria, 2nd edit., Appendix, p. 539, 1859.
— — *Id.* Morris's Catal., 2nd ed., p. 110, 1854.
— — *Id.* Catal. Mus. Pract. Geol., p. 5, 1865.

Il. (*Ill.*) *modicus,* $2\frac{1}{2}$ *uncias longus, latè ovatus, depressus, thorace caudâque ejusdem longitudinis ferè. Caput convexum* (*in juniore*) *glabellâ angustâ, sulcis axalibus brevibus; oculis distantibus posticis. Thoracis axis pleuris longè latior est, his fulcro tenus planis, dein subitò valdèque reflexis, deflexis fulcro distante. Cauda planata, transversa, oblonga, subquadrata, angulis externis longè truncatis; axe magno trientem caudæ efficiente et ultra, sulcis axalibus subparallelis. Fascia lata concava.*

A species well figured in Gen. Portlock's very unfortunate plates; but he referred it, apparently without much consideration, to the *Illænus crassicauda* of Dalman, from which it differs in every particular except the number of body-rings, a sub-generic character. It is of a flattened shape, with short wide triangular tail, and with pleuræ more abruptly bent down than in any other species. *Il. Portlockii* is, perhaps, one of the most distinct of the British *Illæni.* It is only yet known in one locality, the famous Caradoc schists of Desertcreat parish, Co. Tyrone; a spot rendered classic by Portlock's excellent work.

The species must have been fully two and a half inches long, by nineteen lines wide; broad ovate, depressed. The thorax is nearly as long as the tail, which is wide and very short, and has the angles broadly and strongly truncate. Our smaller specimen has only a very imperfect head, which is very convex, probably even gibbous about the base of the distinct glabella; the axal furrows are broad, short, and rather deep: the eyes remote, placed near the posterior margin, apparently close to it, and so far apart as to be more than half the width of the glabella remote from it. Head-angles (probably) obtuse.

Thorax of ten rings, depressed; the axis considerably wider than the pleuræ, and strongly, not deeply, divided from them, scarcely tapering backwards, but more so in the young than the adult.

Pleuræ flat as far as the fulcrum, which is placed far outwards, at about halfway out in all the rings; in the first rings nearly as far outwards as in the hinder ones, a character by no means common. Beyond the fulcrum the pleuræ are abruptly bent downwards, and very much backwards, almost at a right angle to the line of the pleuræ; indeed, more so than in any British species, except, perhaps, *Il. Thomsoni.*

Tail quadrate transverse, the posterior margin elliptical, the front edge slightly sinuated by the arch of the axis,—and with the angles so abruptly truncated beyond the fulcrum as to give an oblong instead of a semicircular shape to the tail. The axis is broader than one third of the width of the tail, marked out by short and rather strong sub-parallel furrows, which reach one third down the tail and are then lost,—at least upon the upper surface. Beneath the crust the anal extremity of the axis is prominent, and forms a narrow sulcus in the cast (fig. 3). This reaches nearly to the margin, and probably indents the broad concave fascia; but our specimen is not quite complete enough to show this. The fascia, however, is slightly convex near the margin, as indicated imperfectly in our figure by a shallow furrow; and extends halfway, or nearly so, up the tail, parallel with the upper surface; above it ends just outside the fulcral point.

Locality.—CARADOC schists of Desertcreat, Co. Tyrone. (Both specimens in Mus. P. Geology; they are Portlock's originals.)

ILLÆNUS (ILL.?) OCULARIS, *Salter*. Pl. XXIX, figs. 7, 8 (9?).

ILLÆNUS OCULARIS, *Salter*. Decade 2, pl. ii, p. 4, 1849.
— — *Id.* Morris's Catal., 2nd edit., p. 110, 1854.
— — *Id.* Catal. Mus. Pract. Geol., p. 5, 1865.

Il. (*Il.?*) *minor, vix unciam latus, lentè convexus; capite* (*solùm adhùc cognoto*) *semicirculato, oculis longis. Caput modicè convexum, insuper depressum, margine frontali gibbo. Glabella genis angustior, sulcis axalibus incurvatis, his dimidium ferè capitis efficientibus.*

Oculi longi, glabellæ vicini, á cervice vix dimidium diametri eorum distantes. Genæ triangulatæ, angulis productis nec acutis, haud spinosis. Thoracis axis pleuris vix latior, pleuris anticis deflexis. Reliqua absunt.

This pretty species may or may not belong to *Illænus* proper. It is more depressed than other Lower Silurian forms, and the front margin is suddenly incurved. The eyes are long, not broad, and gently curved, and with the lentiferous area convex. They are placed so near to the posterior margin as to be about half their length away from it; but being of considerable size for the genus, they reach about halfway up the head. The eye-lobe is depressed, and there is no furrow under the eye itself. The glabella is less than a third the width of the head; the axal furrows, sigmoid in shape, rise up to the level of the front of the eye, and do not converge so much as our figure makes them. Our enlarged figure 8 *a* is still a little further defective, as representing the eye too forward. Fine, concentric, wavy, continuous lines ornament the whole head, but are most conspicuous and coarse around the inflected front margin.

Only parts of three front body-rings are present. Their axis is a little wider than the recurved pleuræ, which are bent down and backwards from the approximate fulcrum, and appear to have rather an attenuated form and oblique apices. They are longitudinally striate.

It is with some doubt that I add fig. 9 to this species; it has a proportionately still larger eye, and a squarer cheek-angle, not so much produced as in the small form. But these may be characters due to age, and there is no other species to compare either with; so I leave it. It is worth figuring, at all events.

Locality.—CARADOC or BALA Limestone, Chair of Kildare, Kildare; quoted as Llandeilo in the 'Decade,' as the Bala Limestone was formerly not known to be the exact equivalent of the Caradoc, but supposed to be the Llandeilo Limestone, until the fossils disproved it. (MUS. P. GEOLOGY. Only these two specimens are known to me.)

ILLÆNUS ROSENBERGII, *Eichwald.* Pl. XXIX, figs. 2—6.

ILLÆNUS ROSENBERGII, *Eichwald,* Geogn. Zool. per Ingriam, Tril. Obs., &c., t. iii, fig. 3 (?), 1833.
— — *Salter.* Mem. Geol. Surv., vol. ii, part 1, p. 338 (not plate), 1849.
— — *M'Coy.* Synopsis Woodw. Mus., p. 172, 1852.
— MURCHISONI, *Salter.* Id., Appendix, pl. i G, figs. 33—35 (not of Decade 2, Geol. Surv., which is the true *Il. Murchisoni*) : see p. 201, 1852.

Il. (*Il.*) 4—5 *uncias longus, gibbus, intús granuloso-striatus, capite valde convexo longitudine caudam rotundatam superante, thorace longo. Caput semiovatum, convexissimum, fronte gibba impendente, sulcis modicis subrectis, tertias longitudinis capitis efficientibus. Genæ majores verticales. Oculi minores, haud distantes, á glabellá dimidio lati-*

tudinis ejus vix sejuncti; á margine postico distantiores. Anguli —? Thorax capite longior, è annulis 10 *latis confectus, quorum axis modicus, sulci axales profundi. Fulcra proxima; pleuræ rectæ, longitudinaliter striato-costatæ, apicibus truncatis. Cauda oblongo-ovata, ad mediam partem gibba, angulis truncatis, axe brevi, angusto, supernè conspicuo, intùs* 3—4-*annulato. Margo caudæ nequidem recurvus; fascia lata concava, striis imbricatis paucis imbricata.*

It is sad patchwork reconstructing this species, though we have plenty of fragments—heads without cheeks, a distorted but complete thorax, and several tail-pieces, all of which show a strong granular character of the inner crust, which is exhibited by short broken lines and puncta in the cast. Even fragments show this character. The species is pretty common in the Coniston limestone; and all who know that rock know what a vast amount of cleavage and distortion the fossils have undergone.

Nor are we quite certain about the name. Eichwald's figure truly represents a Scandinavian form distinct from the common *Il. crassicauda,* and very much like ours, with broad thorax-rings, and the eye placed near the furrows, and rather forward, compared with such species as *Il. Bowmani,* which occurs with it, and has no granular surface.

But in the 'Memoirs Geol. Survey,' vol. ii, pt. 1, when describing this species which I had already distinguished and figured for Prof. Sedgwick's book (the figures quoted are drawn by me), I unfortunately included with it and figured for it the *Il. Murchisoni* described at p. 201. I gave it the same name in the Appendix to the 'Woodwardian Synopsis' above quoted, and thereby vitiated both descriptions. I must, therefore, go over the ground again, as we usually have to do when in a hurry; the Coniston limestone species is the one intended, and should it prove distinct, as I think it will, from Eichwald's fossil (which seems not to have occurred to Prof. Angelin in his Swedish collections), I propose to name this conspicuous Westmoreland fossil *Il. Marshalli,* after the gentleman who has done so much to help forward the study of the Westmoreland fossils.

It is larger than *Il. Bowmani,* and nearer five than four inches long; of a lengthened shape, both the head and tail semioval, with blunt extremities; gibbous, especially in front; deeply trilobate, and marked all over the cast with short wavy impressed lines and puncta. The glabella-furrows are more direct, longer, and converge more than in *Il. Bowmani* (we need not compare it with *Il. Murchisoni,* with which I formerly confounded it). The shape of the head is not very clear, but it is nearly as long as wide, very gibbous forwards, and overhanging like that of *Il. crassicauda,* which species it resembles.

The glabella-furrows are sigmoid, but only slightly converging, and bent out above; they extend forwards twice as far from the posterior margin as the place of the eyes, more than one third, but not half-way up the head (our figure has them rather too much curved). The small eyes are placed about as far out from the glabella as half its width (in *Il. Bowmani* they are two-thirds its width away), and they are placed once and a half their own length from the hinder margin, while in the kindred species they are only their own length in front thereof.

Thorax deeply trilobate; the arched rings broad, and flat from front to back; the pleuræ are equal to the axis in width, and are arched forwards instead of at all backwards; their fulcrum is extremely close to the axal line, seemingly about one fourth out, and from thence the pleuræ are covered by longitudinal lines, not very close-set; their ends are truncate.

The tail (in the figured specimen from the Woodwardian Museum) is compressed; but it was evidently convex; with a narrow axis, defined by broad and rather deep axal furrows, which extend, however, but a short distance, and do not much converge. The axis shows four annulations within the crust (our figure has too many). The sides slope evenly down to the thickened margin, which is not at all flattened or recurved; and the internal fascia is concave, broad, and coarsely and remotely striated with imbricate sculptured lines.

I have described the original specimen, but believe the caudal shield fig. 5, which does not show the axis ribbed, to be the same, and to show the true form of the tail.

Locality.—Coniston; Sunny Brow; and Horton in Ribblesdale, Westmoreland. In CARADOC or BALA LIMESTONE (figs. 2, 4, Woodw. Mus.); fig. 5, Mr. Wyatt-Edgell's Cabinet.

ILLÆNUS (ILL.) MURCHISONI, *Salter.* Pl. XXVI, fig. 1, and Pl. XXX, fig. 7.

ILLÆNUS ROSENBERGII, *Salter.* Mem. Geol. Surv., vol. ii, pt. 1, pl. v, figs. 6—8, 1848. Not of Eichwald.
— MURCHISONI, *Id.* Sedgwick's Synop. Woodw. Mus., Fasc. 1, Appendix, p. iv, 1852; description only. (Only exclude figures of Westmoreland specimens, pl. i G, figs. 33—35, which are *Il. Rosenbergii.*) See p. 199.
— — *Id.* Decades Geol. Survey, No. 2, Art. ii, p. 4, 1849.

Il. (Illænus) magnus, 6-7 uncias longus, ellipsoideus, obtusus, convexissimus, capite et thorace ejusdem longitudinis, caudâ vix minori. Caput pars quarta sphæræ est, sulcis axalibus brevibus, rectis, haud ultra oculos tractis. Oculi majores, vicini, à margine postico diametrum suum distantes. Anguli capitis quadrati. Thorax longus, axe latissimo pleuris subrectis longè latiori, ad fulcrum proximo. Cauda convexa semicirculata, angulis haud truncatis, margine obtuso; axe obscuro lato, fasciâ angustâ.

More desirous to identify the closely related forms of this genus than to multiply species, I too hastily (in 1852) united the fossil from Llandeilo that is figured in the 'Survey Memoirs' with one which is common in beds of the same age in Westmoreland. The latter will be found in p. 199, and I believe it to be the true *Il. Rosenbergii* of Eichwald. The Llandeilo species, however, is clearly distinct, as pointed out by myself in the Appendix to Prof. M'Coy's 'Woodwardian Catalogue ;' and, as there are two species, I must restrict the name to the fossil first figured. That was a crushed specimen; but the finest possible examples have since been obtained from the same locality; and my friend Mr. J. E. Lee, of Caerleon, possesses the best,—our figured specimen. Except the Barr Trilobite, which it much resembles, we have no species of equal bulk in Britain. It is a Bala Limestone fossil; the Grug and Birds Hill quarries being in that rock.

A large and extremely convex species, generally about six or seven inches long. The form a long ellipse, the length being to the breadth as 11 : 5. Head about equal to the thorax, or slightly longer, and the tail a little shorter than either. The head is truly a quarter of a sphere, with square head-angles, and very slightly trilobed by shallow axal furrows, which converge but little, are sigmoid, and reach to the top of the eyes.[1] The glabella thus marked out is not wider than the cheeks. The eyes are large for *Illænus* proper; narrow, with no furrow beneath them, and near to the axis—scarcely a third of the glabella's width away from it. The cheeks steeply decline towards the eyes, and are thence nearly vertical to the blunt margin. The angles, though called square in the diagnosis, are more acute than a right angle, and the corner is rounded off; still the aspect is unusually square for the genus. The shape of the free cheek, from the very outward curve of the facial suture above the eye, is a trapezoid; the side nearest the eye being about one half that opposite to it—viz., the margin.

The convexity of the head is such that a line taken transverse to the eyes would be a semicircle, and that from vertex to front a quarter of a circle. On the underside, the coarsely striate[2] rostral shield (epistome) is as wide as the glabella, and is itself about twice as wide as long. It is narrowed on either side, more so than in *Il. Bowmanni*, and abruptly broader in the middle. And in this form it differs again from the more oval shield of *Il.* (*Bumastus*) *Barriensis*, between which and this species there are many points of resemblance that strike the eye at first glance. Indeed, *Il. Murchisoni* leads from the true *Illæni* to that subgenus, as may be seen by what follows.

The trilobation of the thorax is not very distinct, the axis being almost as broad as that of *Bumastus*, but much more pronounced than in that subgenus.

The thorax of ten rings is about five sixths the length of the head. The greatest breadth of the axis is in the seventh and eighth body-rings, where it is double that of the pleuræ; in the first ring, as the axis is spindle-shaped, the breadth to that of the corresponding pleura is 4 : 3; in the last ring it is 3 : 2. The fulcrum is placed at one third in the first ring, and at one half in the eighth, ninth, and tenth. The pleuræ, flat as far as the approximate fulcrum,[3] thence slightly bent down, and inclining backwards.

[1] Within the crust, on the cast, they show strongly the glandular? depressions opposite the eye, as in *Bumastus Barriensis*, but not so large. These impressions are as yet very problematical. Can they be muscular attachments? I think not.

[2] The striæ are coarsely imbricato-striate, as in *Il. Barriensis*. They are parallel in front with the anterior edge, and behind with the posterior edge; in the middle, nearly direct across.

[3] When the axis is broad, the fulcrum must be always approximate (for mechanical reasons connected with the rolling up), except in very flat Trilobites.

The tail is convex, but not uniformly so, the anterior part and the line of the axis generally being rather flattened. Thence, towards the margin, it is bent down abruptly. The breadth of the tail exceeds the length by about one half the latter; the posterior margin is a semicircle, and the anterior a slightly wavy line (Pl. XXX, fig. 7), on which the broad axis is marked out by shallow indentations, and the angle at which the facet starts a very obtuse one, so that the angles of the tail are scarcely more truncated than in the "Barr Trilobite." There are but slight indications of very shallow axal furrows, but the anterior lateral furrow of the tail beneath the fulcrum is quite distinct. The fascia is slightly convex, and has coarse striæ; it is very narrow. The inner surface of the tail rough, with small points or tubercles.

Locality.—CARADOC or BALA Limestone, at Grug and Birds Hill, close to Llandeilo. (Mus. P. Geol. and Mr. Lee.) Chair of Kildare (Mr. Wyatt-Edgell's Cabinet).

Bumastus follows next in order of affinity; for though technically *Ectillænus* and *Illænopsis* are in the trilobed group, they are so different structurally, on account of the position of the eye, that we may well believe they will form distinct genera by-and-bye. I shall not follow, therefore, the order of the numbers, but place the above-named two sub-genera last. And we now come to the highest, if not the most typical[1] *Illæni*, viz. the Upper Silurian forms of *Bumastus,* distinguished by the great solidity and fine sculpture of the shell or crust, the convexity of the body, the full number of body-rings, largely developed eyes, &c. In this group the trilobation of the tail, all but lost in some of those above described, is altogether absent, and that of the thorax nearly so.

Subgenus 5.—BUMASTUS, *Murch.*, 1837.

ILLÆNUS (BUMASTUS) BARRIENSIS, *Murchison.* Pl. XXVII, figs. 1—5.

"A new species of Trilobite," *F. Jukes.* Ann. Mag. Nat. Hist., vol. ii, p. 42, figs. 8, 9, 10, 1829. "Probably Isotelus," *J. De C. Sowerby,* ib. p. 45. *Silliman,* Amer. Journ. of Science, vol. xxiii, i, p. 203, 1833.

BUMASTUS BARRIENSIS, *Murchison.* Sil. Syst. (description, but not figures), p. 656, pl. vi bis, 1839.

— — *Emmerich.* Dissert., p. 33, 1839.

NILEUS? (BUMASTUS) BARRIENSIS, *Burmeister.* Org. Tril., p. 120, 1843. Ib., Ray edition, p. 104, 1846.

---

[1] The highest and lowest forms of a group are not typical: there is in most natural groups a central form combining all the chief characters. At least that is my idea of type. *Asaphus* is typical of the *Asaphidæ;* it is not the highest form. *Acaste* is the typical *Phacops;* and so forth.

BUMASTUS BARRIENSIS, *Hall.* G. R. New York, p. 102, No. 10, fig. 4; No. 19, fig. 3, 1843.
— — *Id.* Pal. New York, vol. ii, p. 302, pl. lxvi, figs. 1—15, 1852.
ILLÆNUS — *Id.* 18th Rep. State Cabinet (Dec. 1864); possibly also *Il. imperator*, id.
— — *Salter.* Decades Geol. Surv., No. 2, pl. iii (excl. fig. 2), and pl. iv (excl. figs. 9—11), 1849.
— — *Id.* Siluria, 2nd ed., p. 123, Foss. 16, fig. 2 (not plate), 1859.
[NILEUS GLOMERINUS, *Dal.* Arsberätt., p. 136, 1828; Hisinger; Leth. Suec., p. 16, 1837]?

Il. (*Bu.*) *magnus, squamoso-striatus, 6-uncialis, obtusè ovatus, valdè convexus nec gibbus, capite semicirculato, caudâ transversâ. Caput obtusum sub-truncatum, marginibus obtusissimis, sulcis axalibus brevibus valdè convergentibus. Oculi longi haud deflexi, pulvillo longo fulti, sulco lato profundo circumdati. Anguli obtusi. Cauda transversa, semicirculata, convexa.*

This far-famed and common Trilobite has many claims to distinction. Of great actual bulk, for it is remarkably convex; and of no mean linear dimensions, being often nearly 6 inches in length by 3½ in breadth; with the remarkable even contour which induced Sir Roderick to maintain its claims to generic distinction; with a definite geological horizon, the Lower Wenlock, and with a range from New York to Sweden,—the species, which is a peculiarly abundant one, is perhaps as good an Upper Silurian type as can be pointed out. And the beauty of the specimens, the finest known of which are, I believe, figured on our plate, renders *Il. Barriensis* a general favourite. The largest perfect individual extant is in the cabinet of Miss Jukes, of Birmingham; and as it was carefully figured sixteen years ago in the Decades of the Survey, we have been contented to copy that figure, improving it from a good cast. This specimen is 4¾ inches long by 3 inches 4 lines broad at the head. But our fig. 1*a*, from the Museum of Practical Geology, indicates a yet larger size,—it must have been nearly 6 inches in extreme measurement.

It may as well, however, be here stated, that the name *Barriensis*, though evidently intended by Sir R. Murchison for the well-known "Barr Trilobite," was applied by himself, and subsequently by many different authors, to another and a very distinct species. The figures given in the 'Silurian System' represent a fossil which I had already distinguished by the name *Il. pomatia,* before I received Prof. Hall's last paper on the fossils of the Niagara group from the Western States. It would be wrong in principle to apply the name intended to designate the common Woolhope Limestone species to another fossil, because that one happened to be figured (in mistake) by its author; and I think I shall have the consent of all palæontologists to keep Murchison's well-known name for the "Barr Trilobite," while Prof. Hall has furnished us with a name for the Dudley one, viz., *Il. insignis* (*Il. pomatia* of our Plate). I subjoin comparative figures of the three Upper Silurian species of this section of the genus, that the distinctive characters may be better appreciated.

From the *Illænus insignis,* next described, and usually confounded with our fossil, *Il. Barriensis* is at once distinguished by the more depressed shape, the head being regularly and gently convex and obtusely margined, while that of *Il. insignis* is pyramidal and the margin recurved. The latter species is abundant in North America, as described by Hall; and I do not know if the huge *Il. imperator*, lately described as a distinct species by that author (Reports, Dec., 1864), be different or not. It may be, for aught in the description, a giant form of this large species.

It will be perceived that these three species differ in form and sculpture. *Il. Barriensis* is depressed in the front view (fig. 53), and has coarse squamate striæ in front. *Il. insignis* (fig. 52), more elevated, and with a deep pit where the obscure glabellar furrows end, has very fine ornamental striæ. *Il. carinatus* (fig. 53), the most gibbous of all in a front view, has the axal pits like *Il. insignis*, and the coarse striæ, less universal than in *Il. Barriensis.*

British Upper Silurian Illæni of the Section Bumastus.

Fig. 51.

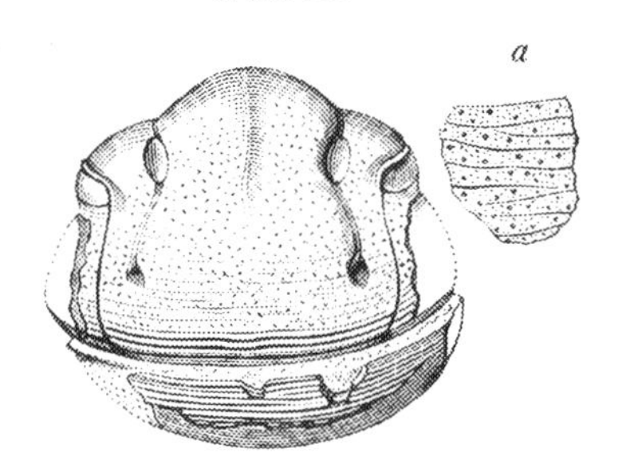

Il. carinatus, Salter.
Wenlock Limestone, Malvern; a rare species.

Fig. 52.

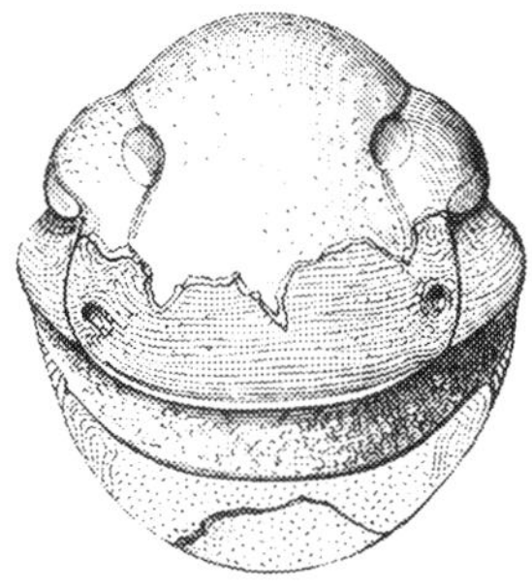

Il. insignis, Hall.
The common Dudley fossil, usually called *I. Barriensis.*

Fig. 53.

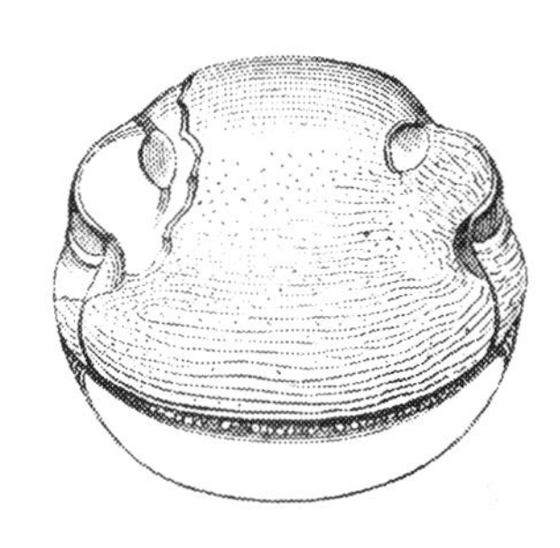

Il. Barriensis, Murch.
Very common in the Woolhope Beds and Wenlock Shale.

Il. Barriensis is broad-oval, with blunt extremities and arched sides. The head measures very nearly an inch and a half long, and is therefore rather less than a third of the whole length. The tail is exactly as long as the head,—the body (of 10 arched rings) occupying one inch ten lines, and being therefore longer than either. The trilobation is extremely faint, and the breadth of the axis as great proportionally as in *Homalonotus.* The fulcral point nearly coincides with the faint axal line (it can scarcely be called a furrow). However, the axal furrows are strongly marked out for a short distance in the head; and we may now describe that portion.

Head forming nearly a quarter of a sphere, and not more pointed in the young than in the old state, the front being obtuse in all ages, and more flattened anteriorly and on the sides than a true semicircle. It is depressed, the vertical height being only half the width; and the convexity being spread over the whole surface, the view of the front edge (woodcut 53) presents a semicircular outline instead of a subpyramidal one. The edge is blunt and convex all round, not at all recurved. The axal furrows converge very strongly above the large eyes, which, in Miss Jukes' specimen, are $2\frac{1}{4}$ inches apart, measuring to their outer edges. The furrows terminate above the eye (but only on the inner surface of the crust) in oval depressed spaces, the nature of which is not yet known, but which probably represent a pair of the curious glands (?) described under the genus *Phacops* (see under *P. caudatus*, p. 52). Beneath the eye, a strong rounded fold or long cushion (*pulvillus*) of the crust supports that organ (the fold is not seen in young specimens), and this cushion is subtended by a deep furrow, that encircles the whole of the lentiferous surface of the eye. The latter is much curved, and quite smooth, and has a thick cornea. The encircling furrow ends in front in a broad deep depression (Pl. XXVII, fig. 16) which indents the head deeply just at the origin of the facial suture above the eye; the suture from thence curves boldly **S**-fashion out to the margin, which it cuts on a line level with the outer edge of the eye. Beneath the eye, the suture again turns sharply outwards, cutting the posterior margin in the same vertical line. The eye-lobe (*palpebra*, Dalman) is very convex and inclines downwards. The inferior fold or eyelid, if so it may be called, is deeply punctate in our specimen, but without any of the wavy lines which cover other parts of the head. (Decade 2, Geol. Survey.) The cheek, outside the eye, is tumid, and the angle quite blunt. The neck-furrow is always obscure in the species of this genus. The whole head is covered with the peculiar squamous lineation characteristic of the family *Asaphidæ;* its peculiarities are more particularly described below.

The epistome beneath the head is broad, and very strongly striated transversely by sharp squamous striæ. (See Decade 2, Geol. Survey, pl. iv, fig. 7.) We have omitted to figure it again in our plate. The rostral shield is very distinct, and often found separate in this and other species of *Illænus;* it is a transverse piece, long-oval in shape in our fossil (many of the species have it rhomboidal) and with pointed ends; the rostral suture is very distinct, and the sutures that separate it from the side-pieces equally so. The labrum should be looked for. It is strange it should not be yet known in collections.

Thorax of 10 segments, arched forward, especially the front ones, and having the three lobes just indicated by a very slight furrow where the fulcrum is placed, the sub-fusiform axis occupying more than two thirds the entire width of the thorax. The fulcrum is, of course, far outwards; and is formed by a short forward bend of each pleura, which then continues in the general direction, is sharpened anteriorly for rolling, and curves forward at its blunt end; the foremost pleuræ are rapidly shortened, the whole of the lateral portions of the great head projecting beyond them (fig. 2).

Tail, in the young state, transverse broad-oval; and in the adult not much more than a semicircle. The upper corners are truncated, the facet strong. There are no indications at all of the axal lobe, and the whole tail is regularly and gently convex from side to side, and from front to back; and is not nearly so hemispherical as in the next species. The incurved portion (caudal fascia) is broad and not so convex as in the next species, *Il. insignis*. It is, however, remote from the inner surface of the crust, leaving a hollow channel, which is conspicuous enough in broken specimens.

The general sculpture of the surface is coarse, and is arranged thus: on the head it lies transversely, and is very strong along the front margin; in old specimens (fig. 2) it appears quite squamous there. For half the length of the head it follows the direction of the front margin (woodcut, fig. 53), and is then met rather abruptly, above the level of the eye, by the more direct and fainter lines of the middle portion, and here the lines inosculate. Behind, the head shows somewhat finer lines, and a central tubercle some way up (see fig. 1*a*). Among the coarse striæ are abundant puncta, wide and strong, the marks, probably, of the bases of short pile or stiff hairs. Similar lines, mixed with puncta, run transversely across the whole thorax, parallel to the course of the rings; both, however, vanish about the fulcral points. The puncta are not closely placed on the caudal shield.

British Localities.—MAY HILL SANDSTONE, Malverns (M. P. G.); Purple shales, Onny River, Shropshire (Mr. Wyatt-Edgell's Cabinet). WOOLHOPE? GRITS of Bogmine, Shelve, Shropshire, rare (M. P. G.). WOOLHOPE LIMESTONE of Woolhope, Herefordshire (abundant); Hay Head, Barr, Staffordshire; Malvern Tunnel (abundant); WENLOCK SHALE of Rushall Canal, near Wallsall (the late Mr. Mushen's Coll.). WENLOCK ROCKS of Llandeilo; Llandovery; Carmarthen; &c., as far as Freshwater, Pembrokeshire (M. P. G.).

Foreign.—NIAGARA LIMESTONE, New York, Wisconsin, &c.

ILLÆNUS (BUMASTUS) INSIGNIS, *Hall.* (*Il. pomatia,* on plate) Pl. XXVII, figs. 6, 7.

ILLÆNUS BARRIENSIS (in part), *Murchison.* Sil. Syst., p. 656, pl. vii bis, fig. 3; pl. xiv, fig. 7, 1839.
— — *Hall.* Geol. Rep., New York, p. 102, No. 10, fig. 4; No. 19, fig. 3? 1843.
— — *Salter.* Decades Geol. Survey, No. 2 (pl. iii, fig. 2; pl. iv, figs. 9—11 only); Siluria, 2nd ed., pl. xvii, fig. 9—11, 1859.
IL. INSIGNIS, *Hall.* 18th Report on the New York State Cabinet for 1865 (printed in advance, Dec. 1864).

Il. modicus, 3-4 uncialis, capite parabolico, caudâ hemisphæricâ. Caput gibbosum anticè ovatum, margine reflexo; sulcis axalibus primùm convergentibus, dein longè divaricatis et in foveas angustas desinentibus. Oculi humiles deflexi, pulvillo nullo fulti. Anguli obtusi. Cauda subhemisphærica, tam longa quam lata fere, margine acuto. Superficies, caput præcipuè, lineis creberrimis.

As above noted, this fine species has all along been confounded with the *Il. Barriensis,* from which it differs in every particular. More convex, with obliquely placed eyes, which have no fold beneath them—deep pits (*foveæ*) at the end of the axal furrows in front of the eyes—a close lineation of the surface; and above all, a highly convex tail-piece, which is subhemispherical and nearly as long as broad,—these are characters which distinguish the species, at a glance, from the much more common *Il. Barriensis.*

It is not nearly so large a fossil as the last, probably never more than 4 inches, but with head and tail so very convex, that when extended only a small portion of the former is seen in a vertical view; while the whole length of the thorax, when not rolled up, is two thirds that of the tail. The pointed or rather parabolic head, when viewed directly in its full measure, is 18 lines long, by 22 broad, and the depth 11 lines (much deeper, therefore, than in *Il. Barriensis*). Of this breadth the glabella, which is strongly marked only in the cast, occupies 14 lines at the base; thence the furrows quickly converge forward to the deep kidney-shaped depression (or gland) above the eye, and thence again diverge to the curious pits[1] near the front of the head. The glabella is most gibbous on the vertex, but is not carinate as in the next species.

From the axal furrows the eye-lobe descends rather rapidly to the *depressed* long lunate eye, which is subtended by a furrow, but by no raised fold of the crust, as in *Il. Barriensis;* nor does this furrow terminate forward in a deep wide depression, as in the larger species. The eye is placed at less than half its length from its margin; the facial suture turns sharply out beneath, but cuts the margin under the eye, not beyond it. Above the eye it bends strongly out, and reaches the margin on a line *outside* the eye. The whole surface is covered by a close sharp lineation which follows the head-margin, and becomes coarse only on the free cheeks, where the puncta are also very strong. The angles are blunt, but rectangular.

The labrum (figured from a Brit. Museum specimen) is very perfectly preserved. It is wide, triangular, half as wide again as long, straight along the expanded base of attachment, and with nearly straight sides, forming thus a right-angled triangle with a rather acute apex. Its centre is very gibbous, and the convexity declines steeply to the sides, where a broad lateral depression, continuous with the marginal furrow, separates a thickened recurved rim all round: this is broadest on the sides. A pair of compressed tubercles occurs at the lower third: they are transverse-ovate, and more than their own diameter apart. The surface is covered with squamous lines (like those on the body and head) curved strongly, and interlined, like the body-sculpture, with short lines.

The thorax-rings, regularly arched, are each convex from back to front, and have the axal line coincident with the fulcrum. The pleuræ curve forward, and are thickened on the hinder edge, and striated along the facets.

The tail, 18 lines long and 21 broad, is all but hemispheric in contour, and most convex at the hinder two thirds. It has lineations mixed with puncta in front, and all the hinder part is covered with punctations only. The edge of the tail is acute, not really recurved, but not at all rounded downwards. The incurved caudal fascia is not broad, but very convex beneath, and is covered closely with granular elevated striæ. This fascia is continued of the same breadth beneath the pleuræ, as in all Trilobites; and a young specimen in Mr. E. Hollier's cabinet shows that it occupies half the breadth of the pleuræ, a very common proportion in the *Asaphidæ*.

[1] These pits represent what are frequently seen in *Calymene, Trinucleus,* and other genera; they are points of attachment for the *alæ* of the labrum.

Young specimens present nearly the same proportions, but the tail is somewhat shorter and the form less convex.

Localities.—Wenlock Limestone of Dudley, and the Malvern Hills, frequent. It is probably common enough in the Wenlock of other places; but I avoid quoting localities which may belong to *Il. Barriensis.*

Foreign.—Niagara Limestone of Illinois and Wisconsin.

[*Illænus orbicaudatus,* Billings ('Catal. Sil. Foss. Anticosti,' 1866, p. 27), is a *Bumastus* so like this that I do not see readily how to distinguish it. It is a Caradoc and Llandovery form, in the Island of Anticosti; but several of our fossils immigrated from the far west in Silurian times; and this may be one of the colonists.]

Illænus (Bumastus) carinatus, n. sp. Pl. XXVII, figs. 8, 9.

[Compare with the kindred species, Il. armatus, *Hall.*[1]]

Il. (*B.*) *modicus* 2—3-*uncialis; capite rotundato gibbo carinato, angulis brevispinosis. Caput gibbosum anticè rotundatum, margine acuto nec recurvo, vertice gibbo carinato; sulcis axalibus primùm convergentibus, dein lentè divaricatis et in foveas desinentibus. Oculi parvi, pulvillo tumido instructi. Anguli in spinas crassas tracti, quæ breves sunt et striatæ.* [*Cauda forsan transversa.*]

This remarkable species, of which we only have two fragments in the Museum of Pract. Geology, and another in Mr. Edgell's fine collection, is one of the rarest of the Wenlock Trilobites. It was detected by myself, after lying for years in the Museum of Pract. Geol. as a variety only of *Il. Barriensis:* a good instance of the desirability, in a public collection, of preserving all the variations that occur, and not reducing the number too greatly. The late Rev. Mr. Dyson, of Malvern, found our specimens in the Winning's Quarry, Malvern. It is closely allied to an American form lately described by Prof. Hall, referred to above; and it may possibly be identical, but differs in the facial suture, and in the proportions. For points of difference from British species, see pp. 204—207.

Fig. 54.

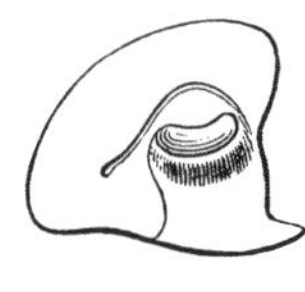

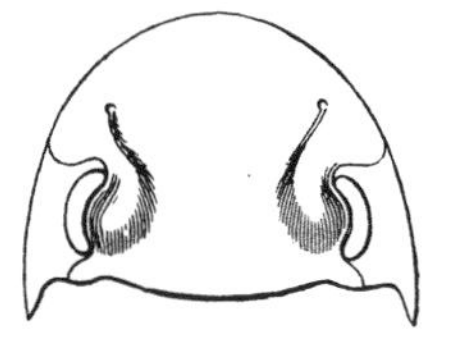

Illænus (*Bum.*) *armatus,* Hall. Niagara group, Wisconsin.

Comparing our specimens, such as they are, with Prof. Hall's smaller species, I find the following differences, that make me hesitate to unite them. Ours is more gibbous on the vertex, has much smaller eyes, set on a more protuberant cushion; and the facial suture in front of them cuts the margin in advance of the eye—not at right angles to it. Prof. Hall's description is, as usual, far too short and general; he gives no note of the

[1] 18th Report New York State Cabinet, printed in advance, Dec., 1864 (see description of *Il. Barriensis*).

sculpture, of the front margin, the depth or shallowness of the axal furrow: nor any characters, other than length and breadth, of the pygidium. Still the two species must be very near indeed. And both are remarkable among the *Bumasti*, for spinous head-angles, a character which we do not know in more than a few instances, even among the typical *Illæni*.

Locality.—Wenlock Limestone, Winnings Quarry, Malvern. (Museum Practical Geol., and the late Mr. H. Wyatt-Edgell's Cabinet.)

Illænus (Bumastus) Maccallumi, n. sp. Pl. XXVIII, fig. 1. Pl. XXX, figs. 2, 3.

Il. (B.) 2 uncias longus, capite obtuso, lunato, caudâ subtrigonâ. Caput gibbosum, oculis majoribus. Thorax subtrilobus. Cauda parabolica, subtrigona fere, lævis: axis ejus latus, ad marginem curvum tantùm conspicuus; latera fulcro valde prominente, axe proximo, lævia, nisi sulco superno profundiore distincta. Fascia caudalis angusta.

A few specimens only of this new species have occurred to the Geological Surveyors in the course of their work at Girvan, Ayrshire; and one or two, more complete, are in Prof. Thomson's cabinet. One of his specimens (Pl. XXX, fig. 3) shows nearly all the body-rings and part of the head. From these we learn that the head was so convex as to be almost hemispherical and blunt-edged, while the tail is very little gibbous, and is pointed—an unusual character.

The head seems to have been transverse, lunate; most gibbous in front, with large eyes proportionally, though they are somewhat less than in *Il. Barriensis*, and placed rather further back. The glabella-furrows curve quite round the eyes in front—incurved and then recurved, as in the larger species. The central punctum on the vertex is strong, and a ridge runs from it; but the head is not otherwise at all keeled. Our figure looks a little as if it were so.

Thorax of ten ? very much arched rings, which show a very broad axis, as usual in the sub-genus, but a more decided axal furrow than in our two Wenlock species. The pleuræ are short; their apices rounded, and curved forward.

The tail is very regularly but not highly convex—a long half-oval. It is one inch long, and as broad at its upper curved margin, which shows distinctly the axal points, distant from each other fully three fifths of the width of the tail, and deeply indented. These are, however, not continued in anywise as furrows down the tail itself. Almost immediately beyond these points the front margin rises abruptly to the prominent fulcrum, and from thence descends to the contracted outer and upper angles; a sharply marked facet is still further defined by the subtending upper furrow, more strongly indented in this than in any other *Bumastus* known.

The surface of the tail is convex, somewhat gibbous down the central line, and this is not due, except in part, to pressure in the arenaceous rock. From this central line the sides slope evenly away to the slightly recurved margin, within which the broken specimen shows the narrow internal fascia.

Locality.—LLANDOVERY SANDSTONE (Lower Llandovery Rock) of Mullock, Girvan, Ayrshire; a locality first described by Murchison, and well searched by the persevering collector, the late Alexander MacCallum (see 'Quart. Geol. Journal,' vol. vii, 1851, p. 170), who discovered most of the species in his own district. *Illænus Thomsoni*, *Trinucleus concentricus*, and *Atrypa hemisphærica* occur in the same deposit of hard yellow sandstone and grit, but no *Pentameri*, such as are found in the overlying "May Hill Sandstone" of Saugh Hill.

Sub-genus 6.—ECTILLÆNUS, *Salter*, 1866.

ILLÆNUS (ECTILLÆNUS) PEROVALIS, *Murchison*. Pl. XXVI, figs. 5—8.

ILLÆNUS PEROVALIS, *Murchison*. Sil. Syst., p. 661, pl. xxiii, fig. 7, 1839.
— — *Salter*. Siluria, 1st ed., pl. iv, figs. 13, 14, 1854. 2nd ed., pl. iv, figs. 13, 14, 1859.
— — *Id.* Morris' Catal., 2nd ed., p. 110, 1854.
— — *Id.* Catal. Mus. P. Geol., p. 5, 1865.
— — *Id.* Mem. Geol. Surv. iii, p. 256, 1866.

Il. (*Ectill.*) *latè ovatus, obtusus, depressus,* 2½ *uncias longus,* 2 *uncias fere latus; capite semicirculato, quam cauda minore, sed thorace brevi paullo longiore; oculis anticis submarginalibus. Caput glabellâ angustâ, haud tertias latitudinis ejus efficiente, sulcis axalibus brevibus rectis. Oculi* (?) *suturaque facialis omnino antici submarginales; anguli obtusi. Thorax bene trilobus, axe angusto subfusiformi, pleuris angustiori. Pleuræ ad fulcrum* (*circiter tertias posito*) *planæ, dein fractæ, subrectæ, apicibus obliquis. Cauda margine antico subrecto, axe brevi conico. Fascia angustior, subconcava.*

This remarkable species, certainly the type of a very distinct sub-genus (probably a new genus), was first obtained by Sir R. I. Murchison from the lower part of the Llandeilo formation, in the Shelve country, west of the Stiper Stones. In the 3rd vol. of the Memoirs of the Survey, I have inadvertently included it in the Arenig group, but wrongly so: see above, p. 182. This I learned by a visit to St. David's, where Mr. Hicks, Mr. Lightbody, and myself obtained a good specimen from the lower part of the true Llandeilo rocks at Abereiddy Bay, near Llanrian (fig. 8); and Mr. Thos. M'K. Hughes, of the Geological Survey, had some time previously obtained a similar specimen (fig. 7). These are larger than Murchison's specimens in the Geological Society's collection, and show more clearly what must be the true position of the eye, *if there be one*. With this Trilobite is found, but rarely, the double Graptolite *Didymograpsus Murchisonæ*, so that its place in the series is unquestionable, for the Graptolite just mentioned is a typical Llandeilo fossil.

We still want to know more about the head of this species, probably not a rare one; and the small free cheek has not yet been found. It must have been quite narrow; very probably it is soldered to the head, and in that case the eye (indicated by a notch at *a*, fig. 8) may be absent, and the species a blind one. Collectors should search for the species in its original locality, between Corndon Hill and Hope Mill, Shelve, Shropshire.

A broad-oval species, two and a half inches long by two inches broad, of which length the semicircular head is rather longer than the thorax, and shorter than the semicircular tail. The axis is well marked out and narrow in the thorax; very slightly so in the head and tail. The semicircular head, with bluntish angles, is very slightly convex, but regularly so; and has short axal-furrows reaching a third up and nearly straight, marking out a glabella so narrow as not to be nearly a third of the whole width. The facial suture is all but marginal, and must cut the exterior edge about half-way up; while the place of the eye, indicated by a notch (the eye has not been found), is fully two thirds up the head. The free cheek, therefore, as before said, must be very small, narrow, and triangular.

Thorax well lobed, the axal furrows sharp, the axis subfusiform, narrower, but not greatly so, than the nearly direct pleuræ, which are flat as far as the fulcrum: this is placed about one third out, and nearly equally distant in all the rings. The outer portions of the pleuræ are neither recurved nor deflexed, but run straight out; and they are oblique at the tips.

Tail semicircular, the front edge straight or a little arched, with a very short narrow conic axis, which indents about one fourth of the length of the tail. The upper angles of the tail are very slightly truncated, and obliquely so. The surface is regularly and only slightly convex. The fascia—seen best in Mr. Lightbody's specimen (fig. 8) and in Sir Roderick Murchison's specimen (fig. 5)—is moderately broad, and of tolerably equal width all round.

Localities.—LLANDEILO FLAGS proper (lower portion); near Hope Mill, north of the Corndon Mountain, Shropshire (Mus. Geol. Soc., figs. 5, 6). Near Llanrian, Abereiddy Bay, Pembrokeshire, (Mus. P. Geology, fig. 7, collected by Mr. Hughes; fig. 8, in Mr. Hicks' Cabinet).

Sub-genus (?) 7—ILLÆNOPSIS, *Salter*, 1865.

I can hardly believe this to be less than a generic group; but prefer to keep it with *Illænus* for the present, as the Canadian sub-genus *Hydrolænus* so much tends to connect it with the more ordinary forms. But while the anterior eye resembles *Ectillænus* just described, the complete axal furrows show a tendency towards *Bronteus*, and the grooved and pointed pleuræ link it more with the ordinary forms of the *Asaphidæ* than is exhibited in any other of the *Illænus* group. Perhaps *Psilocephalus* is the nearest genus we can compare it with. And it is instructive to observe that *Psilocephalus* preceded *Illænopsis* in point of time; and *Illænopsis* and *Hydrolænus*, with complete furrows and forward eyes, preceded *Ectillænus;* and so far as I am aware, this latter sub-genus was rather an earlier one than *Illænus* proper. The group closed with the highest and most perfect form *Bumastus*, in which the characters of the group may be said to culminate. As only one species is known, it is desirable to unite the generic and specific characters in description.

ILLÆNOPSIS THOMSONI, *Salter*. Pl. XX, fig. 1.

ILLÆNOPSIS THOMSONI, *Salter*. Mem. Geol. Survey, vol. iii, pl. 11 B, figs. 1, 2, pp. 256, 316 (not of the May Hill, or Upper Llandovery list in p. 360, which is a misprint for *Illænus Thomsoni*), 1866.

Il. modicus, vix 2 uncias longus, $1\frac{4}{10}$ *latus, ovatus; capite convexo obtuso, sulcis conspicuis; oculis anticis. Caput semicirculatum, convexissimum. Glabella urceolata, genis multò latior, sulcis axalibus profundis supernè divaricatis et ad marginem deductis. Oculi parvi antici, marginales, angulo glabellæ insidentes. Pleuræ cuspidatæ, sulcatæ. Cauda (huíc speciei haud dubiè referta), semicirculata, axe longo distincto ; sulco superno conspicuo. Fascia angusta, axem attingens.*

I have shortly described this form as "ovate, the head very convex, with widely divergent axal furrows reaching to the front; technically, the axal furrows are *complete.* Fixed cheeks large; free cheeks small and anterior; the eye placed close to the front margin; facial suture ending on the outer margin; pleuræ pointed, grooved." [Tail margined, and with complete axis.] See 'Mem. Geol. Surv.,' vol. iii, 1866.

Head semicircular, very convex; rather more so in front, but not gibbous. The glabella, much wider than the cheeks, is marked out by two strong deep furrows, parallel below, but curving out so much above as to give a sheaf-like look to the glabella. The axal furrows continue quite to the front edge, as strong as at the base; and at the front they overhang the small eye, which is close to them. The eye-lobe is not raised. The eye is placed in a rather deep notch, and so close to the front margin that the free cheek (separable, certainly, for it is absent) must have been quite minute and the facial suture all but marginal. Below, the wide fixed cheeks have rather a strong neck-furrow, which does not, however, extend beneath the glabella at all.

Of the thorax we have only the two upper pleuræ, but these are characteristic. Much shorter than the width of the cheek, they present a triangular-pointed outline, the tips being acute and a little curved forwards; a strong oblique groove traverses most part of their length.

The tail found with this head can hardly belong to any other species. It is semicircular, with very slightly truncated front angles; and has the axis distinct for fully four-fifths the length of the tail, and marked sharply out all round (another reason for regarding *Illænopsis* as a genus intermediate between *Illænus* and *Asaphus* or its allies). The axis is narrow, rather more than a fourth the whole width of the tail, and with the first ring strongly marked out, at least in the cast. The sides are smooth, except the rather strong upper furrow. The fascia complete, narrow, and equal all round. I know no fossil that can be a very close ally of this, unless it be the *Illænus conifrons* of Billings, figured by that able palæontologist from the Trenton group of Canada. Perhaps, as suggested by me in the 'Survey Memoirs,' vol. iii, *Il. conifrons* may be of this genus; but it differs in so many respects that I have even ventured to constitute it a distinct sub-genus. *Il. clavifrons*, Billings, and also *Il. arcturus*, Hall, from the Lower Silurian rocks of New York and Canada, have the eye forward, but not nearly so much so as ours. I beg to dedicate the species to my accomplished friend and fellow-worker Prof. Wyville Thomson, whose help to me, both as a thorough naturalist and the owner of a choice cabinet, is invaluable.

Locality.—Arenig Rocks; Ritton Castle Mine Works, near Bogmine, Shelve, Shropshire (Mus. P. Geology); in a lead-coloured soft shale (not slate), which abounds in well preserved and rare fossils.

More profitable ground to work than this locality I hardly know in Britain. Two species of *Æglina*, the rare *Trinucleus Murchisoni*, *Agnostus Morei*, an undescribed *Asaphus* or *Ogygia*, and a new *Cheirurus*, are found there. With them occur the Twin-graptolites of two species; the characteristic *Obolella plumbea;* bivalve Shells; a *Theca;* two species of *Orthoceras; Bellerophon*, &c. This and the neighbouring mine-work of Cefn Gwynlle are the best localities for well preserved fossils in the Arenig or Skiddaw group; and it was by finding fossils there that I was enabled to establish the existence in 1857 of a distinct fauna for this formation, which had been previously named by Prof. Sedgwick. It is a more extensive group than the Llandeilo proper, and has its representative abroad in the fossil-bearing slates of Brittany (Angers, &c.), so celebrated for their great Trilobites. The fossiliferous Lower Silurian schists of Oporto are probably of the same age. Collectors of Trilobites may therefore expect to find in the lead-bearing rocks of the Shropshire district some of the curious Trilobites described by De Verneuil, Sharpe, Rouault, and Barrande,—such as *Placoparia Zippei* and *Pl. Tournemini.*

I think I shall be doing good service to our science by pointing out where a rich harvest of new forms may be expected, and I hope our friends will send me (for examination at least) some of the first-fruits.

The shales and sandstones west of the Stiper Stones; the slates *under* the Arenig and Cader Idris porphyries; those of Ty-obry in the pleasant Tremadoc district; the lower and larger half of the Skiddaw Slate; and the prolific vertical shales of White-sand Bay and Ramsey Island, St. David's, should all be thoroughly worked by those who wish to add new forms to our Trilobitic faunæ.

I take this opportunity also to include in the Arenig formation the puzzling Budleigh Salterton pebble-bed. In former pages of this work it was provisionally referred to the Llandeilo rocks. This pebble-bed (the record of vanished rocks hard by in the old time) and the quartz-rocks of Gorran Haven, S. Cornwall (which are relics of the same old land) will be worth searching for the fossils of the 'Arenig or Skiddaw group,' for many years to come.

To finish the description of this large genus, I may add figures of one or two doubtful forms, respecting which some information is desirable. Two of them are referred to at p. 186.

Fig. 55 represents the original specimen of Prof. M'Coy's *Illænus latus*, from the Wrae limestone of Peeblesshire. I regard it as only a pressed and shortened head of *Il. Bowmanni*, altered by cleavage action so as to destroy the proportions. The glabella-furrows, however, are rather shorter than usual; and unless we had the body and tail, it would be difficult to prove it to be that species.

Locality. Bala Limestone; Wrae quarry, Peeblesshire.

FIG. 55.

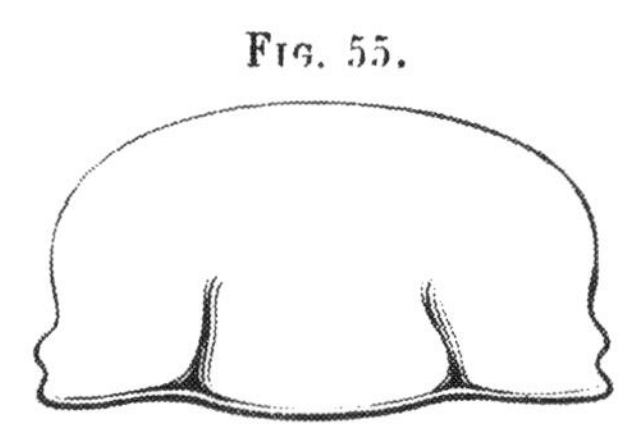

Illænus Bowmanni ?
(*Il. latus*, M'Coy: Peeblesshire).

Fig. 56 is a figure of the species found in the true Llandeilo Limestone of Knockdolian, Ayrshire. I should have little hesitation in describing it as *Il. crassicauda*, which I have before said, p. 193, has never certainly yet occurred in Britain; but the greatly truncated angles of the tail are sufficient to make me for the present quote it with doubt. The shape of the tail is identical with that figured by Dr. Volborth as the common Russian species, and the caudal fascia is of a like breadth: it is very much broader than any British species with which it can be compared, except *I. Portlockii*, pl. xxvi, fig. 3; but the less breadth of the axis, and the more remote fulcrum, as compared with that Caradoc species, are characters in which it resembles *Il. crassicauda*, which comes from Llandeilo rocks.

FIG. 56.

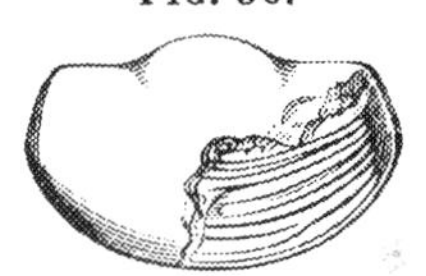

Illænus crassicauda, Wahl. ?
Llandeilo Limestone of Knockdolian, Ayrshire.

Fig. 57 is a small imperfect head of a *Bumastus* from the Bala (or Hirnant?) Limestone near Chirk. It is a solitary specimen, and only shows enough to prevent our identifying it with either of the four species of *Bumastus* figured in our plates. The head is not at all carinate, as in *Bumast. Maccallumi*, pl. xxx, nor are there any glabella-furrows visible. But this may be because we have here the outside crust, which we do not know in the Ayrshire species. It is worth figuring, to incite collectors to search for it in its locality, a prolific spot, and one which contains many rare fossils.

FIG. 57.

Il. (Bumastus), sp.
Mynydd Fron Frys, near Chirk.

Locality. Upper BALA (or Hirnant) Limestone of Mynydd Fron Frys, near Chirk (presented to the Woodwardian Museum by the late Mr. Bowman in 1841).

Perhaps other portions of these species will arrive in time for our promised Appendix. Meantime we have their *disjecta membra*.

We next take up the *Bronteidæ*, a small group in Britain, but a very large one on the Continent. They are closely allied to the *Ogygides* through *Barrandia*.

FAMILY—*BRONTEIDÆ*, BARRANDE, 1852.

If M. Barrande did not actually add the family termination to the name, he showed clearly that *Bronteus* constituted a family group distinct from any other. Originally confounded with *Asaphus* (or rather *Illænus*) by Brongniart, the generic rank of *Bronteus* was perceived by Emmerich, and seized upon by the diligent Dr. Goldfuss, who named the genus in 1839. The name *Brontes* had, indeed, been previously applied to one of the Coleoptera, and it is against all rule to admit a double employ of the same name in zoology. Yet Dr. Koninck's proposition to honour the Bonn Professor by the term *Goldius* has met with no concurrence, and the original name is still preserved.

Dr. Beyrich, who described several of the species first, was the earliest to perceive the necessity of subdividing the cumbrous genus. His view of its relations was not so clear, for he endeavoured to unite it with the *Cheiruridæ*, to which, indeed, as Barrande has pointed out (and, indeed, he seems half inclined to follow the Dr.'s suggestion), the analogies are close. The granulated eye, the short axis and radiated pleuræ of the tail-piece, no doubt suggested this comparison to the naturalist who was the first to accurately describe both genera. But there is, I think, no real affinity between them.

The true relations of *Bronteus* are, as Prof. M'Coy perceived, with the *Ogyginæ*, but it is through forms which were not accessible to Beyrich or Emmerich. I shall not notice the fanciful and absurd division of the genus by Corda; but follow Beyrich, as Barrande has done, in arranging the species according to the number of segments combined in the pygidium. This is not usually a part of much consequence in classification; but by tracing the variation of the tail-piece in *Bronteus*, through *Bronteopsis* to *Barrandia*, the relations of the group to *Ogygia* are clearly seen. Otherwise we should really have had to look for its affinities to *Cheirurus*, as Beyrich has done. But the rostral shield and ungrooved pleuræ are as much characters of *Illænus* as they are of *Cheirurus*. The granulated eye is found, it is true, in *Æglina;* but it is very rare in the *Asaphidæ*. The shape and lobes of the glabella are much like those in *Barrandia;* and, as before said, the shortened tail-axis of the last genus points to the still shorter axis in *Bronteopsis*, and the almost obsolete one in *Bronteus*.

We may define the forms in this family as follows:

Broad oval, generally depressed, sometimes very convex, with nearly equal head and tail portions; lineated (rarely granulated) all over. Glabella distinct and wide above, with three unequal lateral lobes and a broad large forehead-lobe; a rostral shield; granular eyes; the facial suture ending on the posterior margin; ten grooveless pleuræ. The tail large, with greatly abbreviated axis and radiant pleuræ.

INDEX*

TO THE

FAMILIES, GENERA, AND SPECIES DESCRIBED IN THE MONOGRAPH OF THE BRITISH TRILOBITES OF THE CAMBRIAN, SILURIAN, AND DEVONIAN FORMATIONS, BY THE LATE J. W. SALTER, A.L.S., F.G.S.

The synonyms are printed in Italics.

* Prepared by the Rev. Prof. T. Wiltshire, M.A., F.G.S., &c., Hon. Sec. Pal. Soc.

PRINTED BY J. E. ADLARD, BARTHOLOMEW CLOSE.

CORRIGENDA.

In page 14, line 18, *for* Emmerich, "1845," *read* "1839."

,, ,, 13, *for* "ost," *read* "lost."

,, 37, at bottom, *add* the following synonym:

"CHASMOPS ODINI, *Hoffmann*. Trilob. Russlands Verhandl. Kaiserl. Miner. Gesellsch zu St. Petersburg, 1858, t. iv, fig. 7.

,, 39, *for* "fig. 7," *read* "fig. 9."

,, 44, line 4, *for* "VII," *read* "VI."

,, 59, ,, 3, *add* synonym, "*Salter*, in Decades of Geol. Survey, No. 2, Article *P. caudatus*, p. 8, (see line 16)."

,, 64, line 13, *omit* fig. 10.

ERRATA.

Page 84, line 11, for 1855, read 1851.
" 95, last line but one, for *brevicapitata*, read *Caractaci*. (See page 96.)
" 97, line 2, for 5—11, read 6—11.
" 97, line 11, for 1855, read 1851.
" 103, line 6, for 'as,' read 'but.'

Plate X, bottom, *Homal. Brongniartii*, for figs. 15, 16, read 15—17.
" XIII, bottom, for *Homal. Winwoodii*, read *H. Vicaryi*.

ADDENDA ET CORRIGENDA.

Pl. XV, figs. 7, 8, for *Ogygia subduplicata* read *Phacops subduplicatus.*

Pl. XVII, fig. 8, for *O. pelatata* read *O. peltata.*

Pl. XX. The heading should be " UPPERMOST CAMBRIAN (TREMADOC)."

Pl. XXII, fig. 6, belongs to *A. peltastes.*

Pl. XXIII, fig. 6, description. Add: (Mr. Edgell's cabinet).

Pl. XXVIII, fig. 15, for *I. æmula* read *I. æmulus.*

Page 129, line 3 from top, and *passim*, for *anticè* read *antè*, and for *posticè* read *ponè.*

Page 130, after description of *Ogygia angustissima*, add: "This may be but a sub-species of *O. Buchii*, with a very narrow long axis. There is a more important var. common at Llandeilo, which has a wider, shorter tail, and might be called var. *convexa.* It has only 11 lateral furrows to the tail; and is characteristic apparently of the Lower, as *O. Buchii* proper is of the Upper beds of the Llandeilo Flags. (Edgell)." Line 10, dele " *Ogygia.*" It is surely a *Phacops* of the section *Chasmops*, as indicated; correct therefore the name on the plate. See above.

Page 132, bottom. Add: Meadow Town, Shropshire: Cabinets of Mr. Morton of Liverpool, and of Mr. H. Wyatt Edgell. See also page 160.

Page 133, *Ogygia scutatrix*, also page 135, *Ogygia peltata.* I must amend these descriptions (p. 133—136), by referring Pl. XVII, figs. 9, 10, which from the width of the axis I had assigned to *O. scutatrix*, to *O. peltata.* In truth, they represent the ♀ form of the latter species, with which they occur at Whitesand Bay. A noble series, lately transmitted to me by Mr. Henry Hicks, show the distinctive characters of *O. peltata* in both the broad (♀) and narrow (♂) forms.

In the broader form the axis of thorax and tail is $\frac{2}{7}$ths the whole width; and the fulcrum of the thoracic rings, which our figure 8 (copied from the yet unpublished memoir on North Wales by Prof. Ramsay and myself) scarcely shows at all, is placed $\frac{3}{4}$ths out from the axis, and terminates a short distinct facet upon the falcate tip, as shown in fig. 9. The tail has the axis reaching $\frac{3}{4}$ths the whole length; and its width in one form is $\frac{2}{7}$ths, in another barely $\frac{1}{4}$th, of the whole width of the tail. The 8 or 9 cross furrows on the axis are only strong on the sides, leaving the central part only faintly ringed. The side-furrows are like our figure, and have but faint intermediate lines or furrows. The *fascia* or inner border is coarsely striate.

Page 133, lines 9 and 12 from top, for *axin* read *axem.* Line 11 from top, for *secondariis* read *secundariis.*

Page 135, line 9 from top, for *secondariis* read *secundariis.*

Page 134, bottom. Add: " LOWER TREMADOC," Carnarvon Road, 1½ mile west of Tremadoc (Mr. Homfray).

Pages 136 and 142, et *passim*, for *brevispinoso* read *brevi-spinoso;* for *retrocurvæ* read *retro-curvæ.*

Page 143, line 8. Also Mr. H. W. Edgell's cabinet.

Page 150, line 25. The labrum figured probably belongs to "*A: peltastes;*" as a synonym for which I should have quoted " Decade 2, Geol. Surv., pl. v, figs. 2, 4." They are reproduced in our Pl. XX, which see.

Page 153, line 29, dele "Abereiddy Bay," and insert Llampeter Felfrey." Line 5 from bottom, for '*cognota*' read '*cognita.*'

Page 154, line 19, for " LLANDEILO FLAGS ?" read "*Caradoc.*" Line 7 from bottom, for '*basin*' read '*basim.*'

Page 156, line 7, for "Shropshire" read 'Horderley;' the species is local there.

Page 158, note. There has since been found reason to suppose the specimen quoted is really an American one, not British.

Page 162, line 22 from top, strike out '*cujus.*'

Page 164, line 8 from top, for '*cujus caput rotundatum*' read '*capite rotundato.*'

Page 168, line 12 from top, for '*Wrac*' read '*Wrae.*'

J. W. S.

March 12th, 1866.

PLATE I.

DEVONIAN AND LOWER SILURIAN.

FIG.

1—4. *Phacops granulatus*, MÜNSTER. From Upper Devonian Rocks. Fig. 1, Newton Bushell. (Mr. Pengelly's cabinet.) Figs. 2, 3, 4, Petherwin, Cornwall. (Mus. Pract. Geology.) Fig. 4 *a* is fig. 4 enlarged.

5—7. „ *lævis*, MÜNSTER. Knowl Hill, near Newton Bushell. (Figs. 5, 6, Mr. Pengelly's cabinet. Fig. 7, Mr. Vicary's collection.)

8. „ *cryptophthalmus*, EMMR.? Newton Bushell. (Mus. Pract. Geology.)

9—15. „ *latifrons*, BRONN. From the Uppermost Devonian or Pilton Group, near Barnstaple. Figs. 9, 10, somewhat distorted. (Mus. Pract. Geology.) (Fig. 11, cabinet of Mr. Townshend Hall.) Fig 11 *a*, eye, with spaces between the lenses marked out into hexagonal areæ on the inner cast. Fig. 12, specimen with very prominent cups in the inner cast, *i. e.* with thickened cornea. Fig. 13, specimen with thinner cornea and less prominent cups. Fig. 14, Baggy Point, N. Devon. (Mus. Pract. Geol.)

16. „ „ Variety with narrower axis and more side ribs. Newton Bushell. (Mus. Pract. Geology.)

17—19. *Phacops punctatus*, STEININGER. Liskeard. (Figs. 17, 18, Mr. Pengelly's cabinet. Fig. 19, Mus. Pract. Geology.)

20. „ *Brongniartii*, PORTLOCK. Montgomeryshire, Lower Silurian. (Mus. Pract. Geology.)

21—25. „ From Tyrone. Portlock's figured specimens. 25 *a*. Magnified.

26. „ „ Variety or ♀ form. (*P. Dalmani*, Portlock, figured specimen.)

27, 28. *Phacops incertus*, DESLONGCHAMPS. Lower Silurian pebbles of Budleigh Salterton, S. Devon. (Mr. Vicary's cabinet.)

29, 30. „ *Jukesii*, SALTER. N. Wales. (Mus. Pract. Geology.) Caradoc Rocks.

31, 32. „ *alifrons*, Id. From N. Wales (copied from figures in the Woodwardian Synopsis).

33, 34. „ „ „ N. Wales (from specimens in the Mus. Pract. Geology). Caradoc.

35. „ *mimus*, Id. Quartzite of Gorran Haven, Cornwall. (Mus. Pract. Geology.) Llandeilo Rocks?

36—38. „ *apiculatus*, SALTER. Horderley, &c., Shropshire. Caradoc. (Mus. Pract. Geology.)

39—41. *Jamesii*, PORTLOCK. Figured specimens, Waterford. Caradoc. (Mus. Pract. Geology.)

41 *a*. „ „ „ Tail of the same, magnified.

42. „ *obtusi-caudatus*, SALTER. Fine specimen in the Woodwardian Museum, from Coniston Flags. Caradoc Rocks.

43—45. „ „ Variously distorted specimens. (43, 44, Brit. Mus. 45, Mr. Edgell's cabinet.)

DEVONIAN TRILOBITES.

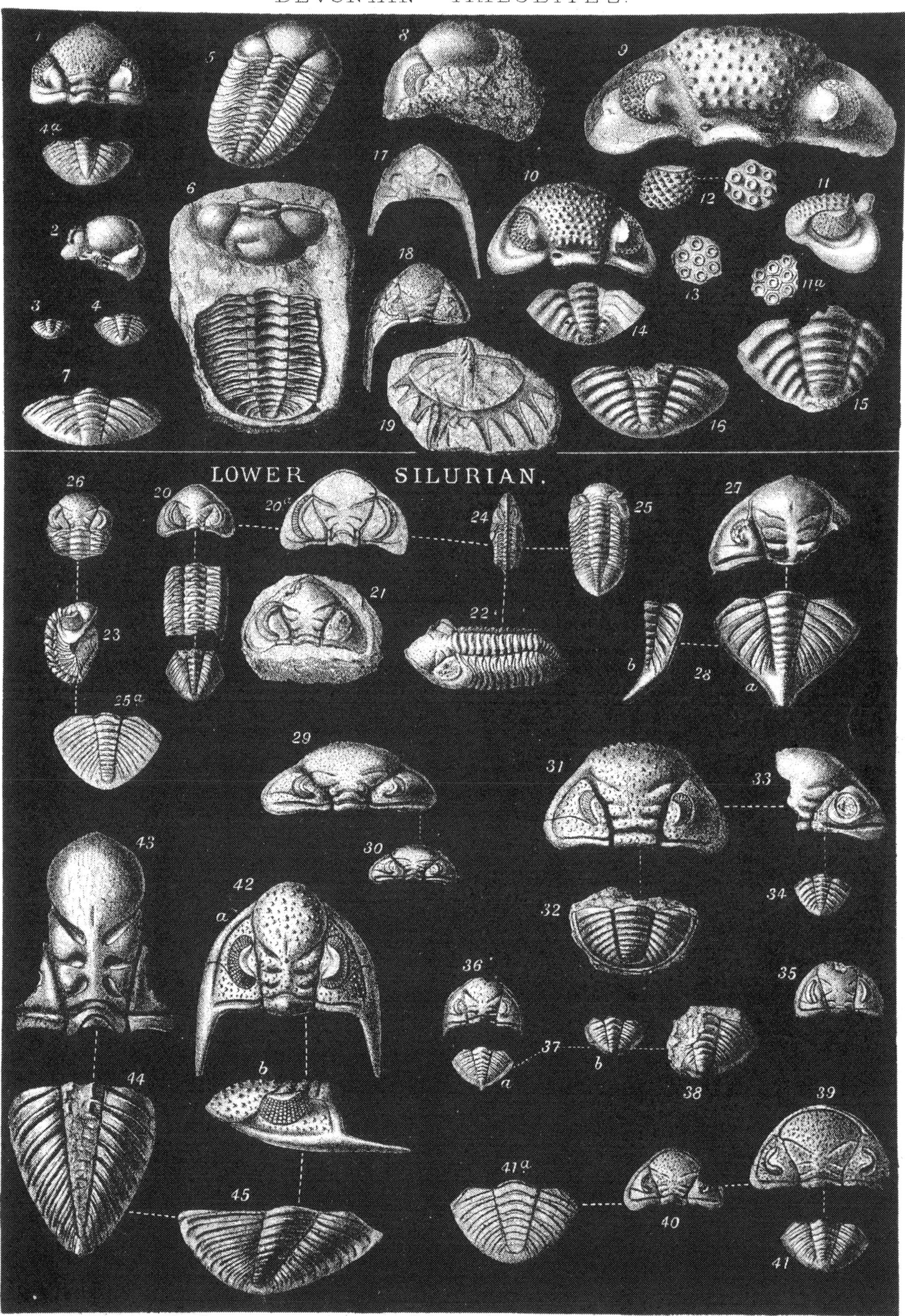

J. W. Salter & A. Gawan, lith. W. West, imp.

1–4. *Phacops granulatus.* 5–7. *P. lævis.* 8. *P. cryptophthalmus.* 9–16. *P. latifrons.* 17–19. *P. arachnoides.* 20–26. *P. Brongniartii.* 27, 28. *P. incertus.* 29, 30. *P. Jukesii.* 31–34. *P. alifrons.* 35. *P. mimus.* 36–38. *P. apiculatus.* 39–41. *P. Jamesii.* 42–45. *P. obtusi-caudatus.*

PLATE II.

UPPER SILURIAN.

FIG.		
1.	*Phacops Stokesii*, MILNE-EDWARDS.	Abberley Hills, Wenlock Shale. (Mus. Pract. Geology.)
2—5.	,, ,,	From Dudley. (Mr. Mushen's cabinet.) 2 *a*. Head, enlarged (3 *a* is 3 magnified). 4. Tail. 5. Eye, magnified.
6.	*Phacops Stokesii*.	Locality uncertain. Wenlock Shale of N. or S. Wales. (Mus. Pract. Geology.)
7—10.	*Musheni*, SALTER.	Of various ages; from the Wenlock Shale, Malvern. 9 *a*. Tail, enlarged. (Cabinets of J. Mushen, Esq., and C. Ketley, Esq.)
11, 12.	,, ,,	Larger specimens, Dudley. (Gray collection, Brit. Mus.)
13, 14.	*Phacops constrictus*, SALTER.	Wenlock Shale, Malvern. 13 *a*. Tail, enlarged. (Dr. Grindrod's cabinet.)
15.	,, ,,	(15 *a*, side view; 15 *b*, magnified.) 15 *c*. Eye of the same species. (Mr. Ketley's cabinet.)
16.	,, ,,	The same species. Wenlock Limestone; Dudley. (Brit. Mus.)
17—25.	*Phacops Downingiæ*, MURCHISON,	var. *a*, *vulgaris*. All from Dudley and Walsall. Fig. 17, a cluster, in the Brit. Mus. Figs. 18, 25, the specimens figured in the original 'Silurian System,' and now in the cabinet of Mr. W. Mathews, junr., of Edgbaston. Fig. 19, half-coiled specimen (Professor Tennant's coll.). Fig. 20, completely coiled (Decades of Geol. Survey). Fig. 21, young ditto (Mr. Mushen's cabinet). Figs. 22, 23, large specimens (Brit. Mus., Gray coll.). Fig. 24, form with large eyes, Malvern, in Wenlock Shale. (Mr. Ketley's cabinet.)
26—29.	,, ,,	Large-eyed variety, *β*, *macrops*. Fig. 26 is copied from the Survey Decades. Fig. 27, large specimen, Malvern, Wenlock Shale. (Mr. Ketley's cabinet.) Fig. 28. Dudley. (Brit. Museum.) Fig. 29, internal cast, from Wenlock Rocks, near Cardiff. (Mus. Pract. Geology.)
30.	,, ,,	var. *γ*, *inflatus*. Malvern Limestone. (Dr. Grindrod's cabinet.)
31—32.	,, ,,	Probably the same variety. Fig. 31, certainly so (glabella only), from Ledbury. Fig. 32, internal cast, same locality. Both are figured in the Decade 7, Geol. Survey, and are in the Mus. Pract. Geology.
33.	,, ,,	Interior cast of tail. Eastnor, near Ledbury. (Mus. Pract. Geology.) Wenlock Limestone.
34.	,, ,,	Dissection of the species. (Decades Geol. Survey.) *a*. Anterior ring, bearing the eyes. *b*. Underside of head (hypostome), bearing the labrum. *c*. Glabella and fixed cheeks. *d*. Magnified eye. *e*. Lenses and interstitial granules, still more magnified. *f*. Casts of the cups which support the lenses. *g*. Thorax-joints. *h*. Underside of the pleuræ of ditto. *i*. Tail of ditto. All more or less enlarged.

UPPER SILURIAN.

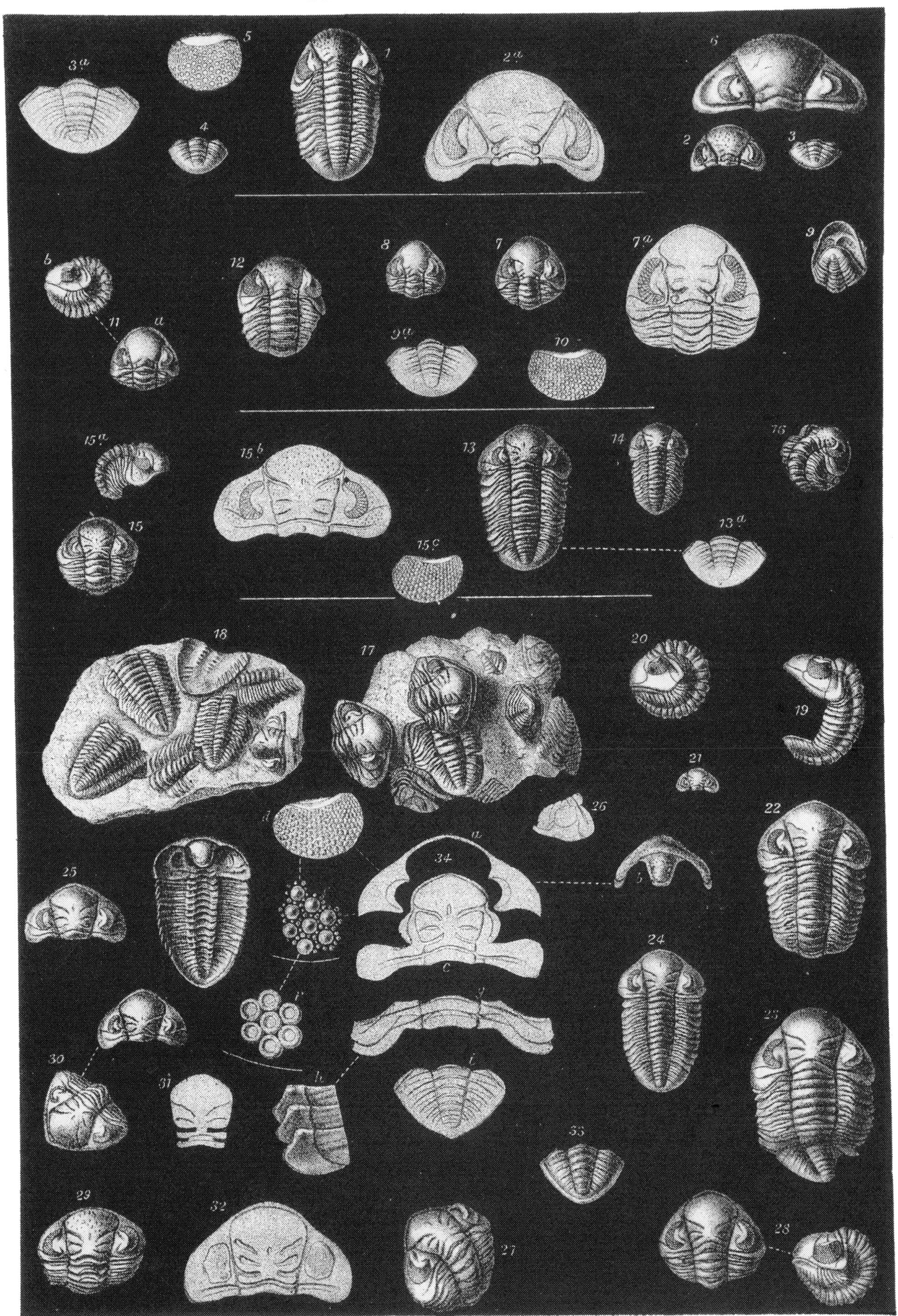

J.W. Salter & A. Gawan, lith. W. West. imp.

1_6. Phacops Stokesii. 7_12. P. Musheni. 13_16. P. constrictus. 17_34. P. Downingiæ, varieties.

PLATE III.

UPPER SILURIAN.

Fig.

1, 2, 3. *Phacops Weaveri*, Salter. Figs. 2, 3, are from (May Hill? or) Ludlow Rocks, Horseshoe Farm, Tortworth. Fig. 1, the head, is added from the May Hill Rock of the neighbourhood to complete the figure of the species. (Mus. Pract. Geology.)

4—14. „ *caudatus*, var. *a*. Various figures of ordinary variety, *vulgaris*, young and old, from Dudley. Fig. 4, Mr. Mushen's cabinet, Rushall Canal. Fig. 5, British Museum. Fig. 6, interior cast, Ledbury. (Mus. P. Geology.) Figs. 7, 8, labrum, from Mr. Mathews' and Mr. Ketley's cabinets. Fig. 9, side view of head, Dudley. (Mus. Pract. Geology.) Fig. 10, eye (Dr. Grindrod's cabinet), Ledbury. Fig. 11, same, magnified. Fig. 12, interior cast, Ledbury (Mus. Pract. Geology), shows the casts of the interior of the gland-like tubercles (see p. 52), such as are seen on the exterior of figs. 16, 17. Fig. 13, large but worn Dudley specimens (in Brit. Mus.), figured by König under the name of *Asaphus myops*. Fig 14, fine interior, in Mr. Mushen's cabinet.

15. „ „ Young (Brit. Mus.), showing the central tubercles.

16, 17. „ „ Dwarf specimens, retaining the characteristic tubercles of the young state; Ledbury. (Fig. 16, from my own cabinet. Fig. 17, Mus. Pract. Geology.)

18. „ „ var. *δ*, *aculeatus*. Wren's Nest, Dudley (Mr. Ketley's cabinet). All the above are from the Wenlock Limestone.

19—28. *Phacops longicaudatus*, Murchison. Figs. 19—21, ordinary variety, *a*. Figs. 22—28, short-tailed variety, *β*. All from Wenlock Shale. Figs. 19, 20, are Burrington specimens. (Mus. Pract. Geology.) Fig. 21, Brit. Mus. Fig. 22—25 are from the Wenlock Shale of the Malvern Railway. (Mr. Mushen.) Fig. 23, Dr. Grindrod. Fig. 26, Brit. Mus. Fig. 27 is from the Rushall Canal. (Mr. Mushen's cabinet.)

MIDDLE SILURIAN ? | UPPER SILURIAN.

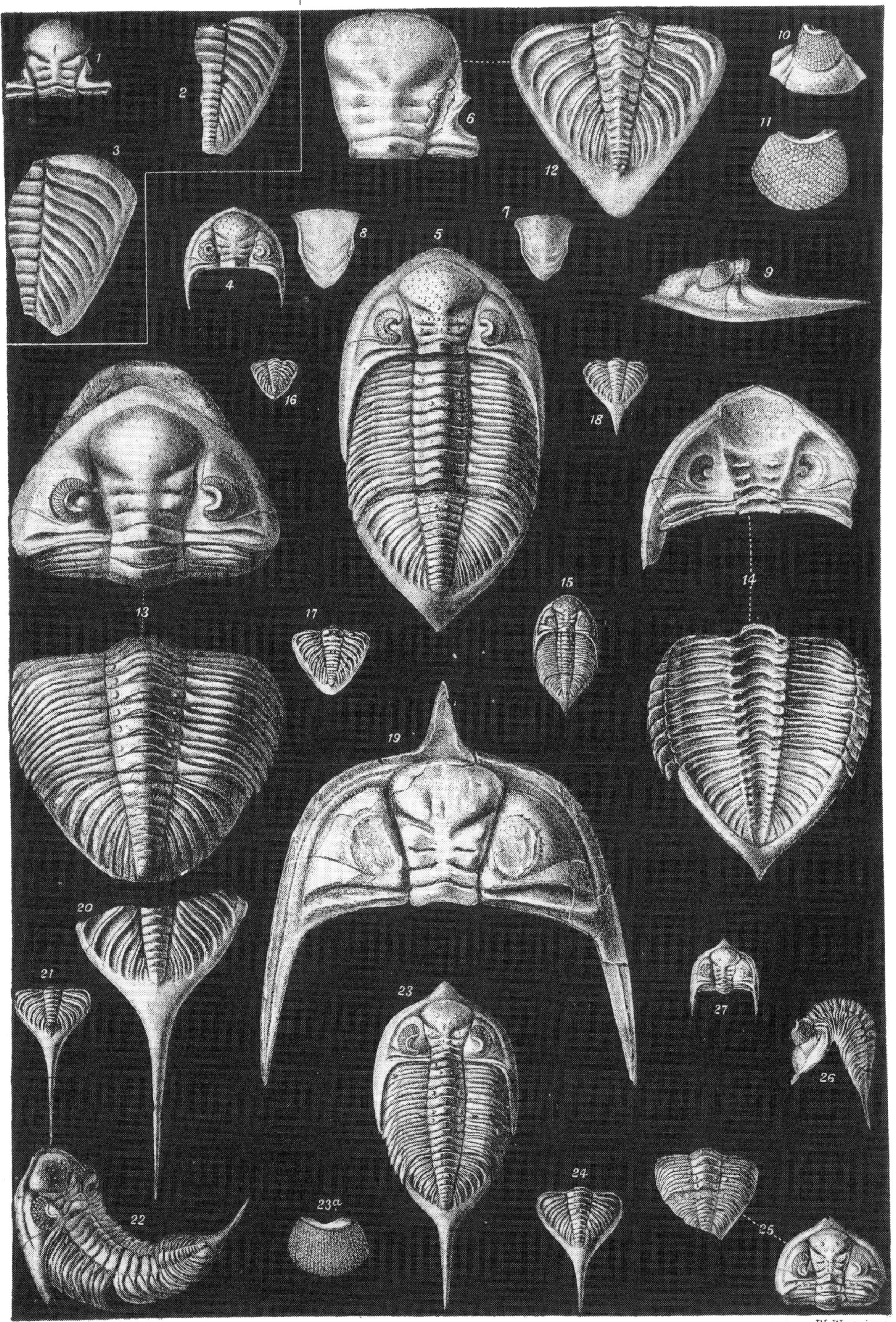

J. W. Salter, & A. Gawan, lith. W. West, imp.

1–3. Phacops Weaveri. 4–17. P. caudatus, v. 18. Var. δ. 19–21. P. longicaudatus, var. α. 22–28. P. longicaudatus. var. β.

PLATE IV.

UPPER AND LOWER SILURIAN.

Fig.

1. *Phacops caudatus*, var. β, *tuberculato-caudatus*. A large and fine specimen in Dr. Grindrod's cabinet, Lower Ludlow, Malvern.

2. „ *caudatus*, junior. Mr. Mushen's cabinet, Dudley.

3, 4, 5. „ „ var. γ, *nexilis*. Vinnal Hill, Ludlow. (Mus. P. Geology.)

6—9. *Weaveri*, SALTER. Fig. 6, imperfect head from Tortworth. (Mus. P. Geology.) Fig. 7, tail, ordinary size, Tortworth. Figs. 8, 9, tails of mucronate variety, May Hill Schist, Marloes Bay, Pembrokeshire. (All in Mus. P. Geology.)

10. *Phacops imbricatulus*, ANGELIN? May Hill Rocks, Presteign. (Mus. P. Geology.)

11, 12. „ *mucronatus*, BRONGNIART. From near Bala. (Mus. Pract. Geol.)

13—15. „ *truncato-caudatus*, PORTLOCK, his original figured specimens. Fig. 13, natural size. Fig. 14, restored figure, reduced size. Fig. 15, labrum. All from the Caradoc of Desertcreat, Tyrone. (Mus. Pract. Geology.)

16, 17. „ *amphora*, SALTER. Grug, near Llandeilo. (Mus. Pract. Geology.)

18—23. „ *macroura*, SJOGREN. From the Caradoc of Acton Scott, &c., Shropshire. Figs. 18, 19, adult head and tail. (Ludlow Museum.) Fig. 20, external cast of head, middle age. (Mus. Pract. Geology.) Fig. 21, labrum (Ludlow Museum), showing the perforation *a*, which marks the place of the ascending process. Fig. 22, perfect tail, and fig. 23, very young tail, pointed. (Ludlow Mus.)

24, 25. „ *conophthalmus*, BOECK? Fig. 24, young heads. (Mr. Lightbody's cabinet.) Fig. 25, internal cast of head from Llansantffraid Glyn Ceiriog, N. Wales. (Woodwardian Museum); for figure of tail from same locality, see Pl. VI, fig. 25.

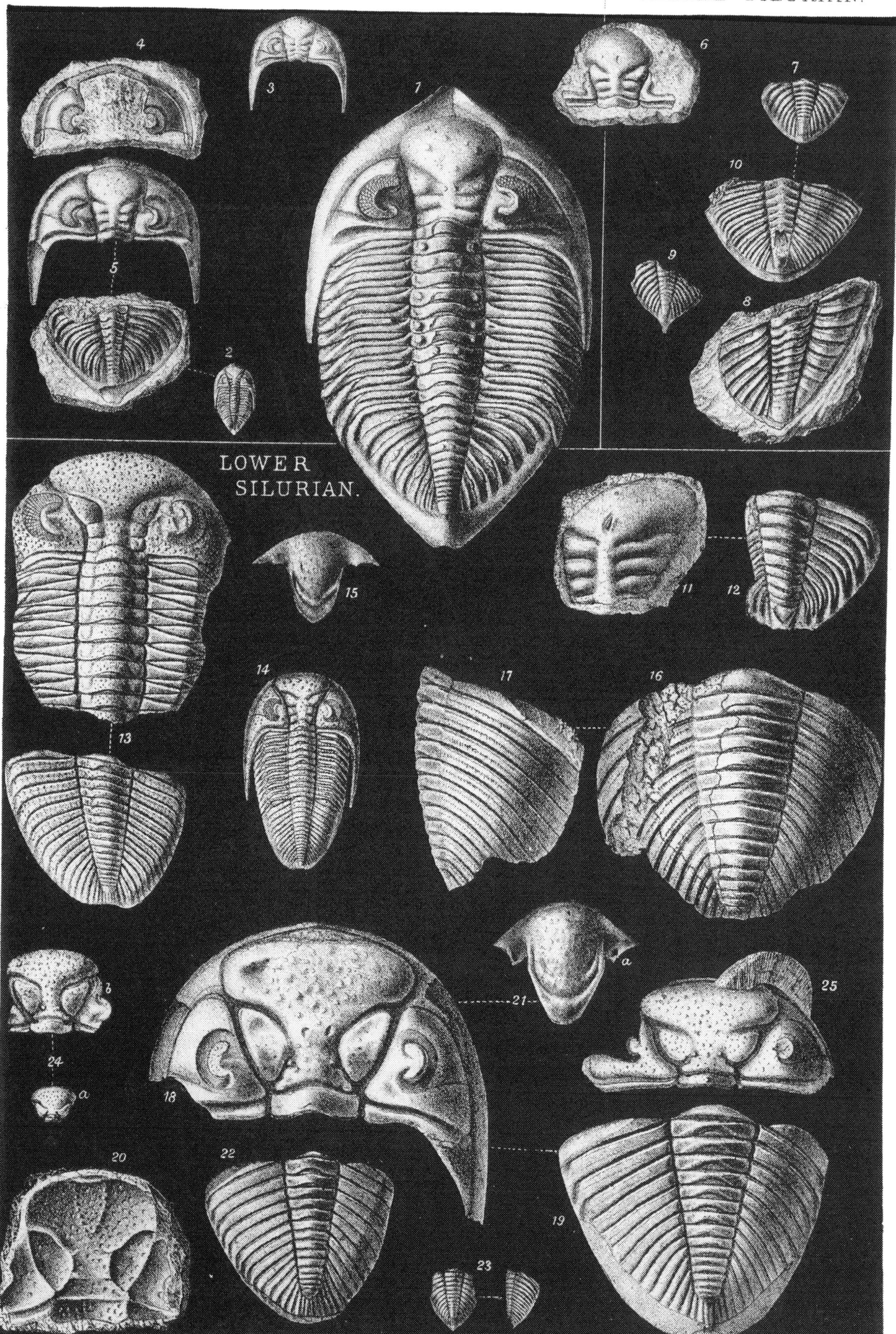

J.W.Salter, & A.Gawan lith. W. West, imp.

1_5. Phacops caudatus, varieties. 6_9. P. Weaveri. 11, 12. P. mucronatus. 13_15. P. truncato_caudatus. 16.17. P. amphora. 18_23. P. macroura. 24, 25. P. conophthalma ? 10. P. imbricatulus ?

TABLE V.

LOWER SILURIAN.

FIG.		
1.	*Cheirurus bimucronatus*, MURCH.	Large internal cast from Caradoc Strata, Sholeshook, Pembrokeshire, also figured in Memoirs of Geol. Survey, vol. ii, pt. 1, pl. vii, fig. 4. (Mus. Pract. Geol.)
2.	,, ,, ,,	Smaller head, external surface, Caradoc Strata, Chair of Kildare. (Mus. Pract. Geol.)
3.	,, ,, ,,	Young. May Hill Sandstone of Norbury. (Mr. W. Edgell's cabinet.)
4.	,, ,, ,,	Young, distorted (*Ceraurus Williamsii*, McCoy, figured specimen), Llandovery Rock, Llandovery. (Woodw. Mus.)
5.	,, ,, ,,	Labrum of large specimen, Chair of Kildare. (Mus. Pract. Geol.)
6—8.	,, *gelasinosus*, PORTLOCK.	Fig. 6, tolerably perfect surface of Ayrshire specimen. (Mus. Pract. Geol.) Figs. 7, 8, Portlock's figured specimens from Tyrone. Both are Caradoc localities.
9—12.	,, *juvenis*, SALTER.	Fig. 9, exterior of glabella, Chair of Kildare. Fig. 10, internal cast, Sholeshook, Pembrokeshire. Fig. 11, side view of same. Fig. 12, tail from Chair of Kildare. Caradoc Beds. (All from Mus. Pract. Geol.)
13.	,, *octolobatus*, McCOY.	Caradoc of Penwhapple, Ayrshire. (Brit. Mus.)
14.	,,	The same, tail from Rhiwlas, Bala. (Mus. Pract. Geol.) (Probably this is the *C. clavifrons*, Dalman ; *C. affinis*, Angelin.)
15.	,, *cancrurus*, SALTER.	Originally described specimen, from the Chair of Kildare. (Mus. Pract. Geol.)
16.	,, ,,	Chair of Kildare. (Sir. R. Griffith's coll.)
17.	,, *Sedgwicki*, McCOY.	Llandeilo Flags of Builth. (Woodwardian Museum.)
18—20.	,, *Frederici*, SALTER.	Upper Tremadoc Slate of Portmadoc, N. Wales. (Fig. 18, Mr. D. Homfray's cabinet. Fig. 19, Mr. Ash's cabinet. Fig. 20, Mus. Pract. Geol.)
21.	,, *Frederici ?*, variety.	Same formation and locality, Garth Hill, Portmadoc. (Mr. D. Homfray's cabinet.)

LOWER SILURIAN.

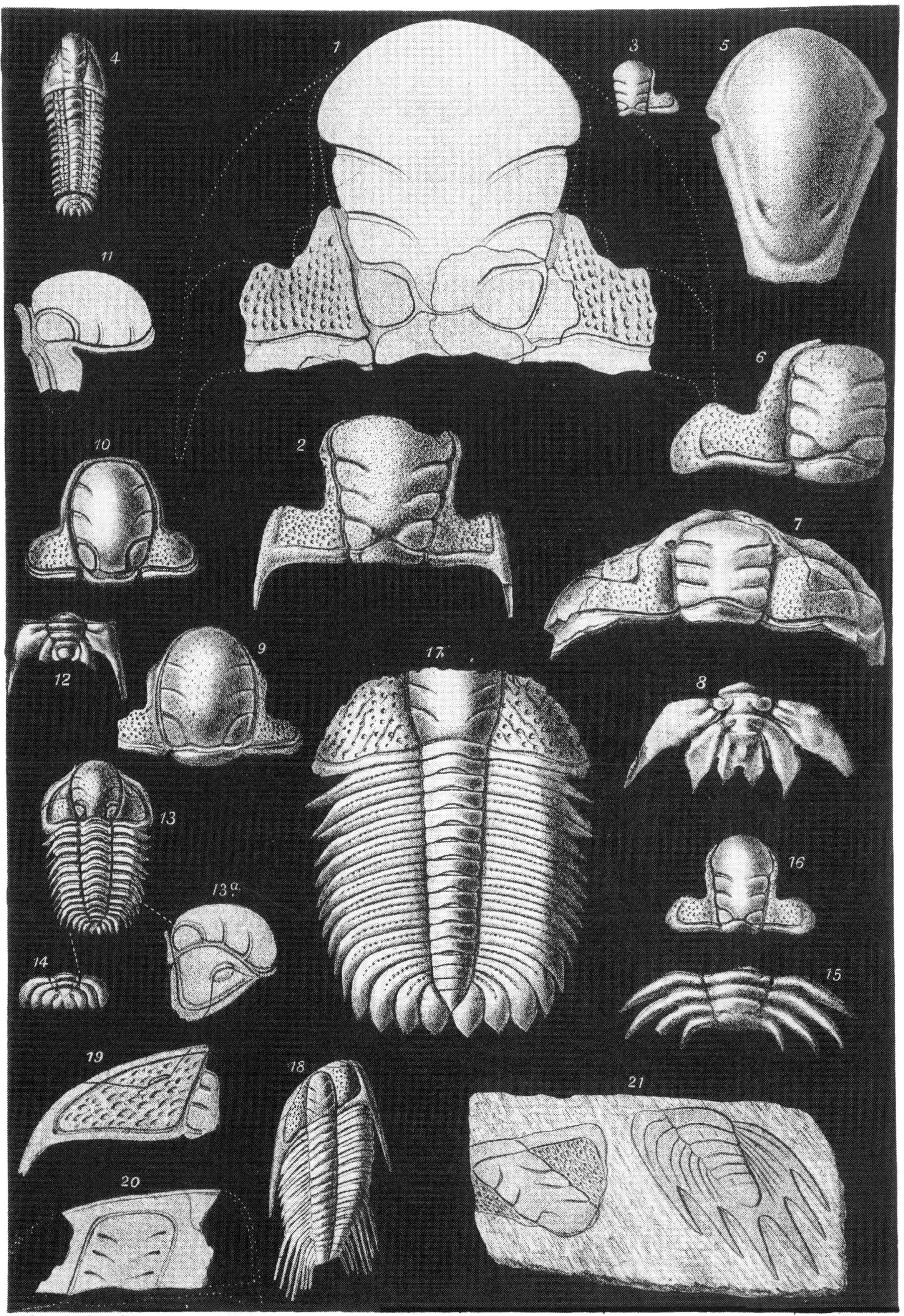

J.W. Salter & A. Gawan, lith. W. West, imp

1_5. *Cheirurus bimucronatus.* 6_8. *C. gelasinosus.* 9_12. *C. juvenis.* 13,14. *C. octolobatus.*
15?, 16. *C. cancrurus.* 17. *C. Sedgwicki.* 18_21. *C. Frederici.*

TABLE VI.

UPPER AND LOWER SILURIAN.

UPPER SILURIAN.

Fig.	
1.	*Sphærexochus mirus*, Beyrich. Dudley. (Mr. Shedden's cabinet.) Young coiled specimen.
2, 3, 4.	,, ,, Of various ages. From the Wenlock Shale of Malvern Tunnel. (Dr. Grindrod's cabinet.)
5.	A coiled-up specimen. (Brit. Mus., Gray coll.) 5 *a*. The head, dissected, front view, showing the anterior glabella-furrows, and at *b* the anterior segment, the cheeks connected by the hypostome, without the intervention of a rostral shield. *c*. The convex eye. *d*. The lentiferous surface, magnified, the lenses removed.
6.	The same viewed dorsally, and showing the strong basal glabella-furrows, the surface finely granular. *b*. The cheeks. A minute tubercle indicates the place of the head-spines. *c*. A body-ring, third or fourth, showing the position of the fulcrum (*d*), and the prominence behind, against which the fulcrum abuts. *e*. Incurved tip of pleura. *f*. Last thorax-segment, the fulcrum near the axis. In this and the preceding figure the pleuræ are represented as flattened out to show their characters; they would appear much shorter on viewing them from above. *g*. The tail, magnified. (Figs. *c* to *g* are from Capt. T. Fletcher's Dudley collection, now in the Fitzwilliam Museum, Cambridge.)
7, 8.	*Cheirurus articulatus*, Münster. A large and a small head from Mr. Pengelly's cabinet. From Lammaton, Torquay, in Middle Devonian Limestone.
9.	,, *bimucronatus*, Murchison, var. *β*. A half-grown but very perfect specimen. (Cabinet of Mr. E. Hollier, junr.) Dudley.
10.	A fine specimen, shortened by pressure, from Mr. Ketley's collection. Same variety,—*β*.
11 *a*, *b*.	A Malvern specimen, cleared by Mr. C. Ketley, and now in his cabinet. It shows the labrum and the central mucro to the tail, var. *α*. 11 *b*. Side view of ditto.
12.	Labrum, also from Mr. Ketley's cabinet.
13.	Shows the interior view of the same organ. At *a*, *a*, the ascending processes, which are attached to the sides of the glabella. *b*, *b*. The incurved triangular plates. There is a hollow space under the ascending processes, answering to the lateral notch on the upper surface (fig. 12).
14.	Pleuræ of *C. bimucronatus*, from a specimen in Mr. Mushen's cabinet. The lower figure is an internal cast of ditto, copied from the Decades of the Survey. *a*. The posterior fulcral tubercle. *b*. The anterior tubercle. *c*. Free tip of pleura. *d*. Cast of vertical ridge on interior surface.
15, 16.	Tails of young specimens, var. *α*. Dudley collections.
17.	Entire young specimen, var. *α*. (Mr. Mushen.) Dudley.
18.	*C. bimucronatus*, var. *β*, *centralis*. Dudley. (Mr. E. Hollier, junr.)
	(Figs. 19—24 are from the Mus. Irish Industry.)
19, 20.	*Phacops nudus*, Salter. *a*, *b*, *c*. Fragmentary heads. Fig. *d* is *c* magnified. *e*. Eye, magnified. *f*. The lenses, magnified. Fig. 20, tails, two specimens. Upper Silurian Limestone, Dingle, Co. Kerry.

LOWER SILURIAN.

21—24.	*Phacops Bailyi*, Salter. Caradoc of Tramore, Co. Waterford. 24 *a*. Interior cast of tail. *b*. Exterior cast.
25.	,, *conopthalmus*, Boeck.? Caradoc, N. Wales. (Cambridge Museum.)
26.	*Sphærexochus mirus*, Beyrich. Caradoc, Chair of Kildare. (Mus. Pract. Geology.)
27, 28.	*boops*, Salter. Caradoc. Haverfordwest. Fig. 27 (Mus. P. Geol.). Fig. 28, Westmoreland. (Cambridge Museum.)
29, 30.	*Amphion pseudo-articulatus*, Portlock. Caradoc of Tramore, Waterford. (Mus. Pract. Geology.)
31.	,, *benevolens*, Salter. Caradoc of Newtown, Waterford. (Mus. Pract. Geology.)
32.	,, *pauper*, Salter. Caradoc of Tramore. (Mus. Irish Industry.)

UPPER SILURIAN. DEVONIAN.

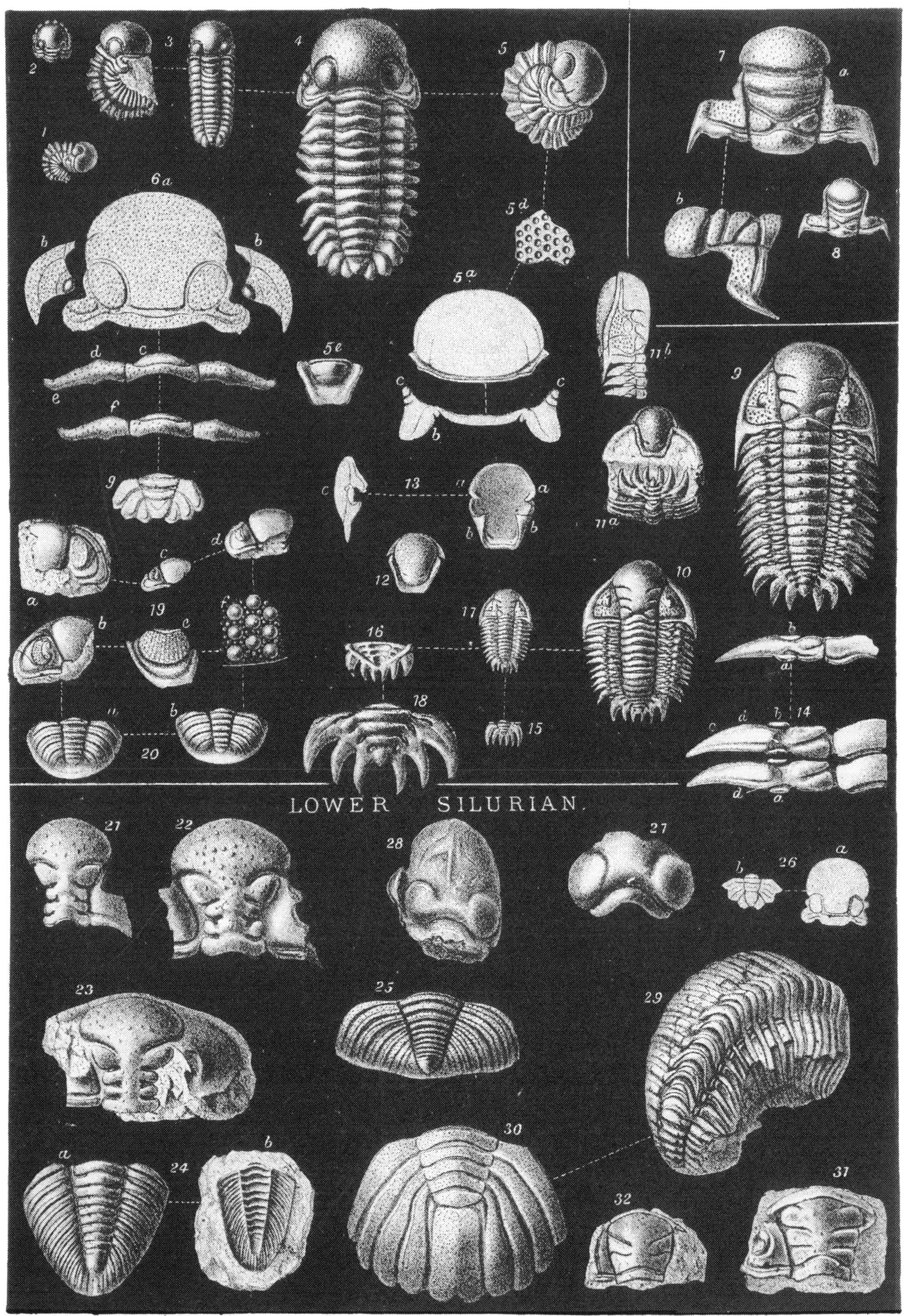

J. W. Salter & A. Gawan, lith. W. West, imp.

1–6. Sphærexochus mirus. 7, 8. Cheirurus articulatus. 9–18. C. bimucronatus. 19, 20. Phacops nudus. 21–24. P. Baïlyi. 25. P. conophthalmus? 26. Sphærexochus mirus. 27, 28. S. boops.– 29, 30. Amphion pseudoarticulatus. 31. A. benevolens. 32. A. pauper.

PLATE VII.

UPPER SILURIAN (WENLOCK).

Figs. 1—12. *Deiphon Forbesii,* BARRANDE. All the principal specimens known.

1. „ „ The most perfect specimen extant; natural size and magnified. It is in the rich cabinet of my friend Dr. Grindrod, of Malvern, and shows the eyes, pleuræ, and tail, to perfection. Malvern Tunnel.

2. „ „ A younger, but nearly perfect, example. Same locality and cabinet.

3. „ „ From Mr. C. Ketley's collection. Same locality.

4. „ „ Young head, from the late Mr. Mushen's collection. Dudley.

5. „ „ Somewhat older. Dudley. (Mr. Mushen's cabinet.)

6. „ „ Largest head known. (Mr. Mushen's collection.) 6 *a*. Front view. 6 *b*. Side view. Dudley.

7. „ „ Portion of glabella, magnified.

8. „ „ Side view of perfect eye and free cheek; from fig. 1.

9. „ „ Showing the supporting folds of the eye (the lentiferous surface lost) from fig. 5. This specimen also shows the curious blunt process on the forward edge of the head-spine.

10 *a*. „ „ Mr. Ketley's specimen, showing hypostome and labrum, and the underside of the first six pleuræ; a beautiful specimen, which also shows the eyes in place. 10 *b*. The same, enlarged. 10 *c*. Is a magnified front view of the labrum; 10 *d*, side view of ditto. From Malvern Tunnel.

11. „ „ Tail, and part of the tenth body-segment. Dudley. (Woodwardian Museum.)

12. „ „ Do., enlarged to twice its size. (Mr. Mushen.) Dudley.

Figs. 13—20. *Staurocephalus Murchisoni,* BARR.

13 *a*. „ A Dudley specimen, from Mr. E. Hollier's cabinet, magnified.

13 *b*. „ Side view of another. (Same cabinet.) Dudley.

14. „ Coiled specimen. (British Museum.) Natural size.

15 *a*. „ Do. a fine example from Dr. Grindrod's cabinet: Wenlock Shale, Malvern. *a*. Natural size. *b*. Magnified.

16. „ Do., showing the border-spines to the head. (Same cabinet.)

17. „ Fragment with the frontal crest of tubercles and the blunt labrum. (British Museum.)

18. „ Middle and last body-segments, and tail; from the Decades of the Survey; partly restored, and magnified.

LOWER SILURIAN.—CARADOC.

19. *Staurocephalus Murchisoni.* Small specimen, figured by M'Coy, and in the Woodwardian Museum. Rhiwlas, Bala, N. Wales.

20 *a*. „ „ Do. (In Mus. Pract. Geology.) Same locality; 20 *b*, enlarged.

Fig. 21. „ *globiceps,* PORTLOCK. The original specimen figured in Portlock's work and in Decade 11, Geol. Survey. *a*. Natural size; *b*, enlarged; the tail is broken off at the apex. Tyrone.

Figs. 22—24. „ *unicus,* THOMSON. A series of fine specimens from Professor Wyville Thomson's cabinet. Others are in the Museum Pract. Geology. Girvan, Ayrshire.

22 *a*. „ „ The same head, front view, nat. size; *b*, viewed sideways. *c*. Fig. 22 *a* magnified; it shows the gibbous glabella, forward eyes, and minute free cheeks.

23, 24. „ „ Body and tail. Nat. size, same locality.

Fig. 25. „ *sp.* (*St. nodulosus,* MS., SALTER), Professor Thomson's cabinet. This shows many differences from *St. unicus,* but is not sufficiently preserved for me to be quite sure of its distinctness from *St. globiceps.* Girvan, Ayrshire.

UPPER SILURIAN.

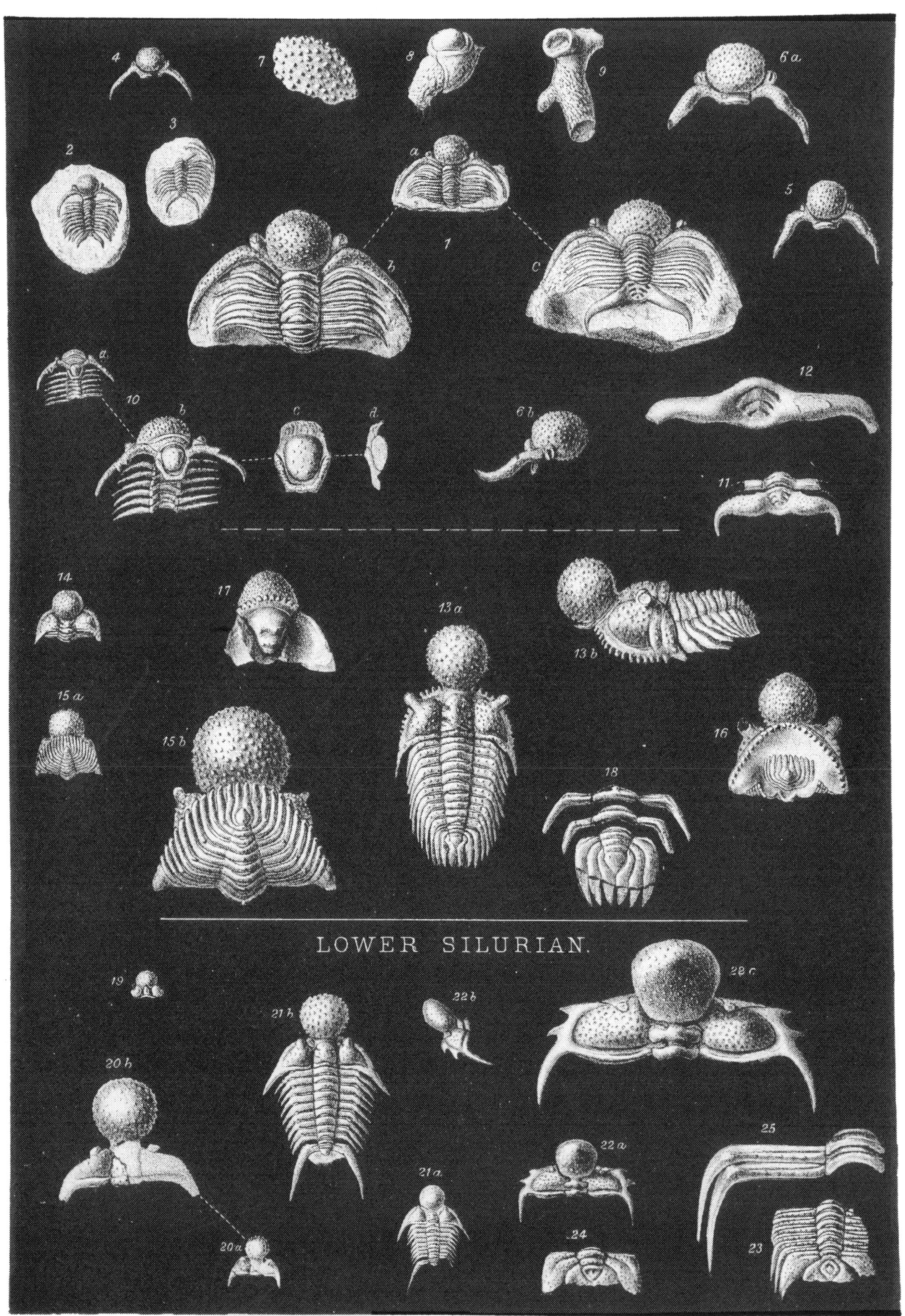

J.W. Salter & A. Gawan, lith. W. West, imp

1–12. Deiphon Forbesii. 13–20. Staurocephalus Murchisoni.— 21. Staur. globiceps. 22–24. S. unicus.— 25. S. ——— sp

PLATE VIII.

UPPER SILURIAN (WENLOCK).

Figs. 1—6.	*Calymene tuberculosa*, SALTER [Mus. Pract. Geology]. Burrington, Shropshire.	
1.	" "	Full grown.
2.	" "	Specimen figured in Decade 2, pl. viii, fig. 1.
3.	" "	Side view of the same.
4.	" "	Coiled specimen, showing the labrum imperfect; this is also figured in the plate of Decade 2, above quoted.
5.	" "	Young specimen.
6.	"	Another specimen, with the labrum *in situ*.
Figs. 7—16.	" *Blumenbachii*, BRONGN. Ordinary Dudley variety.	
7.	" "	The largest specimen known, the same that is figured in the original 'Silurian System,' and in 'Siluria,' both editions. (Cabinet of W. Matthews, Esq., junr., Edgbaston, Birmingham.)
8 *a*, *b*, *c*,	" "	Outlines of characteristic head and tail, from a Dudley specimen in Mr. E. Hollier's cabinet, Dudley.
9.	" "	Labrum of the species, from the specimen figured in Pl. IX, fig. 2. (Mus. Pract. Geology.)
10.	" "	Very young Dudley specimen (*C. pulchella*, Dalman ?), from Dudley. In the Gray collection, Brit. Museum.
11.	" "	Young Dudley specimen, same collection.
12.	" "	Ordinary half-grown specimen, Dudley. (E. Hollier, Esq., junr.)
13 *a*, *b*.	" "	Under view and side view of a fine specimen in Prof. Tennant's cabinet. Dudley.
14.	" "	Specimen with coarse granulation. (Mus. Pract. Geology.)
15.	" "	Large granulated specimen of the full size. (Cabinet of E. Hollier, Esq., junr.) Dudley.
16 *a*, *b*.	" "	Two views of very large coiled specimen, the granules much worn off. (Cabinet of E. Hollier, junr.)

UPPER SILURIAN.

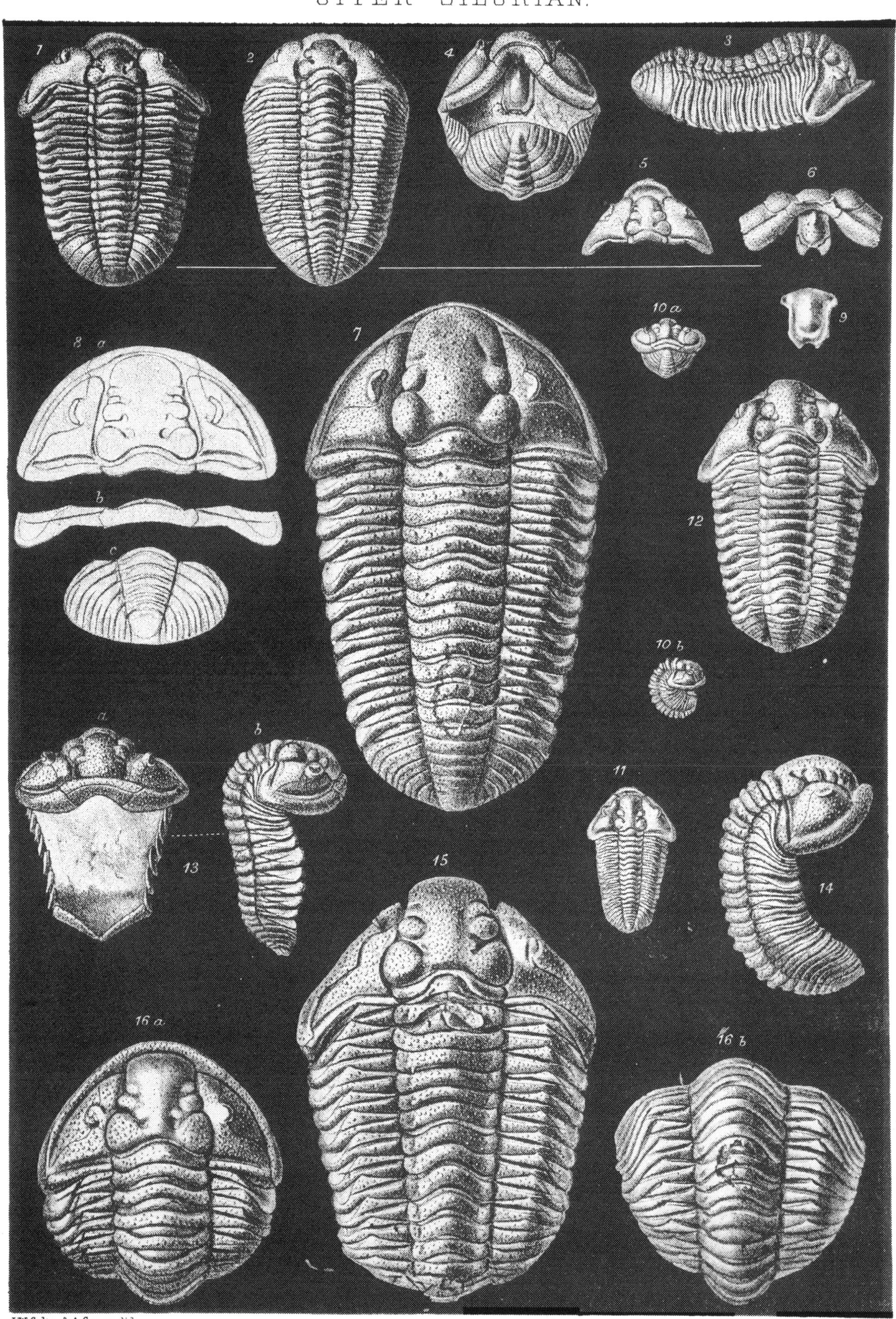

J.W. Salter & A. Gawan, lith.

W. West, imp.

1-6. *Calymene tuberculosa.* 7-16. *C. Blumenbachii.*

PLATE IX.

MIDDLE AND LOWER SILURIAN.

Figs. 1—5.

1. *Calymene Blumenbachii*, BRONGN, var. *a*, *auctorum*. Interior cast, from the Woolhope or Lower Wenlock Grits of Bogmine, Shelve, Shropshire. (Mus. Prac. Geology.) The crust of this specimen is much thickened.

2. „ Same species and variety. Llandovery Rock, Mullock, near Girvan, Ayrshire. (Mus. Pract. Geology.) [The labrum figured in Pl. VIII, fig. 9, is from the same locality.]

3. „ var. *Caractaci*. Fine specimen from a large slab in the Geological Society's Museum. Acton Scott, Shropshire.

4. „ Do. A head from Shropshire. (Mr. Edgell's cabinet.)

5. The same variety, from Gen. Portlock's collection in Mus. Pract. Geology. (*C. brevicapitata*, Portl., in part.)

Figs. 6—11. „ *senaria*, CONRAD.

„ „ (Original figured specimens of *Calym. brevicapitata*, Portl.) from Caradoc Rocks, Tyrone. Interior cast. (Mus. Pract. Geol.)

8. „ „ A head of same species from Caradoc Rocks of Dolbenmaen, near Pwllheli, Caernarvonshire. Interior cast. (Mus. Pract. Geol.)

9. „ „ A specimen with external crust from the blue Trenton Limestone of Ohio, N. America. (Mus. Pract. Geology.)]

10. „ „ Young, perfect (the specimen figured in the Memoirs of the Geol. Survey, vol. ii, plate 1, plate 11). 10 *a*. Tail magnified. Bala, N. Wales.

11. „ „ Head from same locality. (Mus. Pract. Geol.)

Figs. 12—14. „ *Cambrensis*, SALTER.

12. „ „ Interior cast of glabella, Llandeilo Flags, Lann Mill, Narberth.

13. „ „ Cast of exterior of perfect specimen. Near Llandeilo, in Llandeilo flags. (In the cabinet of J. E. Lee, Esq., Caerleon.)

14. „ „ Same species. (M'Coy's figured specimen of *C. brevicapitata*. Woodwardian Museum.)

Figs. 15—18. „ *Tristani*, BRONGN.

15, 16. „ „ Llandeilo or Arenig Rocks; Gorran Haven, Cornwall. (Mr. Edgell's cabinet.)

17. „ „ Tail of same species, from the Budleigh Salterton pebble-bed. (See Quart. Geol. Journ. for 1864, vol. xx, pl. xv, fig. 5, where the same specimen is figured.)

18. „ „ [Outline head from Nehou, Normandy, for comparison.]

Figs. 19—24. *duplicata*, MURCHISON. 19. Small specimen, var. *a* (*mas*), Pen Cerrig, Builth. (Mus. Pract. Geology.) 20. Dissected head, same locality and collection. 21. Large tail, same locality. (Mr. Griffith Davies's cabinet.) 22. Labrum, natural size and twice magnified; Builth. (Mus. Pract. Geology.) 23. Nearly complete specimen from Builth (Mr. Edgell's cabinet); the front of the head is added in outline from other specimens. 24. Variety (*fœmina*), probably the ♀ form. Trecoed, Builth. (Mus. Pract. Geology.)

Fig. 25. „ *parvifrons*, SALTER. Arenig group (Lower Llandeilo). Tai Hirion, Bala (Woodw. Museum).

Figs. 26—28. „ *id.* VAR. *Murchisoni*. Same formation, west of Stiper Stones, Shropshire. Fig. 28 is partly restored in shaded outline, derived from several specimens, to show the probable shape of the entire tail.

MIDDLE AND LOWER SILURIAN.

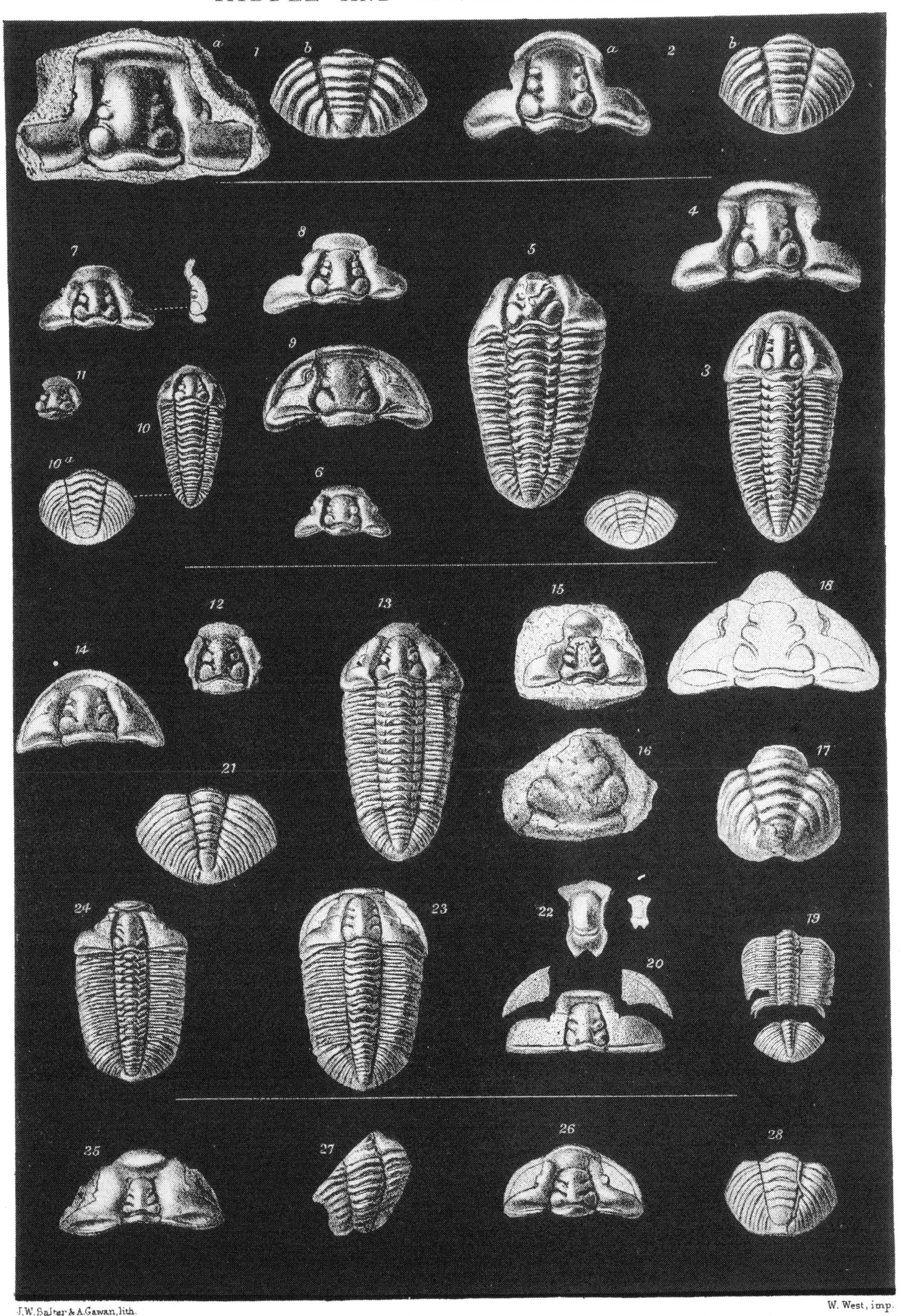

J.W. Salter & A. Gawan, lith.

W. West, imp.

1, 2. *Calymene Blumenbachii.* 3–5. *C. Blum. var. Caractaci.* 6–11. *C. senaria.*
12–14. *C. senaria var. cambrensis.* 15–18. *C. Tristani.* 19–24. *C. duplicata.*
25. *C. parvifrons.* and 26–28 *var. Murchisoni.*

PLATE X.

LOWER DEVONIAN.

Figs. 1, 2. *Homalonotus elongatus*, SALTER. Front and side views of a unique specimen in the cabinet of Mr. Townshend Hall, of Pilton Rectory, N. Devon. Lowest Devonian; Meadsfoot, Torquay.

LOWER SILURIAN.

Figs. 3—10.[1] ,, *bisulcatus*, SALTER. From the Caradoc of Shropshire.

3. ,, ,, Tail, from Wittingslow, Shropshire. (Museum of the Geological Society.)

4. ,, ,, Body and tail. Same locality and museum.

5. ,, ,, Choice young specimen. Acton Scott. (Mr. Edgell's cabinet.)

6. ,, ,, Specimen from Bala, N. Wales, elongated by cleavage. (Mus. Pract. Geology.)

7, 8. ,, ,, Young specimens. North Wales. Showing a more conical glabella than the adult. (Woodwardian Museum.)

9. ,, ,, Fine head. (Mr. Lightbody's cabinet.) Onny River.

10. ,, ,, Very young head (possibly of the next species.) Horderly. Mr. Edgell's cabinet.

Fig. 11. ,, *Edgelli*, SALTER. Tail from Acton Scott. (Same cabinet.)

Figs. 12—14. ,, *rudis*, SALTER. CARADOC of North Wales and Shropshire.

12. ,, ,, Giant specimen. Nantyr, near Llanarmon, in the Berwyn Mountains, N. Wales. (Mus. Pract. Geology.)

13. ,, ,, Probably same species, showing nodular rings to the axis. Caradoc Grits of Cressage, Shropshire. (Mus. Pract. Geology.)

14. ,, ,, Coiled body (*a*), and tail (*b*), from Cader Dinmael, Denbighshire. (Mus. Pract. Geology)

Figs. 15—17 ,, *Brongniartii*, DESLONGSCHAMPS. From the Lower Silurian pebbles, Budleigh Salterton, S. Devon. (Mr. Vicary's cabinet.) 15. Head, wanting the free cheeks. 16, *a*, *b*. Body ring; *a* shows the blunt end of the pleura, *b* the axal point. 16. Is an edge view of the ring. 17. Tail, with serrate edges.

Fig. 18. ,, — *sp*. Same locality and cabinet. The last two are also figured in the Quart. Geol. Journ., vol. xx.

[1] Wrongly 2—10 at bottom of plate.

1, 2, Homalonotus elongatus. *2–10. H. bisulcatus.* *11. H. Edgelli.* *12–14. H. rudis.*
15, 17. H. Brongniarti. *18. H. ——— sp.*

PLATE XI.

UPPER SILURIAN (WENLOCK).

Figs. 1—11. *Homalonotus delphinocephalus*, GREEN.

1. ,, ,, From Dudley: the finest known. This is the specimen figured in the 'Silurian System,' and is in Mr. Blackwell's possession. 1 *a*. Is a side view of the same.

2. ,, ,, A smaller specimen. (Mr. E. Hollier's collection.)

3. ,, ,, A very young one. (Mr. Edgell's cabinet.) Invaluable, as showing the eyes. 3 *a*. The same, magnified. Fig. 6. The magnified eye. Dudley.

4. ,, ,, Somewhat older, and also showing the lentiferous surface perfectly. (From the cabinet of Mr. E. Hollier, junr.)

5. ,, ,, The angle of the cheek, with the squamate granules. (From Mr. Blackwell's specimen.)

6. ,, ,, Eye and free cheek of fig. 3, magnified, to show the eye-lenses.

7. ,, ,, Labrum, interior view. (Mr. E. Hollier, junr.)

8. ,, ,, A thorax-ring, the fulcrum close to the puncta which mark out the axis.

9, 10. ,, ,, Pleuræ enlarged, to show the squamate character of the granulation.

11. ,, ,, Tail from large specimen. (E. Hollier, junr.)

All the above are from the Dudley Limestone.

Fig. 12. *Homalonotus cylindricus*, SALTER. See p. 116, figs. 27, 28 (woodcuts), for a more complete specimen. Woolhope Limestone, Woolhope. (Cabinet of Mr. H. Edgell.)

J. W. Salter & A. Gawan. lith.

W. West, imp.

1-11. Homalonotus delphinocephalus. *12. H. cylindricus.*

PLATE XII.

UPPER SILURIAN (LUDLOW ROCK.)

Fig. 1. *Homalonotus (Koenigia) ludensis*, SALTER, (not of 'Sil. System'). Nat. size. So far as known, a unique specimen. (Mr. Edgell's cabinet.)

Figs. 2—10. „ *Knightii*, KOENIG.

2. „ „ Perfect head, showing the sinuated front; at *b*, the attachment for the labrum. (Mr. G. Cocking's collection.)

3. „ „ Fine specimen, from Ludlow Museum, with the crust preserved (all the others are casts).

4. „ „ Interior cast, showing the depressed spaces at the base of the cheeks.

5. „ „ Half-coiled specimen, copied from the 'Silurian System.' Ludlow.

6. „ „ Half-grown specimen, the original from which Dr. Koenig's figure in the 'Icones Sectiles' was taken. Probably from Radnorshire.

7. „ „ The largest specimen known: from the Upper Ludlow Rock of Malvern. (Dr. Grindrod's cabinet.)

8, 8 *a*. „ „ Side view of the same. 8 *a*. Two of the pleuræ magnified, to show the granules (not squamæ) in this species.

9. „ „ Side view and underside of tail; showing the strong broad fascia, granulated and angularly bent along its middle. Ludlow. (Mr. Cocking's cabinet.)

10. „ „ Labrum of an Upper Ludlow specimen. Ludlow Museum. (Mr. Edgell's cabinet.)

Fig. 11. „ *Johannis*, SALTER (see also Pl. XIII). Wenlock Rocks of Carmarthenshire. (Mus. Pract. Geology.)

UPPER SILURIAN.

J.W. Salter & A. Gawan, lith.

W. West, imp.

1. Homalonotus ludensis. 2_10. H. Knightii. 11. H. Johannis.

PLATE XIII.

UPPER SILURIAN.

Figs. 1—7. *Homalonotus Johannis*, SALTER. A new species, from the Wenlock Shale of Usk. (Cabinet of Mr. John E. Lee, Caerleon.)

1. „ „ Largest head known, interior cast. The oval spaces at the base of the cheeks are well seen.

2. „ „ Showing the rostral shield, and the granular cheek-border—interior cast.

3. „ „ Imperfect body, and pointed tail, interior.

4. „ „ Exterior cast, in gutta-percha, of the body-rings.

5. „ „ The pleuræ from fig. 7, enlarged, to show the emarginate tips, and the spinose granules on the border.

6. „ „ Large but less perfect specimen. Same cabinet.

7. „ „ Most complete one known, showing the rostral shield, and the deep circumscription of the glabella and axis of the tail. The crust is preserved in part, and is magnified in fig. 5.

Fig. 8. „ *Knightii*, MURCHISON. The original specimen figured in the 'Silurian System.' (Mus. Geol. Society.)

LOWER SILURIAN.

The following are from the Budleigh Salterton pebbles in Mr. Vicary's cabinet.

Fig. 9. *Homalonotus Brongniartii* (see Pl. X), showing the serrated border and vertical margin. Nat. size.

Fig. 10. „ *Vicaryi*, n. sp. 10 *a* (marked *H. Winwoodii* on plate). A small Budleigh specimen. [10 *b*. A Normandy specimen for illustration. Mus. Geol. Soc.]

All the figures except fig. 5 are of natural size.

UPPER SILURIAN.

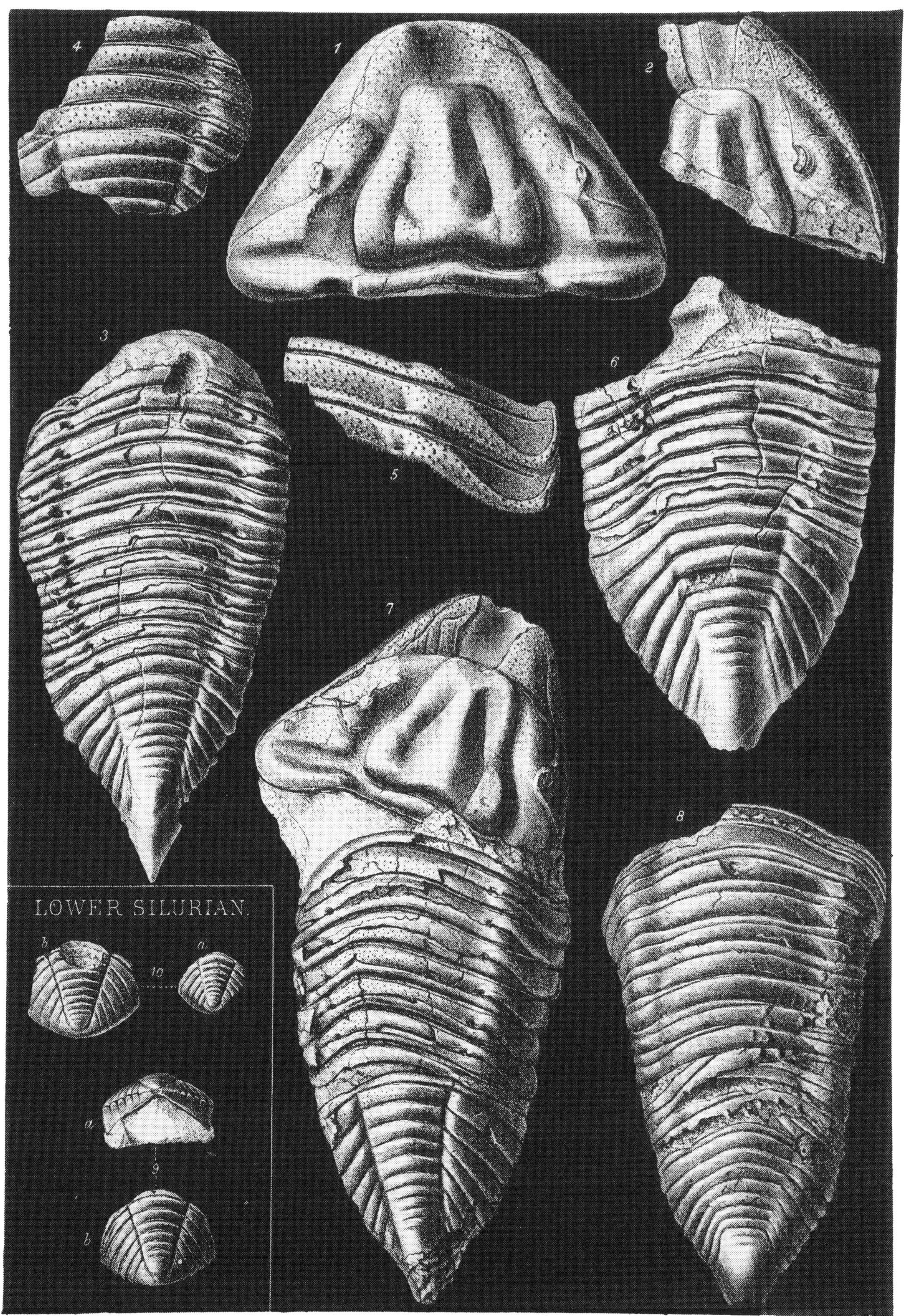

J.W. Salter & A. Gawan, lith.

W. West, imp

1_7. *Homalonotus Johannis.* 8. *H. Knightii.* 9. *H. Brongniartii.* 10. *H. Winwoodii.*

PLATE XIV.

LLANDEILO FLAGS.

Figs. 1—7. *Ogygia Buchii*, Brongniart. A series of fine specimens from Builth, in the cabinet of Mr. Griffith Davies, Cloudesley Street, Islington.

1. ,, ,, Group of the female form, not fully grown.

2. ,, ,, The same variety, or ♀ form.

3. ,, ,, Large head of the ♂ form, showing the lineation of the surface, and the labrum shifted from its natural place.

4. ,, ,, ♂ form, not fully grown; the glabella-furrows strongly marked.

5. ,, ,, Young specimen.

6. ,, ,, Young specimen; the original of *Trinucleus asaphoides* figured in the 'Sil. System.'

7. ,, ,, Eye, natural size. *a*. Lenses of the same, magnified.

Figs. 8, 9. *Ogygia angustissima*, Salter. From the same cabinet and locality.

LOWER SILURIAN.

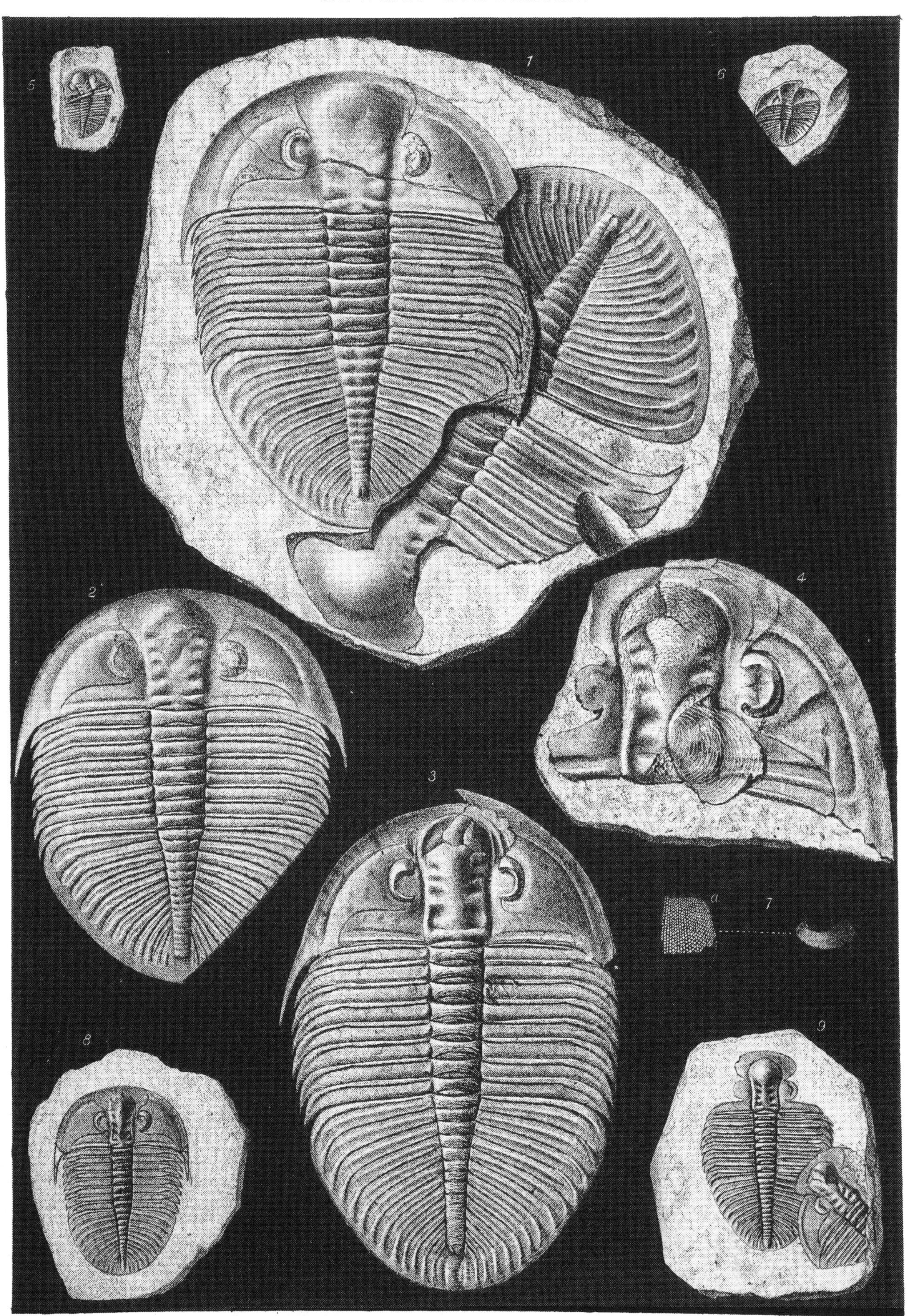

J. W. Salter & A. Gawan, lith.

W. West, imp.

1–7. Ogygia Buchii. 8, 9. O. angustissima.

PLATE XV.

LLANDEILO FLAGS.

Figs. 1—6.	*Ogygia Buchii*, Brongniart.	
1.	,, ,,	Male form ♂, Builth (Mr. Davies' cabinet), of the true convexity (most of the specimens are flattened).
2.	,, ,,	Specimen showing the labrum *in situ*. (Same cabinet.)
3.	,, ,,	Labrum. (Mus. Pract. Geology.)
4.	,, ,,	Portion of the striated upper border of the tail.
5.	,, ,,	Female form ♀; a variety with the dorsal tubercles conspicuous. (Mus. Pract. Geology.)
6.		Female form; a fine external cast. Presented to the Mus. Pract. Geology by R. Banks, Esq., of Kington.
Figs. 7, 8.	*Ogygia ??* (or *Phacops*) *subduplicata*, Salter.	Llandovery Rocks of Haverfordwest. (Cabinet of Mrs. Breawell, Brighton.)

LOWER SILURIAN.

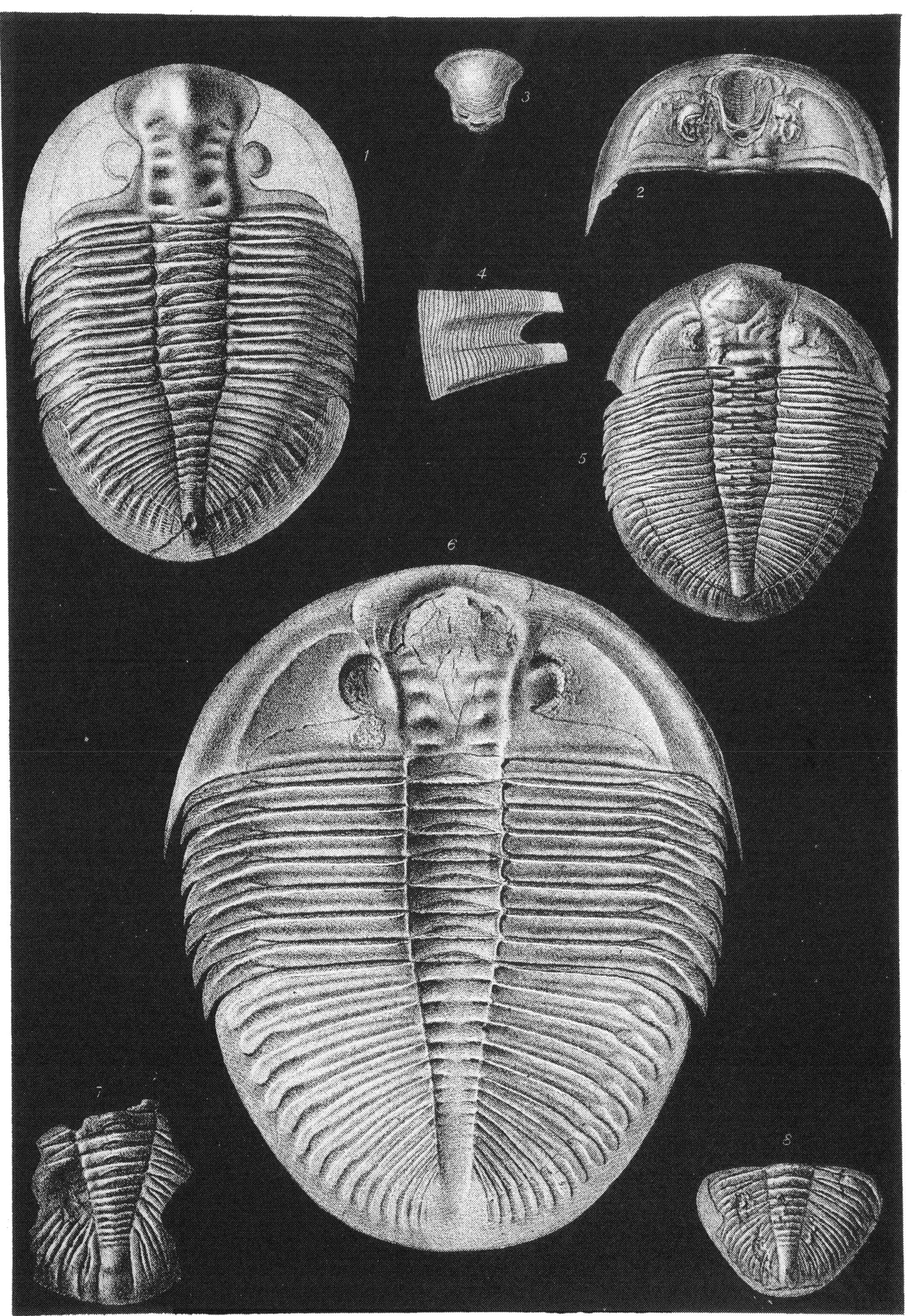

J.W. Salter & A. Gawan. lith.

W. West. imp

1_6. *Ogygia Buchii.* 7, 8. *O. subduplicata.*

PLATE XVI.

LLANDEILO FLAGS OF BUILTH, RADNORSHIRE.

Figs. 1—14. *Ogygia Corndensis*, (MURCHISON.) Chiefly from the cabinet of Mr. Griffith Davies, of Cloudesley Street, Islington. The smaller specimens (figs. 2—6) are from the Shales of Trecoed, North of Builth, and are in the Museum Pract. Geology.

1. „ „ Male ♂ form? with pointed head. Gilwern, near Llandrindrod.

2. „ „ Young specimens, but with the full number of rings. Trecoed.

3, 4. „ „ Distorted young specimens. Trecoed, Builth. The number of rings not quite certain.

5. „ „ Very young state, with only six body-rings.

6. „ „ Somewhat older, with seven rings.

7. „ „ Half-grown specimen, the head has slipped over three of the body rings, and only five are visible.

8. „ „ The originally figured specimen, from the Corndon Mountain, Shelve. (See 'Siluria,' pl. iii, fig. 4.)

9. „ „ Very large specimen, showing impression of labrum, and the caudal fascia below. Gilwern.

10. „ „ Young specimen, with labrum *in situ*. Gilwern.

11. „ „ Full-grown female ♀ form, rounded at both ends.

12, 13, 14. „ „ Half-grown specimens. Gilwern.

LOWER SILURIAN.

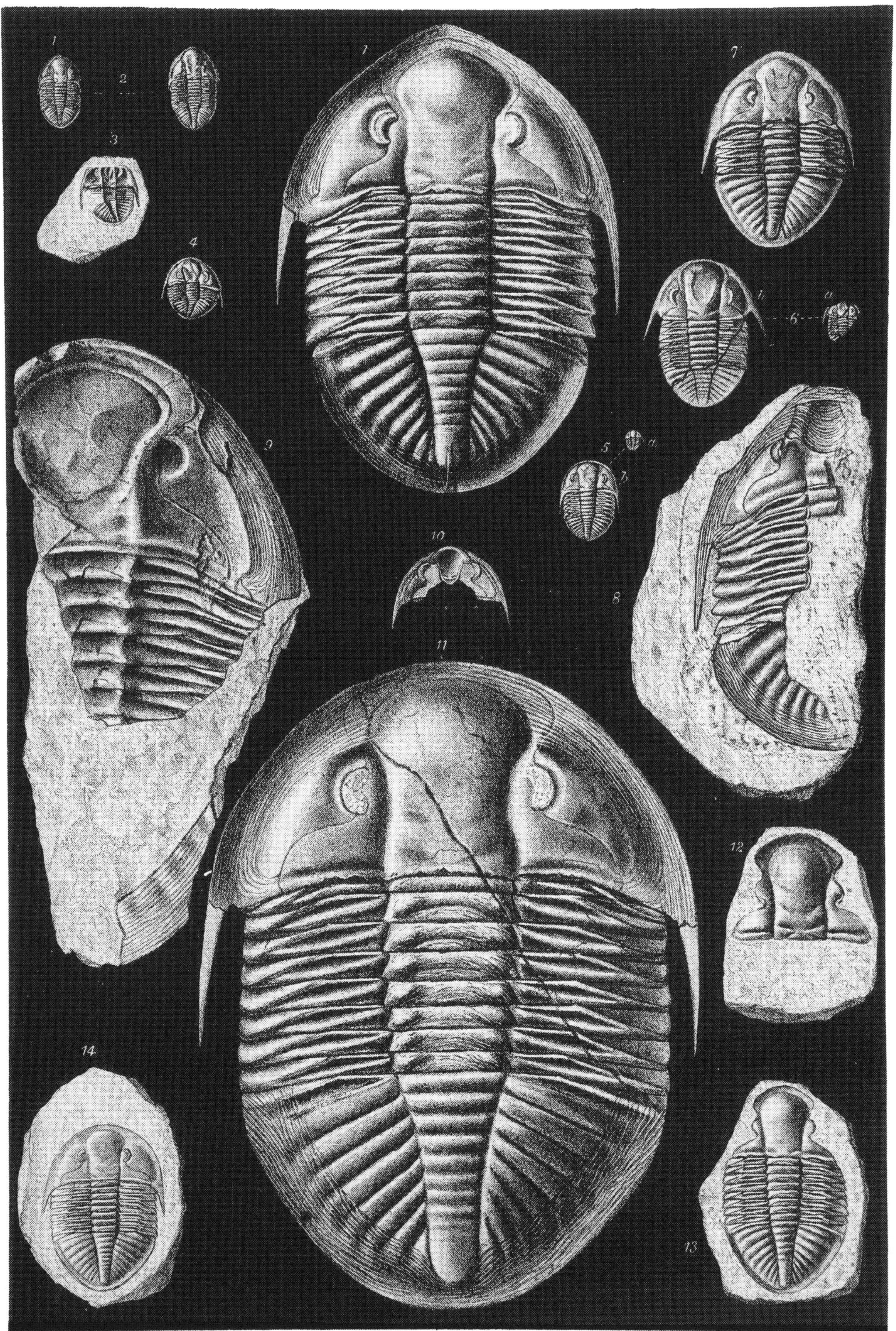

J. W. Salter & A. Gawan. lith. W. West, imp

Ogygia Corndensis.

PLATE XVII.

LOWEST SILURIAN (ARENIG GROUP).

Figs. 1—7. *Ogygia Selwynii*, SALTER.

1. „ „ Restored figure (not quite correct in proportions, the glabella should be broader), chiefly from specimens in the Museum of Practical Geology.

2. „ „ Glabella. (Mr. Lightbody's cabinet.) White Grit Mine, Shelve.

3. „ „ Compressed. From Llanfaelrhys, Aberdaron, South Carnarvonshire. (Mus. Pract. Geology.)

4. „ „ Large caudal shield, White Grit Mine. (Mus. Pract. Geology.)

5, 6. „ „ Thorax-joint and caudal shield. Same locality and cabinet.

7. „ „ Labrum. Same locality. (Mr. Lightbody's cabinet.)

Figs. 8—10. *Ogygia peltata*, SALTER. From Whitesand Bay, St. David's, Pembrokeshire.

8. „ „ A figure partly restored, and made too narrow in the axis, and the eye wrongly placed. Please to correct this by fig. 10.

9, 10. „ „ Parts of one fine specimen, pressed by cleavage-action into a shorter form than ordinary. (Cabinet of J. E. Lee, Esq., Caerleon.) Fig. 9 should show faint traces of the intermediate furrows of the tail-ribs; and fig. 10 shows the labrum broken at the tip, so as to look like an *Asaphus*. It is, however, incomplete, and must have been pointed, as in *Ogygia*.

TREMADOC ROCKS (UPPERMOST CAMBRIAN).

Figs. 11—13. *Ogygia scutatrix*, SALTER. Fig. 11, the specimen originally found at Garth, in Upper Tremadoc. Fig. 13, labrum, imperfect, from Portmadoc. (Mus. Pract. Geology.) Fig. 12, tail from West of Penmorfa, Lower Tremadoc. (Mr. Ash's cabinet.)

ARENIG ROCKS.

J. W. Salter & A. Gawan, lith.

W. West, imp.

1_7. Ogygia Selwynii. 8_10. O. peltata. 11_13 O. scutatrix.

PLATE XVIII.

CARADOC AND LLANDEILO ROCKS.

Figs. 1—5. *Asaphus radiatus*, SALTER.

1. " " Largest specimen known. Berwyn Mountains. (Collection of Mr. Blunt, of Shrewsbury.)

2. " " Small but perfect specimen, figured by M'Coy as *A. laticostatus*. Bala, North Wales. (Woodw. Mus.)

3. " " Tail, compressed. Same locality. (Mus. Pract. Geology.)

4, 5. " " Caudal shields, shortened by pressure. Caradoc Slates of Louth, Ireland.

Fig. 6. *Asaphus laticostatus*, M'COY. The only specimen known. Llandeilo flags, Maen Goran, Builth. (Woodwardian collection.)

Figs. 7—10. *Stygina latifrons*, PORTLOCK. (Chiefly specimens figured in Decade II of Geol. Survey.) Tyrone.

7. " " Compressed laterally; best specimen known. (Mus. Pract. Geology.)

8. " " A little shortened by pressure. (Mr. J. E. Lee's cabinet.)

9. " " Epistome and labrum. (Prof. W. Thomson's cabinet.)

10. " " Dissected head, from various specimens. It shows there is no rostral shield.

Fig. 11. *Stygina Murchisonæ*, MURCHISON. Head and tail from the only locality known. Mount Pleasant, Carmarthen.

LOWER SILURIAN. (Caradoc and Llandeilo.)

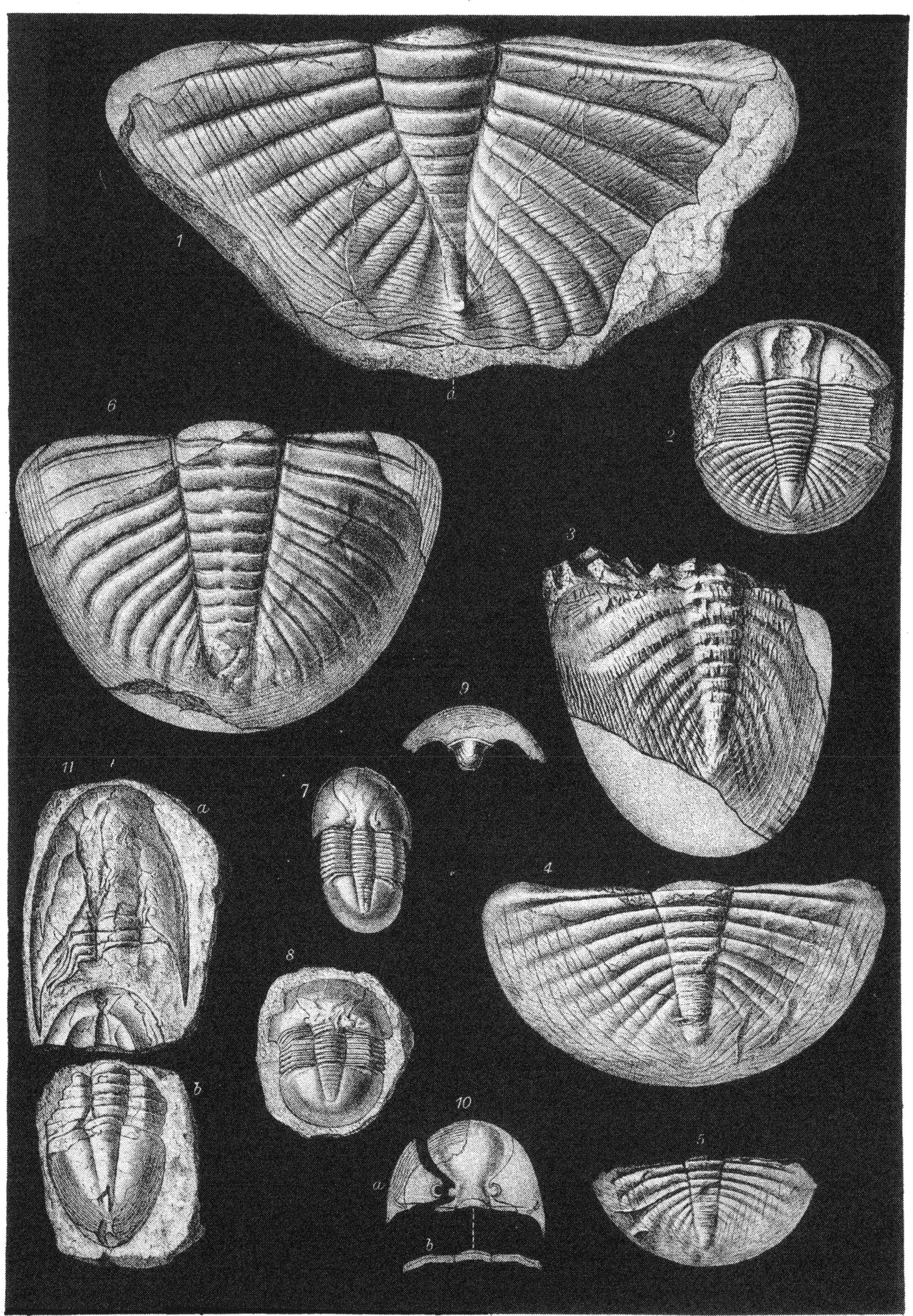

J.W. Salter & A. Gawan, lith

W. West, imp

1_5. Asaphus radiatus. 6. A. laticostatus. 7_10. Stygina latifrons. 11 S. Murchisonæ.

PLATE XIX.

LLANDEILO FLAGS.

Figs. 1—4. *Barrandia (Homalopteon) radians*, M'Coy. In various stages of growth.

1. " " " Young, with only four thorax-rings developed. *a*, natural size; *b*, magnified. Wellfield, Builth.

2. " " " Somewhat older, with seven thorax-rings. Pencerrig, Builth.

3. " " " Not quite half grown, the eighth thorax-ring partly developed. (All the above in the Mus. Pract. Geology.)

4. " " " Full-grown tail. Gwernyfyd, Builth. (Mr. J. E. Lee's cabinet.)

Fig. 5. *Barrandia (proper) Cordai*, M'Coy. The only specimen known. Pencerrig, Builth. Once and a half natural size. The true length is represented by the line beside it. 5 *a*. A thorax-ring, enlarged. (Woodwardian Museum.)

Figs. 6—10. *Barrandia (Homalopteon) Portlockii*, Salter.

6. " " " Large specimen, partly restored. 6 *a*. One of the thorax-rings; diagram.

7. " " " Young specimen.

8. " " " Full-grown caudal portion.

9. " " " Labrum, with one side and the tip broken off.

10. " " " Head, showing glabella-furrows very perfectly. The free cheeks are wanting in all our specimens.

Figs. 3, 4, 6—10 are natural size; figs. 1, 2 magnified; fig. 5 enlarged one half.
Figs. 6—10 are from the Mus. Pract. Geology.

LOWER SILURIAN. (Llandeilo Flags.)

J.W. Salter & A.Gawan lith.

W. West, imp.

1_4. *Barrandia radians.* 5. *B. Cordai.* 6_10. *B. Portlockii.*

PLATE XX.

LOWER TREMADOC SLATES AND ARENIG GROUP.

ARENIG ROCKS.

Figs. 1, 2. *Illænopsis Thomsoni*, SALTER. From the mine-works at Ritton Castle, Shelve, Shropshire. (Mus. Pract. Geology.)

LOWER TREMADOC SLATES.

Figs. 3—12. *Niobe Homfrayi*, SALTER. Penmorfa Church, near Tremadoc. (In the cabinet of Mr. David Homfray.)

3. ,, ,, Nearly full-grown, the free cheeks only wanting.

4. ,, ,, Head, showing more or less completely the glabella-furrows.

5. ,, ,, Tail, showing the caudal fascia well. (My own cabinet.)

6. ,, ,, Complete head; the free cheeks obtuse, not pointed. (Probably ♂.)

7. ,, ,, Young specimen (♂ form), the general shape and the axis being narrower than usual.

8—11. ,, ,, Outline diagrams from actual specimens, showing the true position of the glabella-furrows.

12. ,, ,, Broad (♀) form, compressed and shortened by cleavage.

Figs. 13—19. *Psilocephalus innotatus*, SALTER. Specimens of all ages and various degrees of compression in the rock. Fig. 16 is the fry; but the metamorphosis has not yet been properly observed.

UPPER CAMBRIAN. LOWER SILURIAN.

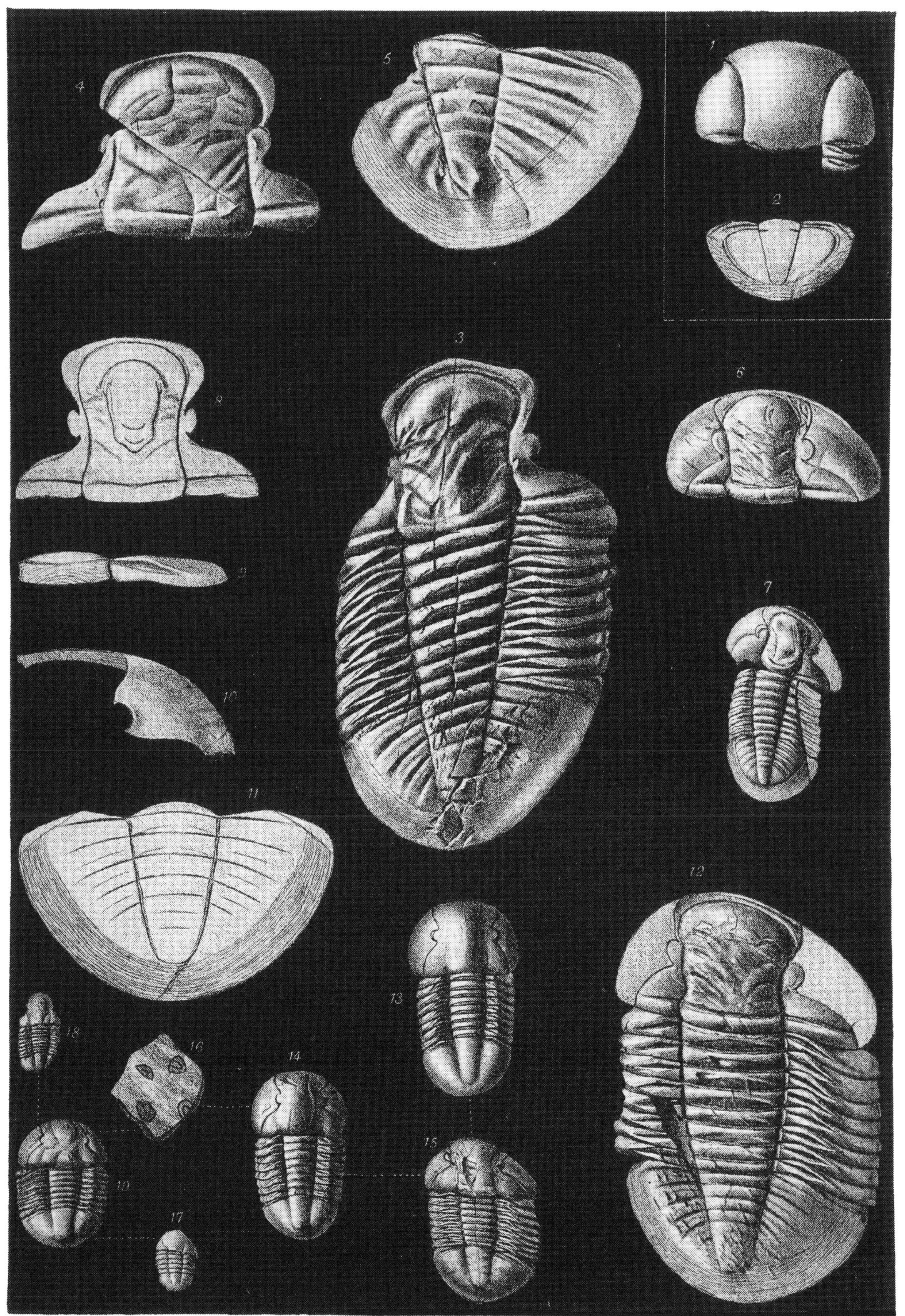

J. W. Salter & A. Gawan, lith. W. West, imp.

1, 2. Illænopsis Thomsoni. *3_12. Niobe Homfrayi.* *13_19. Psilocephalus innotatus.*

PLATE XXI.

LLANDEILO FLAGS.

The large and fine specimen of *Asaphus tyrannus* figured in the 'Silurian System,' pl. xxiv, as a variety, *ornatus*.

[We have, as usual in these plates, restored missing portions in a light tint. No inconvenience can arise from this practice so long as care is taken to distinguish accurately between the real and the missing parts. In the present instance one rib too many is added to the tail, which never shows more than twelve side-ribs.]

This is not the largest specimen known. *A. tyrannus* grows very nearly a foot in length.

LOWER SILURIAN (Llandeilo Flags.)

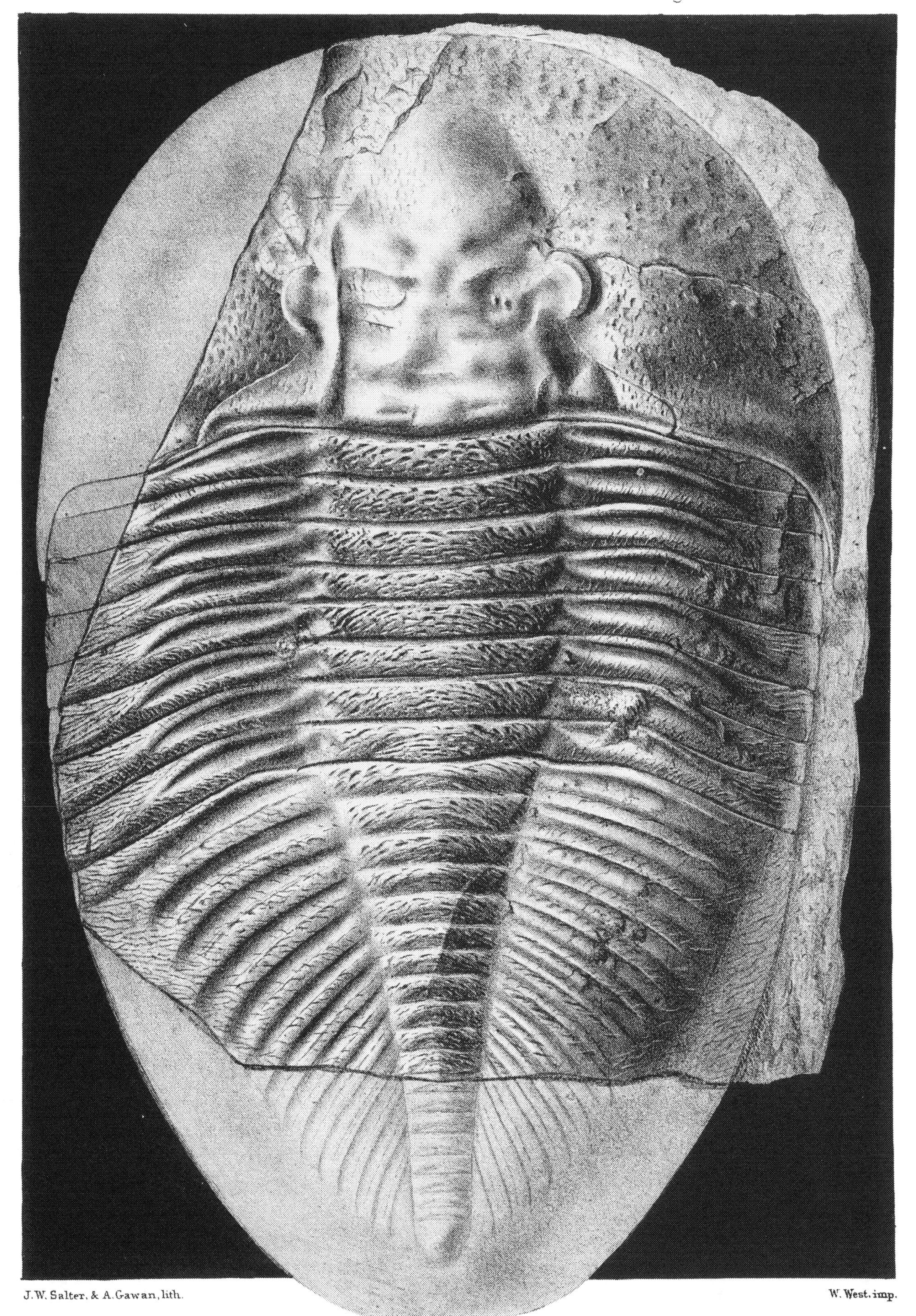

J. W. Salter, & A. Gawan, lith.

W. West. imp.

Asaphus Tyrannus, Murchison!

PLATE XXII.

LLANDEILO FLAGS.

Figs. 1—4. *Asaphus peltastes*, Salter.

1. „ „ Fine young specimen, showing the lateral lobes of glabella and long head-spines. (Mus. Pract. Geology.) Llandeilo.

2. „ „ Free cheek and eye. The smooth cornea is removed, and shows the minute lenses, 4000 or more in a single eye. (Mus. Pract. Geology.)

3. „ „ Head of young specimen. Llampeter Felfrey. (Mus. Pract. Geology.) It is a little shortened by pressure.

4. „ „ Tail of nearly full-grown specimen. Llandeilo.

Figs. 5—12. *Asaphus tyrannus*, Murchison. Old and young. Llandeilo.

5. „ „ Head of very perfect specimen. (In the cabinet of Mrs. Breawell, of Brighton.)

6. „ „ Labrum from Llampeter Felfrey. (Mus. Pract. Geology.)

7. „ „ Magnified eye-lenses.

8. „ „ End of large thorax-ring. (Mus. Pract. Geology.)

9. „ „ Fine caudal shield. (Brit. Mus.) The nodes on the sides of the axis are the internal glands mentioned in p. 52.

10, 11, 12. „ „ Young caudal shields. The ribs are shorter and more numerous than in corresponding specimens of *A. peltastes*. (Mus. Pract. Geology.)

LOWER SILURIAN (Llandeilo Flags.)

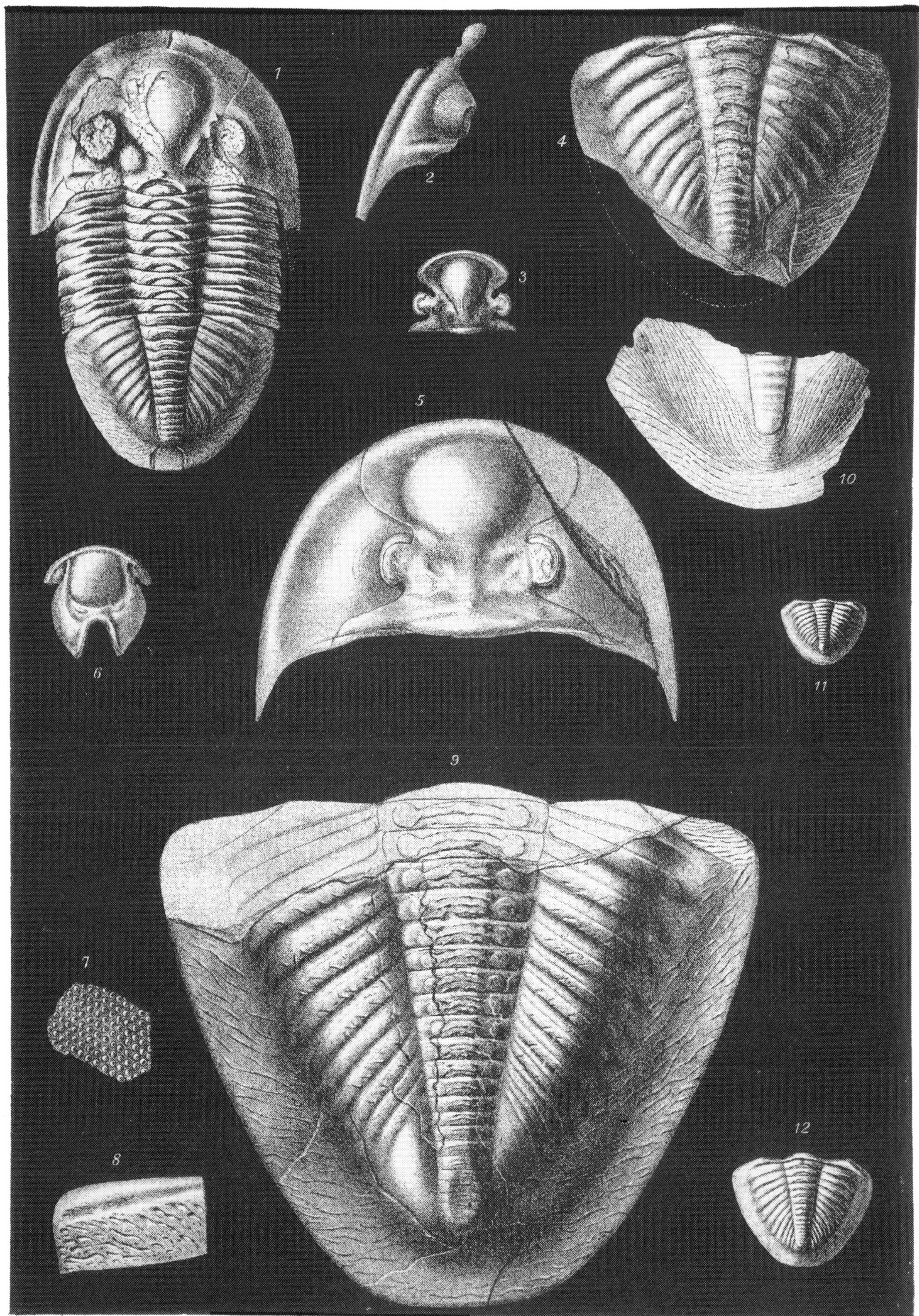

J.W. Salter & A. Gawan, lith.

W. West, imp

1_4. Asaphus peltastes. *5_12 A. tyrannus.*

PLATE XXIII.

LOWER SILURIAN (CARADOC).

Fig. 1. *Asaphus* (*Basilicus*) *Marstoni*, SALTER. Natural size; 1 *a*, magnified. *b* Is a free cheek of a larger specimen. Shales of Horderly. (Mr. H. W. Edgell's cabinet.)

Figs. 2—7. *Asaphus* (*Basilicus*) *Powisii*, MURCHISON. Chiefly from the Mus. of Pract. Geology. Figs. 3, 4, 6, are from Mr. H. Wyatt Edgell's cabinet.

2. " " " Restored figure, from many specimens. The head is not so perfectly known, and is only given here in outline. *a* Is a free cheek from Mr. Edgell's cabinet.

3. " " " Tail of young specimen. (Mr. Griffith's cabinet.)

4. " " " Somewhat older. Horderly.

5. " " " Large specimen from Waterloo Bridge, Conway. (Mus. Pract. Geology.)

6. " " " Labrum, not quite perfect at its base.

7. " " " Specimen from Dinas Mowddwy, North Wales. (Mus. Pract. Geology.)

Figs. 8, 9. *Asaphus* (*Basilicus ?*) *hybridus*, SALTER. From the Caradoc Shales (called in the text "Llandeilo flags"?) Henllan Amgoed, near Narberth, Carmarthenshire.

LOWER SILURIAN. (Caradoc.)

J. W. Salter & A. Gawan, lith. W. West. imp.

1. *Asaphus Marstoni.* 2–7. *A. Powisii.* 8, 9. *A.? hybridus.*

PLATE XXIV.

CARADOC ROCKS.

Figs. 1—5. *Asaphus* (*Isotelus*) *gigas*, DEKAY. From the Caradoc Rocks of Tyrone. (Mus. Pract. Geology.)

1. „ „ „ Young specimen, showing labrum. (Portlock's *I. planus*, pl. viii, fig. 1.)
2. „ „ „ Larger head. (Portlock's specimen, *I. sclerops*, pl. x, fig. 2.)
3. „ „ „ Tail of a large specimen.
4. „ „ „ Quite young. (*I. ovatus*, Portlock, pl. viii, fig. 5.)
5. „ „ „ Free cheek, showing eye and cast of lateral tubercle (*a*). *b* Is a cast in sealing-wax of the same outer portion of the cheek, showing the pit for the reception of the pleuræ in rolling up. (Portl., pl. vii, fig. 4.)

UPPER TREMADOC ROCKS.

Figs. 6—12. *Asaphus* (*Isotelus ?*) *Homfrayi*, SALTER. (From Mr. D. Homfray's cabinet, and Mus. P. Geology.)

6. „ „ „ Elongated and compressed specimen.
7. „ „ „ Young individual.
8. „ „ „ Labrum and epistome; the latter less perfect.
9. „ „ „ Labrum, pressed out of shape.
10. „ „ „ Tail, shortened by pressure, and showing the rudimentary rings on axis and sides.
11. „ „ „ Another specimen, similarly compressed.
12. „ „ „ Specimen shortened obliquely.

All the Tremadoc Slate specimens are from Garth, Portmadoc.

Figs. 13, 14. *Asaphus* (*Isotelus ?*) *affinis*, M'COY. Young and older specimen. (Mr. Ash's cabinet.)

LOWER SILURIAN (Caradoc Rocks.)

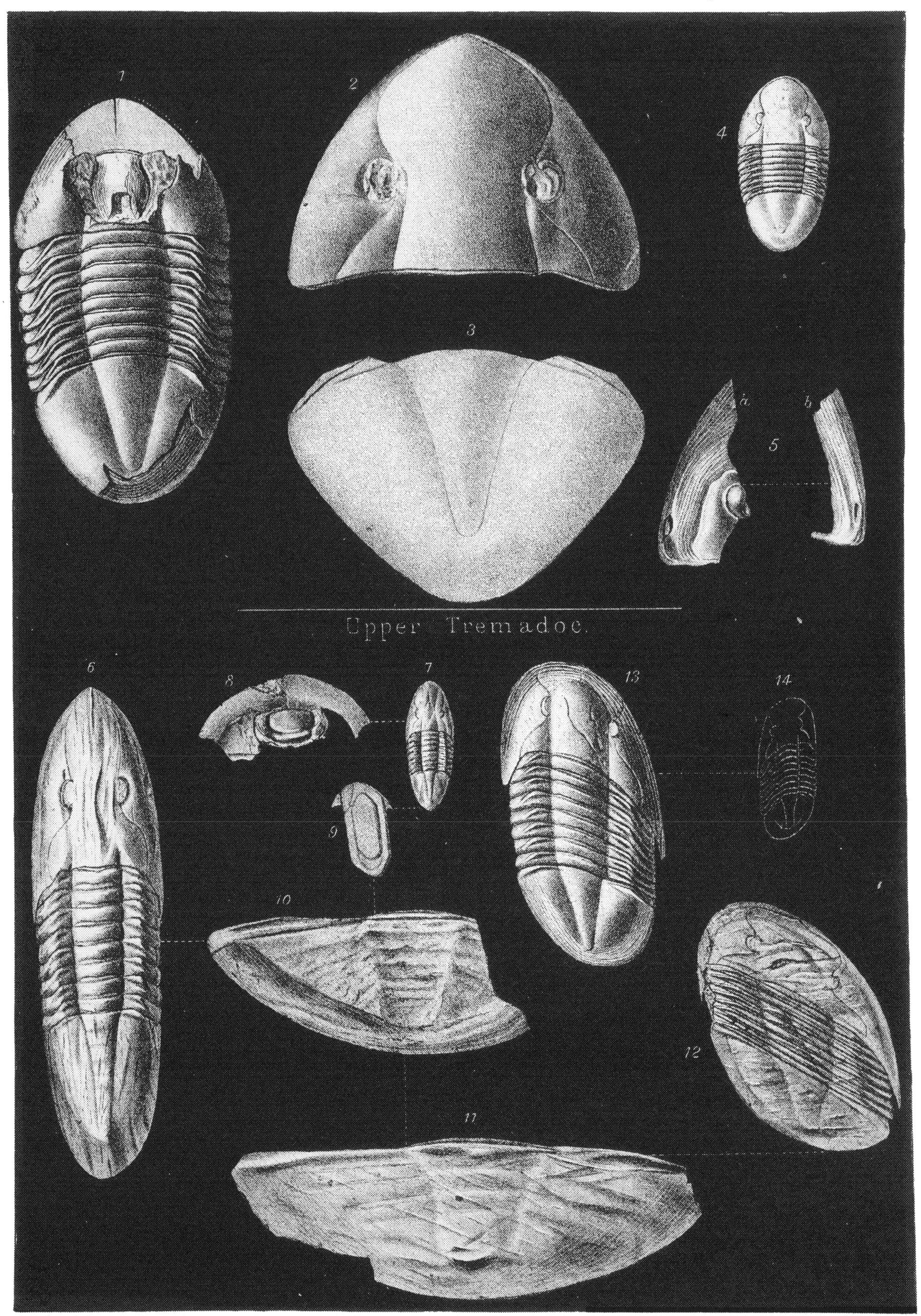

J.W. Salter & A. Gawan, lith. W. West, imp.

1–5. Asaphus gigas. 6–12. A. Homfrayi. 13, 14. A. affinis.

PLATE XXV.

CARADOC ROCKS.

Fig. 1. *Asaphus gigas*, DEKAY, variety. This is from Portlock's specimen of *I. intermedius,* from his pl. ix, fig. 5.

Figs. 2, 3. *Asaphus* (*Cryptonymus*) *scutalis,* SALTER.

2. ,, ,, ,, Head (Portl., pl. ix, fig. 4) Tyrone. (Mus. Pract. Geology.)

3. ,, ,, ,, Body-rings and tail (labelled *I. intermedius* by Portlock). (Same collection.)

Fig. 4. *Asaphus,* sp. undetermined. From the Caradoc? Rocks of Mount Pleasant, Carmarthen. (Mus. Geol. Society.)

Figs. 5—10. *Asaphus* (*Brachyaspis*) *rectifrons,* PORTLOCK.

5. Under surface of the head, showing the entire epistome. Portl. *I. planus,* pl. viii, fig. 3. Tyrone. (Mus. Pract. Geology.)

6. ,, ,, ,, Side of head. Ditto. Portl., pl. ix, fig. 1 *b*.

7. ,, ,, ,, Glabella only. Ditto. Ib., fig. 1 *a*.

8. ,, ,, ,, Body-rings and tail. Ditto. *I. arcuatus,* Portl., pl. ix, fig. 2.

9. ,, ,, ,, Fascia of underside of tail. Ditto. Ib., pl. ix, fig. 2.

10. ,, ,, ,, Under margin of cheek, showing the pit for the reception of the pleuræ.

All these are in the Mus. Pract. Geology.

LOWER SILURIAN. (Caradoc.)

J.W. Salter & A. Gawan, lith. W. West, imp.

1. Asaphus gigas. var. 2, 3. A. scutalis. 4, 5. A. — sp 6.10. A. rectifrons.

PLATE XXV*.

(To aid Plate XVII.)

ARENIG ROCKS (LOWEST SILURIAN).

Figs. 1—4. *Ogygia peltata*, SALTER. All natural size. From the vertical black slates of Whitesand Bay, St. David's, Pembrokeshire. (See Pp. 133—135, and 177.)

1. ,, ,, Intaglio of young ♂ specimen. (Woodwardian Mus., Cambridge, case G*h*.)

2. ,, ,, Very perfectly preserved ♀ form, showing the place of the eye, facets of the pleuræ, and the short interlining furrows of the tail, which differ from *O. scutatrix*. (Same Museum.)

3. ,, ,, Wide labrum, broken at the tip. (Cabinet of the late H. Wyatt-Edgell, Esq.)

4. ,, ,, Full grown ♀ form (Woodwardian Museum, case G*h*.)

Fig. 5. *Ogygia bullina*, SALTER. Same Locality and Museum. P. 178.

All these specimens were collected by H. Hicks, Esq., M.D., St. David's.

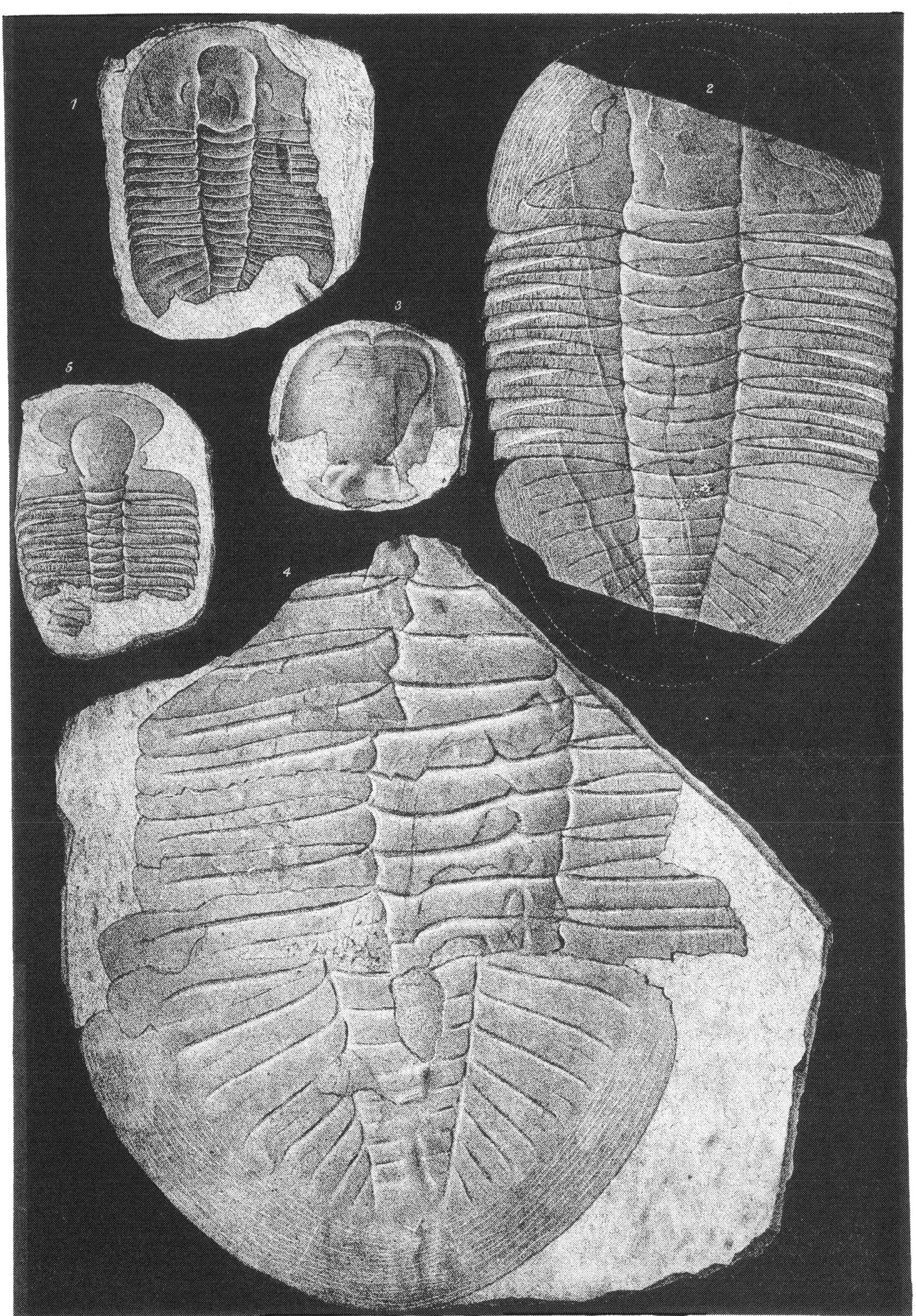

J. W. Salter & A. Gawan lith

W. West, imp.

1_4. *Ogygia peltata.* 5. *O. bullina.*

PLATE XXVI.

LOWER SILURIAN (LLANDEILO AND CARADOC).

Fig. 1 *a*, *b*. *Illænus* (*Ill.*) *Murchisoni*, SALTER. Front and side views of a very fine specimen in Mr. J. E. Lee's collection; from Grug Quarry, Bird's Hill, Llandeilo; in Caradoc or Bala limestone. P. 201.

1 *c*, 1 *d*. „ „ Direct and front views of the head, same specimen.

Fig. 2. *Illænus* (*Panderia*) *Lewisii*, SALTER. Nat. size and magnified. Moelydd, Oswestry, in Caradoc rocks. (Cabinet of Rev. D. Phillips Lewis, of Guilsfield.) P. 183.

Figs. 3, 4. *Illænus* (*Ill.*) *Portlockii*, (SALTER.) Caradoc schists of Desertcreat, Co. Tyrone. (Portlock's originals, Museum P. Geology.) P. 197.

Figs. 5—8. *Illænus* (*Ectillænus*) *perovalis*, MURCHISON. Llandeilo rocks of Shropshire and South Wales, P. 211.

5, 6. „ „ „ Specimens figured in the 'Silurian System.' (Museum of the Geol. Society.) From near Shelve.

7. „ „ „ More fully grown specimen, without the head, which is restored in outline from fig. 8. Llanrian, near Abereiddy Bay, Pembrokeshire. (Mus. P. Geol.)

8. „ „ „ Nearly perfect specimen, with head. We were compelled only to outline the head, as the plate was already finished ere this specimen was found.) At *a* the place of the eye is indicated. (Cabinets of Mr. Lightbody and Dr. H. Hicks.)

LOWER SILURIAN. (Llandeilo and Caradoc.)

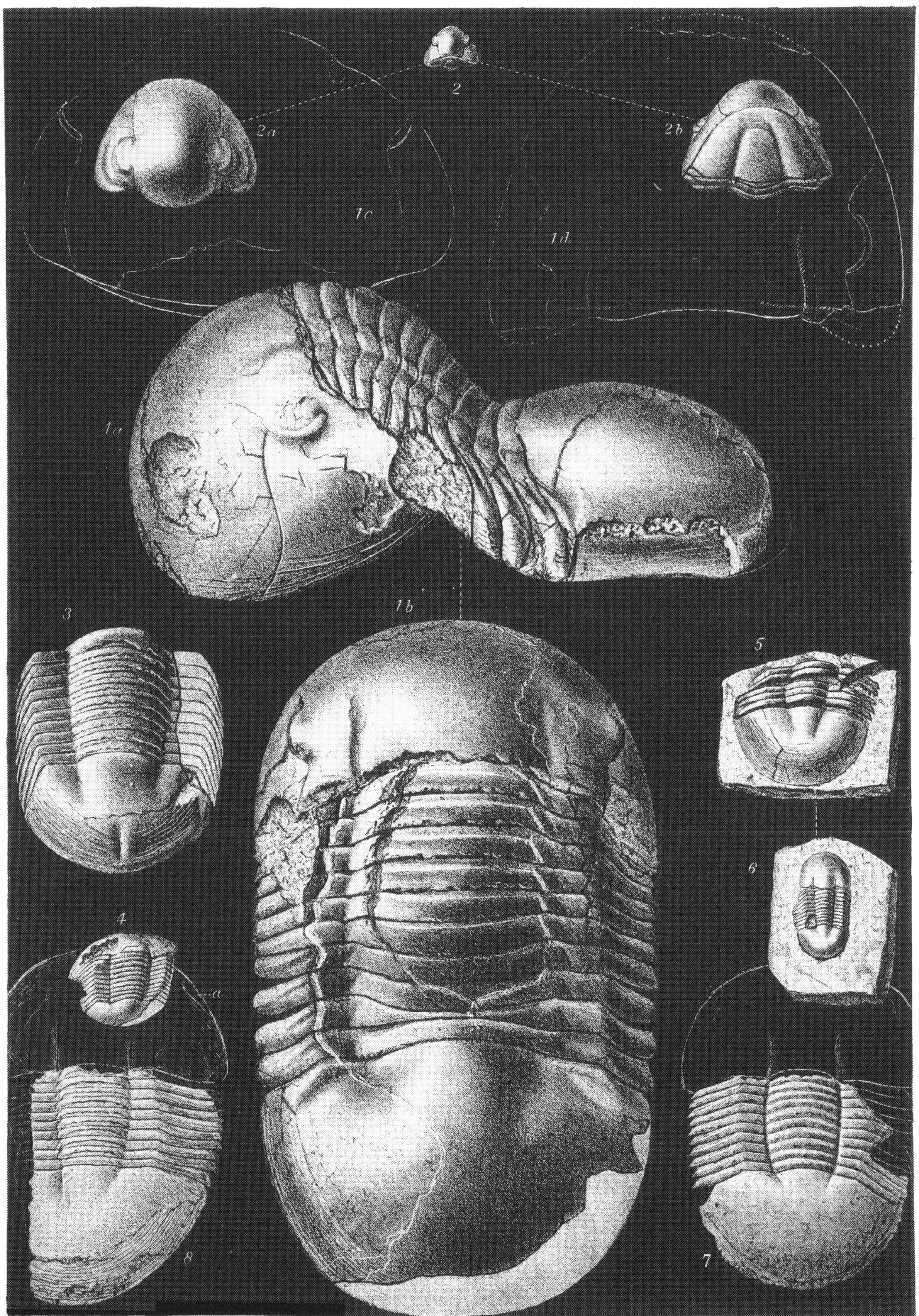

J.W. Salter & A. Gawan, lith.

W. West, imp.

1. *Illænus Murchisoni.* 2. *I. Lewisii.* 3, 4. *I. Portlockii.* 5–8. *I. perovalis.*

PLATE XXVII.

WOOLHOPE AND WENLOCK ROCKS (UPPER SILURIAN).

Figs. 1—5.	*Illænus* (*Bumastus*) *Barriensis*, MURCHISON.	
1 *a*, 1 *b*.	,, ,, ,, (Pp. 203—207.)	Direct and side views of the fine head figured in Decade 2. pl. iv, 'Mem. Geol. Survey.' Woolhope. (Mus. P. Geol.) It shows well the great glands (or muscle-spots?) over the eyes, and the strong sculpture, even on the cast.
2, upper figure	,, ,, ,,	Large head, wanting the free cheeks. Woolhope shales of Malvern Tunnel. (Dr. Grindrod's Collection.)
2, lower figure	,, ,, ,,	Miss Jukes' original specimen. From Barr, Staffordshire. An unequalled example, of which casts are in all museums. Woolhope limestone.
3.	,, ,, ,,	Young specimen, front view, to show flatness of head. (Dr. Grindrod's Collection.) Malvern Tunnel.
4.	,, ,, ,,	Younger still. Woolhope. (Mus. P. Geol.)
5.	,, ,, ,,	Side view of young head. (Dr. Grindrod.) Malvern Tunnel.
Figs. 6, 7.	***Illænus*** (***Bumast.***) ***insignis***, HALL (*Il. pomatia* on plate).	
6.	,, ,, ,, (P. 207.)	Fig. 6 is Dr. Grindrod's exquisite specimen from Ledbury, Herefordshire. WENLOCK LIMESTONE.
7.	,, ,, ,,	Labrum, from Dudley. (Gray's Collection, British Museum.)
Figs. 8, 9.	***Illænus*** (***Bumast.***) *carinatus*, SALTER. (P. 209.)	From the Wenlock limestone of Winnings Quarry, Malvern. (Presented to the Mus. P. Geol. by the late Rev. F. Dyson.)
8.	,, ,, ,,	Free cheek, eye, and broken spinous head-angle.
9.	,, ,, ,,	Upper view of head, showing carinated form of same, and the axal puncta.

UPPER SILURIAN.

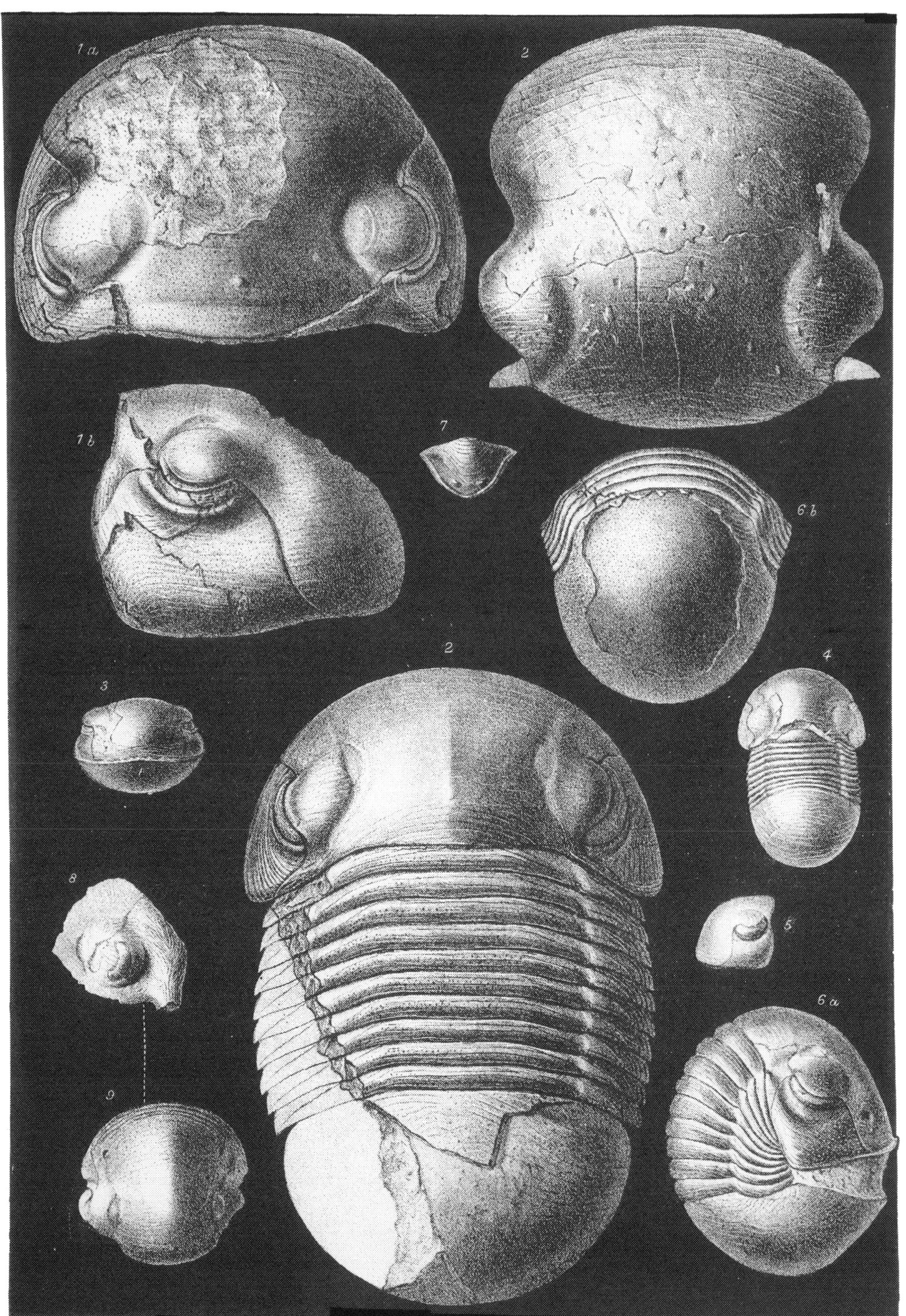

J. W. Salter & A. Gawan, lith.

W. West, imp.

1–5. *Illænus Barriensis.* 6, 7. *I. pomatia.* 8, 9. *I. carinatus.*

PLATE XXVIII.

MIDDLE SILURIAN (LLANDOVERY AND MAY HILL ROCKS).

Fig. 1. *Illænus* (*Bumastus*) *Maccallumi*, SALTER. Mullock, Girvan, Ayrshire. (Mus. P. Geol.) P. 210.

Figs. 2—4. *Illænus* (*Dysplanus*) *Thomsoni*, SALTER. P. 188.

2. " " " Internal cast, the same figured by Murchison, 'Quart. Geol. Journ.' From Mullock, Girvan. (Mus. P. Geol., presented by Sir Roderick I. Murchison.)

3. " " " Tail of ditto. Same Locality and Museum.

4. " " " Obscure young specimen. Purple shales (Tarannon shale) of Onny River, Shropshire. (Mus. P. Geol.)

Fig. 5. *Illænus* (*Dysplanus*) *æmulus*, SALTER. May Hill sandstone of Norbury. (Mus. P. Geol.) P. 187.

Fig. 6. *Illænus* (*Dysplanus*) *Bowmanni*, SALTER. Same Locality and Museum as the last. P. 185.

LOWER SILURIAN (CARADOC).

Figs. 7—13. *Illænus* (*Dysplanus*) *Bowmanni*, SALTER. P. 185.

7. Half-grown. Desertcreat, Tyrone. (Mus. P. Geol.) Portlock's original.

8. " " Young. Same Locality and Museum.

9. " " " Young tail, very short. Shoals Hook, near Haverfordwest. (Mus. P. Geol.)

10. " " " Half-grown fine specimen from the Bala slates of Llanwddyn, Montgomeryshire. (Woodwardian Museum; M'Coy's figured specimen of '*Il. centrotus.*')

11. " " " Variety, with longer and more connivent head-furrows. (Mus. P. Geol.)

12. " " " Outline of large head. Shoals Hook, Pembrokeshire. (Mus. P. Geol.)

13. " " " Young head. (Old Red ? conglomerates, derived from the) Coniston or Bala limestone, Troutbeck. The species shows the attachment of the lower (mandibular?) muscles. (Mus. Wyatt-Edgell.)

Fig. 14. *Illænus* (*Ill.*) *Bailyi*, SALTER. Dunabrattin Head, Waterford. In limestone. (Mus. Irish Survey.) P. 192.

MIDDLE SILURIAN. (Llandovery and May Hill Rocks.)

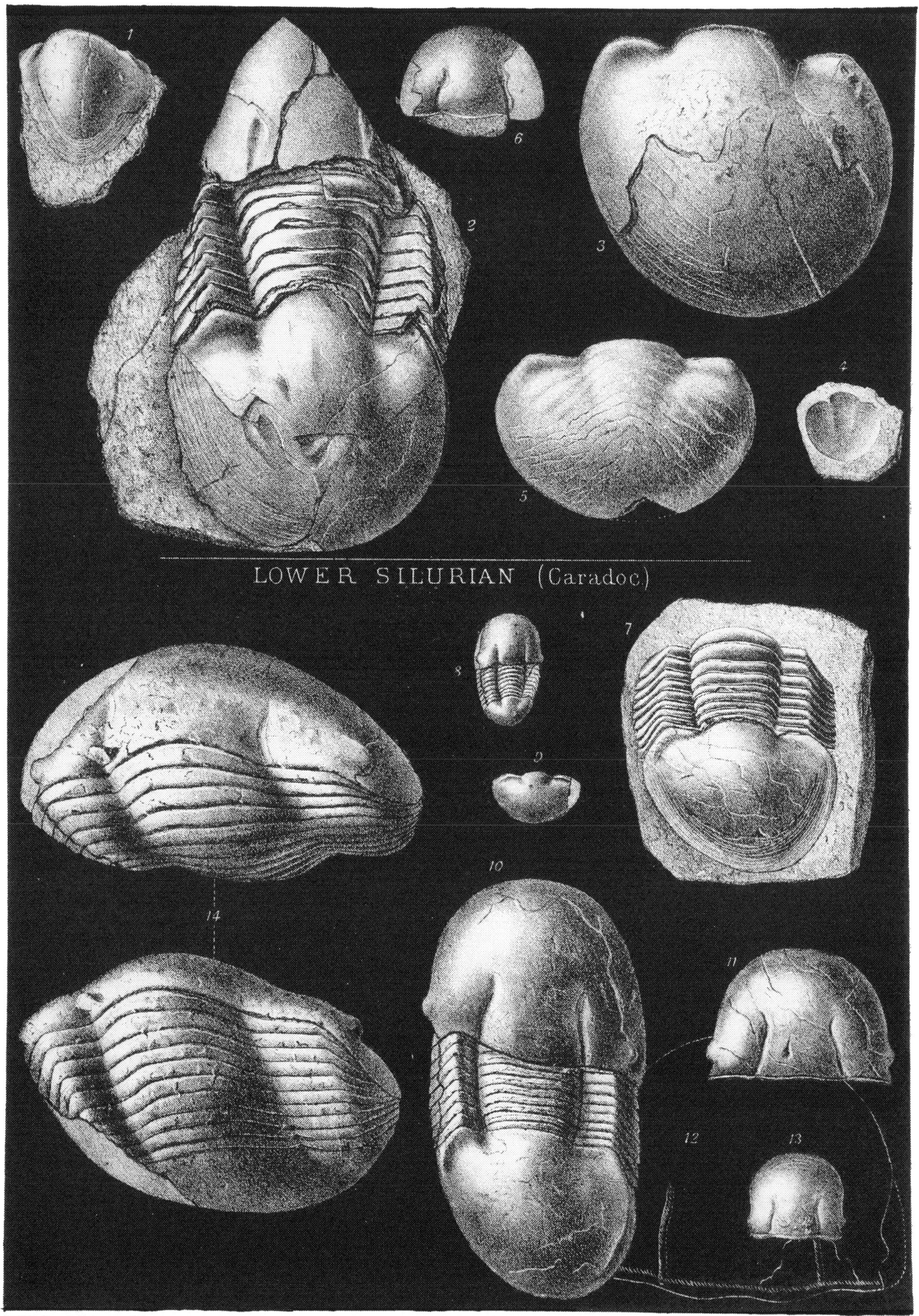

J.W. Salter & A.Gawan, lith. W. West, imp.

1. Illænus Macallumi. *2–4. I. Thomsoni.* *5. I. æmula.* *6–13. I. Bowmanni.* *14. I. Bailyi.*

PLATE XXIX.

LOWER SILURIAN (CARADOC).

Fig. 1.	*Stygina? Musheni*, SALTER. (P. 174.)	Boulder (probably of Caradoc), from the Drift, Buildwas, Shropshire. (Brit. Mus.)
Figs. 2—6.	*Illænus* (*Ill.*) *Rosenbergii*, EICHWALD? (P. 199.)	From the Coniston or Bala limestone of Coniston and Troutbeck. (Except fig. 5, which is Mr. Wyatt-Edgell's specimen, all are in the Woodwardian Museum.) Figs. 2, 3, 4, are figured in Prof. Sedgwick's 'Synopsis,' and show the sculpture well.
Figs. 7—9.	*Ill.* (*Illænus?*) *ocularis*, SALTER. (P. 198.)	7, Side view of head; 8, head, nat. size; 8 *a*, magnified; 8 *b*, sculpture enlarged. CARADOC or BALA limestone, Chair of Kildare. (Mus. P. Geol.)
9.	,, ,,	? A larger fragment of the side of the head, probably the same species. Same Locality and Museum.
Figs. 10—16.	*Ill.* (*Illænus*) *Davisii*, SALTER. (P. 194.)	CARADOC or BALA limestone.
10.	,, ,,	The finest known. Figured in the Woodwardian 'Synopsis' by M'Coy and Salter. (Woodw. Mus., Cambridge.)
11.	,, ,,	Ordinary form at Rhiwlas, Bala. (Mus. P. Geol.)
12, 13.	,, ,,	Restored figure, from Decade 2, Geol. Survey.
14.	,, ,,	Epistome or rostral shield. (Mus. P. Geol.)
15.	,, ,,	Side view. Same locality and cabinet.
16.	,, ,,	Young tail, from Mr. Wyatt-Edgell's cabinet, showing a nearly complete axis.

LOWER SILURIAN (Caradoc.)

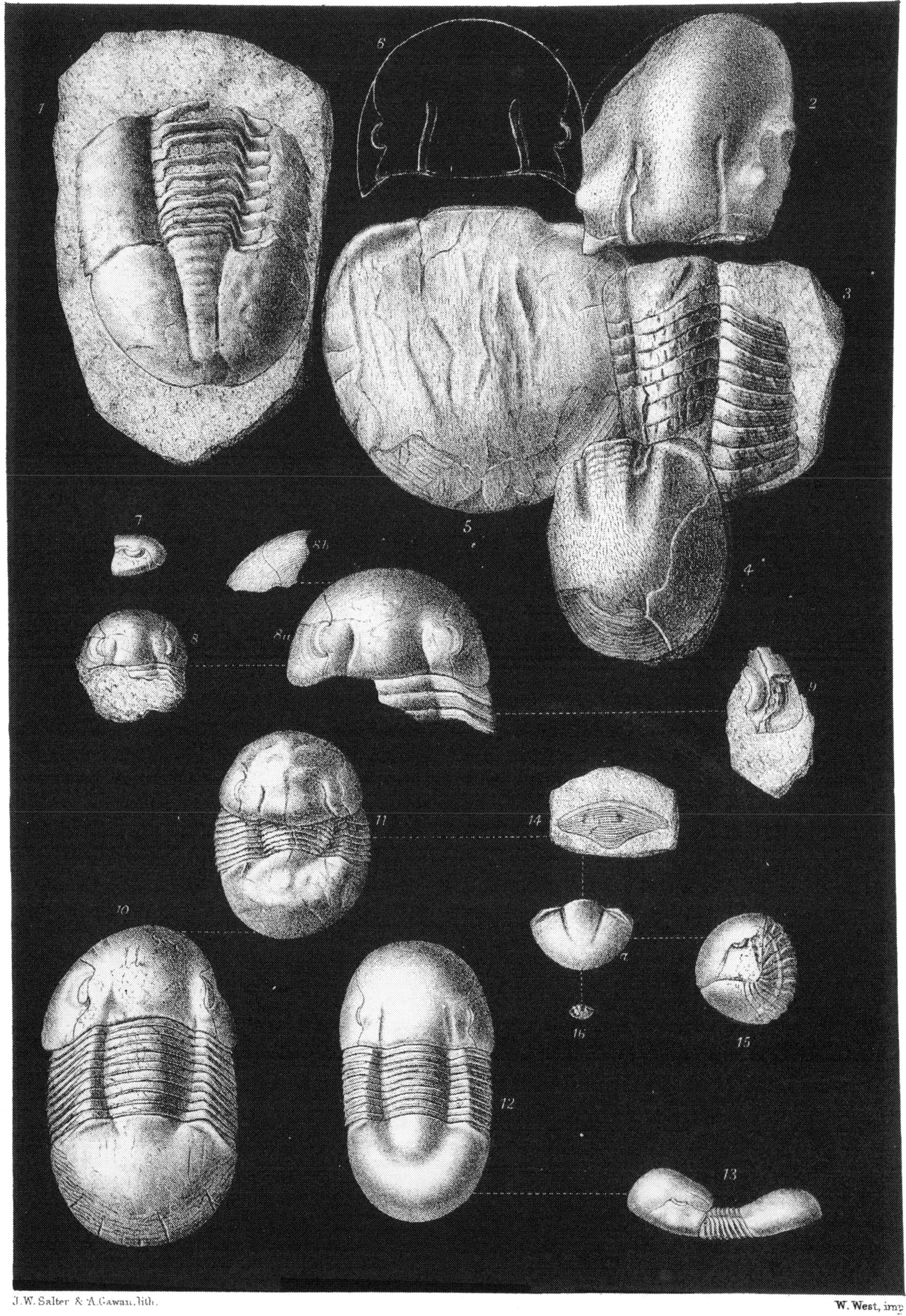

J. W. Salter & A. Gawan, lith. W. West, imp.

1. Stygina Musheni. 2_6. Illænus Rosenbergii. 7_9. I. ocularis. 10_16. I. Davisii.

PLATE XXX.

LLANDOVERY ROCKS (MIDDLE SILURIAN).

Fig. 1. *Illænus* (*Bumastus*) *Barriensis*, MURCHISON. A solitary specimen from the Purple (Tarannon) shales of the Onny River, Shropshire. (Mr. Wyatt-Edgell's Cabinet.) P. 203.

Figs. 2, 3. *Ill.* (*Bumastus*) *Maccallumi*, SALTER. From the Llandovery (not the May Hill) rocks of Girvan, Ayrshire. (See also Pl. XXVIII, fig. 1.) Fig. 2, tail, full-grown; 3, young specimen, dismembered. The free cheeks are yet wanting; possibly they are spinose. (Mus. Prof. Wyville Thomson.) P. 210.

Figs. 4, 5. *Ill.* (*Dysplanus*) *nexilis*, SALTER. Llandovery rocks of Drummuck, near Girvan, Ayrshire. (Prof. Wyville Thomson's cabinet.) Figs. 4 and 5 show the broad epistome, unlike that of *Il. Bowmanni.* Mr. Wyatt-Edgell's Cabinet.) Fig. 4 shows the fascia, which is much bent in this species. P. 190.

Fig. 6. *Ill.* (*Dysplanus*) *Bowmanni*, SALTER. Llandovery mudstone of Mullock, Girvan, Ayr. (Cabinet of Prof. Wyville Thomson.) P. 185.

Figs. 8, 10. *Ill.* (*Dysplanus*) *Thomsoni*, SALTER. Llandovery rock. P. 188.

8. ,, ,, Large intaglio in yellow sandstone, from Mullock, near Girvan, Ayrshire. (Prof. Wyville Thomson's cabinet.)

9. ,, ,, Side of head, showing eye and free cheek (very rarely found). The Gas Works, Haverfordwest. (Mr. Wyatt-Edgell's Cabinet.)

10. ,, ,, Tail, from Mullock, Girvan, Ayrshire; showing truncate angles. (Woodw. Mus.)

LOWER SILURIAN.

Fig. 7. *Ill.* (*Illænus*) *Murchisoni*, SALTER. CARADOC limestone of Chair of Kildare, Ireland. (Cabinet of Mr. Wyatt-Edgell.) This should have been in Pl. XXVI. P. 201.

MIDDLE (AND LOWER) SILURIAN. (LLANDOVERY &c) Pl. 30.

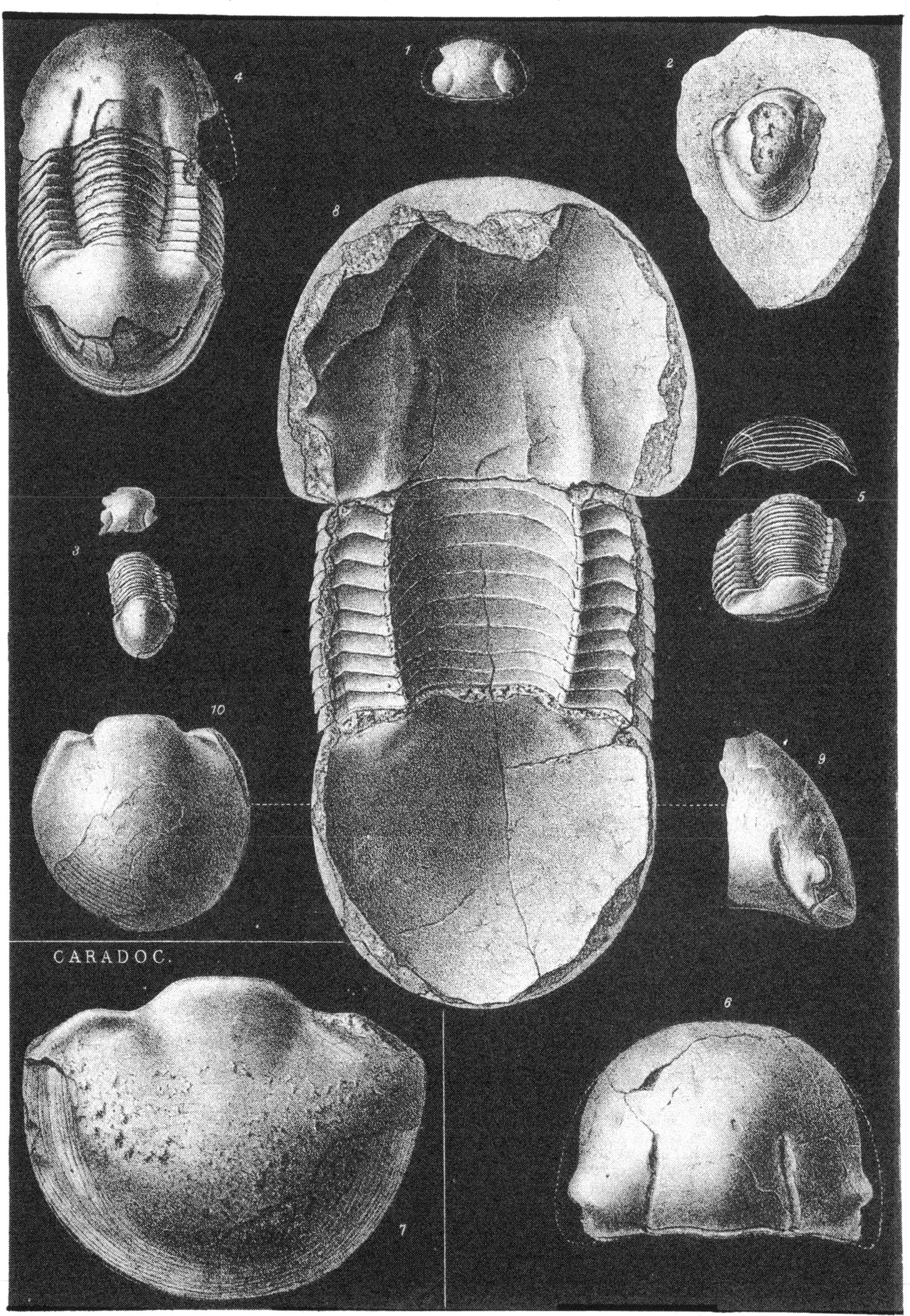

J W. Salter & A. Gawan lith. W. West imp.

1. *Illænus Barriensis* 2.3. *I. Mc Callumi.* 4.5. *I. nexilis* *I Bowmanni.* 7 *I Murchisoni.*
8_10. *I Thomsoni*

Printed in the United States
By Bookmasters